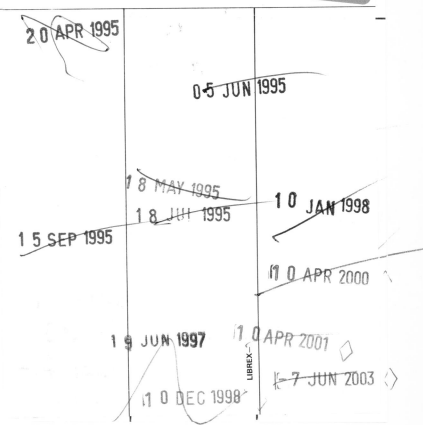

Surveys in Industrial Wastewater Treatment
Vol. 1 Food and Allied Industries

VOL I

Surveys in Industrial
Wastewater Treatment

Food and Allied Industries

Edited by
D Barnes
University of New South Wales

C F Forster
University of Birmingham

S E Hrudey
University of Alberta

Pitman Advanced Publishing Program
Boston · London · Melbourne

PITMAN PUBLISHING LIMITED
128 Long Acre, London WC2E 9AN

PITMAN PUBLISHING INC
1020 Plain Street, Marshfield, Massachusetts 02050

Associated Companies
Pitman Publishing Pty Ltd, Melbourne
Pitman Publishing New Zealand Ltd, Wellington
Copp Clark Pitman, Toronto

First published 1984
© Pitman Publishing Limited 1984

Library of Congress Cataloguing in Publication Data
Surveys in Industrial Wastewater Treatment
   Includes bibliographical references and index.
   Contents: v. 1. Food and allied industries.
   1. Factory and trade waste—collected works.
   2. Sewage—purification—collected works.
   I. Barnes, D.   II. Forster, C. F. (Christopher F.)
   III. Hrudey, S. E. (Steve E.)
   TD897.S85   1984   664'096   84-1898

   ISBN 0-273-08586-7 (v. 1)

British Library Cataloguing in Publication Data
Surveys in industrial wastewater treatment.
   Vol 1. Food and allied industries
   1. Water—Purification
   I. Barnes, D.   II. Forster, C. F.
   III. Hrudey, S. E.
   628.1'62   TD430

   ISBN 0-273-08586-7

All rights reserved. No part of this publication may be reproduced, stored in a retrieval system, or transmitted, in any form or by any means, electronic, mechanical, photocopying, recording and/or otherwise, without the prior written permission of the publishers. This book may not be lent, resold or hired out or otherwise disposed of by way of trade in any form of binding or cover other than that in which it is published, without the prior consent of the publishers.

Text set in 10/12 pt Linotron 202 Ehrhardt,
Printed in Great Britain at The Pitman Press, Bath

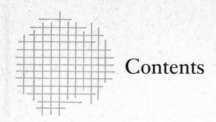

Contents

Preface xi

Introduction xiii
D Barnes, C F Forster and S E Hrudey

1 The treatment of wastes from the sugar industry 1
K E McNeil

1.1 Introduction 1

1.2 Production and refining of raw sugar—sources of wastewater 2
    1.2.1 Processing operations 2
    1.2.2 Water use and sources of wastewater in a beet sugar factory 9
    1.2.3 Water use and sources of wastewater in a cane sugar factory 17
    1.2.4 Sources of cane sugar refinery wastewater 26

1.3 Manufacture of sugar—treatment and disposal of wastewater 29
    1.3.1 Irrigation 29
    1.3.2 Treatment in ponds or aerated lagoons 31
    1.3.3 Activated sludge plants 35
    1.3.4 Biofiltration or trickling filters 43
    1.3.5 Anaerobic methane-generating processes 44

1.4 Fermentation of molasses and sugar cane juice—sources of distillery wastewater 46

1.5 Distillery wastewater—treatment, disposal and utilization 47
    1.5.1 Reduction of effluent levels by process modification 47
    1.5.2 Land disposal of stillage 48
    1.5.3 Evaporation, incineration and potassium recovery 48
    1.5.4 Biological treatment 50
    1.5.5 Production of yeast or fungal biomass 55

1.6 Conclusions 56

Acknowledgements 57

References 57

2 **The treatment of wastewater from the beverage industry** 69
  H M Rüffer and K-H Rosenwinkel

  2.1 Introduction 69

  2.2 Juices—production processes, the wastewater and its treatment 75
    2.2.1 Juice production 76
    2.2.2 Wastewater characteristics 76
    2.2.3 Wastewater load reduction 81
    2.2.4 Wastewater treatment 83
    2.2.5 Examples 85

  2.3 Mineral waters and soft drinks—the wastewater and its treatment 86
    2.3.1 Wastewater characteristics 86
    2.3.2 Wastewater load reduction 88
    2.3.3 Wastewater treatment 88
    2.3.4 Example 89

  2.4 Beer—production processes, the wastewater and its treatment 90
    2.4.1 Beer production 90
    2.4.2 Wastewater characteristics 93
    2.4.3 Wastewater load reduction 97
    2.4.4 Wastewater treatment 100
    2.4.5 Examples 101

  2.5 Wine—production processes, the wastewater and its treatment 103
    2.5.1 Wine production 103
    2.5.2 Wastewater characteristics 103
    2.5.3 Wastewater load reduction 110
    2.5.4 Wastewater treatment 111
    2.5.5 Example—combined treatment of winery wastewater and municipal sewage 111

  2.6 Spirits—distillation, the wastewater and its treatment 111
    2.6.1 Distillation 112
    2.6.2 Wastewater characteristics 116
    2.6.3 Wastewater load reduction 118
    2.6.4 Wastewater treatment 118
    2.6.5 Examples 122

  2.7 Conclusions 122

  References 123

3 The management of wastewater from the meat and poultry products industry 128
S E Hrudey

3.1 Introduction 128
    3.1.1 Relevant water pollutant parameters 128
    3.1.2 Regulatory controls for the industry 130

3.2 Industry characterization 131
    3.2.1 Red meat products 132
    3.2.2 Poultry products 139
    3.2.3 Independent rendering 146

3.3 In-process waste control 151
    3.3.1 General concepts 151
    3.3.2 Red meat products 153
    3.3.3 Poultry products 157
    3.3.4 Independent rendering 160

3.4 External treatment 161
    3.4.1 Equalization 162
    3.4.2 Coarse solids removal 164
    3.4.3 Settleable solids and FOG removal 167
    3.4.4 Soluble organic carbon removal 171
    3.4.5 Disinfection 194
    3.4.6 Land treatment 197

3.5 Summary 202

References 203

4 The treatment of waste from the fruit and vegetable processing industries 209
J R Harrison, L A Licht and R R Peterson

4.1 Introduction 209

4.2 Waste characteristics 210
    4.2.1 Pollutants 210
    4.2.2 Waste survey 212

4.3 Aims of waste treatment 216
    4.3.1 Economic considerations 216
    4.3.2 Meeting discharge standards 217

4.4 Pretreatment   219
   4.4.1 Flow measurement and sampling   220
   4.4.2 Screening   220
   4.4.3 Silt removal   223
   4.4.4 Neutralization   228
   4.4.5 Oil and grease removal   229
   4.4.6 Flow equalization   229

4.5 Primary treatment   230
   4.5.1 Gravity clarifier   231
   4.5.2 Flotation clarifier   233
   4.5.3 Inclined plate separator   234
   4.5.4 Centrifugal concentrators   235
   4.5.5 Complete primary system   235

4.6 Anaerobic treatment   236
   4.6.1 Ponds   238
   4.6.2 Filters   238
   4.6.3 Contact process   239
   4.6.4 New anaerobic processes   240

4.7 Secondary treatment   240
   4.7.1 Ponds and lagoons   248
   4.7.2 Trickling filters   253
   4.7.3 Activated sludge   256
   4.7.4 Biofilter activated sludge   258
   4.7.5 Pure oxygen activated sludge   261
   4.7.6 Rotating biological contactors   263
   4.7.7 Pilot testing   263

4.8 Tertiary treatment   271
   4.8.1 Nitrification   271
   4.8.2 Chemical precipitation and sedimentation   272
   4.8.3 Filtration   272

4.9 Land treatment   274
   4.9.1 Process variations   274
   4.9.2 Constraints   276
   4.9.3 Operation and management   278

4.10 Solids concentration and disposal   279
   4.10.1 Conditioning   280
   4.10.2 Thickening   281
   4.10.3 Dewatering   282
   4.10.4 Disposal   284

4.11 Byproduct utilization 287
    4.11.1 Conversion of waste to livestock feed 288

4.12 Treatment costs 290
    4.12.1 Present-day costs 290
    4.12.2 Future trends 292

References 293

5 The treatment of wastes from the dairy industry 296
K R Marshall and W J Harper

5.1 Introduction 296

5.2 Milk 297
    5.2.1 Composition of milk 297
    5.2.2 History of commercial processing 298
    5.2.3 Milk production 299

5.3 Milk products 300
    5.3.1 Dairy product manufacture 301
    5.3.2 Process equipment 309

5.4 Dairy plant wastes 310
    5.4.1 Sources of dairy plant wastes 310
    5.4.2 Composition of dairy fluids and products 311
    5.4.3 Composition and characteristics of dairy plant wastes 316

5.5 Waste reduction 323
    5.5.1 Use of effluent data to compute losses and yield 324
    5.5.2 Waste monitoring 325
    5.5.3 Control programme 329
    5.5.4 In-plant control 330
    5.5.5 Achievable levels of losses 334

5.6 Treatment of unavoidable wastes 334
    5.6.1 Pretreatment 336
    5.6.2 Discharge to natural waterways 338
    5.6.3 Municipal treatment 340
    5.6.4 Land treatment 341
    5.6.5 Aerobic biochemical treatment 347
    5.6.6 Anaerobic treatment 363
    5.6.7 Chemical precipitation 364
    5.6.8 Tertiary treatment methods 364
    5.6.9 Sludge disposal 364

5.7 Conclusion 370

Acknowledgement 370

References 371

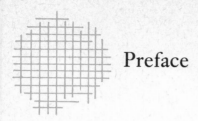

Preface

THE CONCEPT of a series of books on aspects of industrial wastewater treatment was developed during 1979. At that time one of the editors (DB) was on study leave at the University of Birmingham and another (SEH) had occasion to make several visits to Britain. All of the editors had some experience of treating industrial wastewaters and all had experienced difficulties in obtaining detailed information both about wastewater characteristics and about the performance of particular processes.

This volume is the first of a series which attempts to provide a detailed review of the treatment of wastewaters from groups of industries. Authors have been selected who can provide a balanced summary based upon their direct experience. Each has been asked to discuss a subject in sufficient depth that readers can obtain a reliable insight into the options available. As such, the individual reviews are not summaries of specific research endeavour: rather they reflect established or proven practice. Equally, the reader has been assumed to have a basic understanding of wastewater treatment, such that the principles of analytical procedures and unit operations need not be explained, except as they have specific features unique to a given industrial wastewater.

In order to provide some cohesion to each volume, related subjects have been grouped together. This first volume is concerned with the food and beverage industries, with chapters on sugar, dairy, beverage, fruit and vegetable and meat and poultry industry wastes. The second volume groups together organic-based wastewaters, with chapters on the dyestuffs, petrochemicals, oil refining and synthetic fuels industries. The third volume, which deals with inorganic wastewaters, covers plating, silver recovery (particularly from the photographic industry), general inorganic chemical industries, chloralkali (particularly the treatment and disposal of mercury sludges) and the steel industry. Subsequent volumes will group topics in a similar manner.

The editors would like to express their thanks to the authors for providing the manuscripts and for tolerating the corrections, threats and browbeatings as editorial deadlines approached and passed. Also we thank the publisher, Pitman, for help and cooperation and our students and professional colleagues who provided the necessary stimulus for such an endeavour. Finally we express our thanks to our wives and families for their help and support.

December 1983

DB *Sydney*
CFF *Birmingham*
SEH *Edmonton*

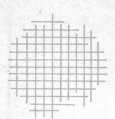

Introduction
Principles of industrial wastewater treatment

D Barnes, C F Forster and S E Hrudey

THE WIDE range of industrial manufacturing processes generates an equally diverse range of wastewaters. Clearly it is inappropriate to assume that similar wastewaters will be generated in, say, a brewery and in the manufacture of steel, or that the wastewaters can be treated by the same unit processes designed to the same criteria. However, some aspects of both waste generation and treatment are common to many industries. This brief introduction attempts to summarize some of these common aspects.

Wastewater characteristics
Industrial wastewaters tend to be characterized by great variability in both flow and composition. Only a small number of industrial plants operate continuously to generate wastewater of non-varying characteristics. It is only very large organizations, such as oil refineries, that attempt to maintain a continuous throughput of consistent quality and so can produce a consistent effluent. The majority of industries are small-to-medium sized, do not operate 24 hours a day and do not attempt to produce either product or effluent continuously or consistently.

Many dairies, for example, work for only 8–12 hours a day so only generate wastewater during that period. Several of the effluent-producing operations, such as tank cleaning and disinfection, are discontinuous and hence give rise to variability of both flow and load for treatment. Often the dairy manufactures several products, such as a range of conventional milks (homogenized, high or low butterfat), condensed or evaporated milk, cream products and more specialized products such as yoghurt and cheese. Each of these manufacturing processes produces wastewaters of different volume and composition. Thus, while the BOD_5 received for treatment may average 3000 mg l^{-1} over a working day, this may include periods when the BOD_5 exceeds 20 000 mg l^{-1} or contains high concentrations of sodium hydroxide with associated high pH values.

This variability of industrial wastewaters is accentuated by the short sewer lengths through which wastewater is conveyed to the treatment plant. Municipal plants receive wastewater that has been subjected to considerable mixing, dilution and attenuation in the sewerage system, whereas industrial plants are served by sewers that rarely exceed 1 km in length, so must tolerate flow and load variations that directly reflect manufacturing operations. Most industrial operations use some chemicals that are toxic, corrosive or of extreme pH, hence it is inevitable that some of these chemicals arrive at the wastewater treatment plant. Small or continuous discharges usually represent only minor problems; however, accidental or deliberate release of such chemicals represents a major disruption for many wastewater treatment processes, particularly biologically based unit operations.

Thus, the industrial wastewater treatment plant must be able to accommodate a range of flow and load variations far greater than that encountered at municipal plants. Preferably data should be accumulated on a continuous or short time-interval basis, to indicate likely variability. A designer should also be aware of probable shock loads in the form of erratic discharges of specific chemicals or debris which may be discarded to the factory drainage system. Hence the pretreatment section of an industrial wastewater treatment plant is likely to include a screening operation and, often, facilities for flow balancing and neutralization.

Importance of wastewater treatment to industries
Centres of industrial activity were generated by the growth of manufacturing industry and the transfer of manpower from agriculture to manufacture in the Industrial Revolution. This process has accelerated and continues with the growth of the chemical and allied industries and the centralized processing of foodstuffs. Wastewaters from industry received relatively little treatment before the second half of the twentieth century. At this time, increased awareness of the limited supply of fresh water initiated a general campaign to reduce the pollutional load imposed upon this resource. As a consequence, regulatory agencies were set up to monitor and control discharges to fresh water and, subsequently, to other bodies of water. This led to a more stringent attitude towards industrial discharges to the sewers in order to control the loads at municipal wastewater treatment plants.

While responsible industrial organizations generally accept the basic premise that 'the polluter pays' and hence the manufacturer installs pollution control equipment, in many cases this represents an additional manufacturing cost. There are several examples of byproduct recovery to offset costs, or even to increase cash flow, but still most manufacturers see pollution control as a drain on resources. The capital cost of the wastewater treatment plant does not improve manufacturing efficiency and the plant incurs additional operating, maintenance and running costs. Therefore, while it is easy to advocate stringent water pollution control measures, these represent an economic disadvantage in many cases. The modification of some taxation principles to permit rapid depreciation of pollution control equipment — and even concessionary allowances on running costs for the equipment — can only be advantageous.

Until recently, industrial wastewater treatment plants were installed on the basis of lowest capital cost and were then expected to operate with minimum attention from unskilled or untrained operators. Such an approach, when combined with the erratic and difficult nature of many industrial wastewaters, has led to poor performance from many plants. For effective wastewater treatment the unit processes of pollution control must be integrated with the unit processes of manufacture and should be considered equally important to the overall efficiency of the industry. This ideal is rarely achievable, particularly in old-established industries where often the site has been reused and modified over several decades of industrial occupation. At such sites it may be difficult even to trace the drainage system or to separate waste streams.

In new industrial developments wastewater treatment should be considered at an early planning stage and fully integrated into the total facility. In some cases this can lead to major conflict between the wastewater treatment aspects and the manufacturing aspects of the industry. For example, in the processing of potatoes the use of a highly alkaline solution to remove skins — lye peeling — produces a strong solubilized waste that can be awkward to treat. More recent potato processing plants have replaced lye peeling, for some applications, with dry non-chemical peeling processes; while wastewater treatment is not the only consideration in making such process changes, it does have a significant influence on potato processing.

Another influence upon industry's attitude to pollution control is the potential volatility of any manufacturing activity. While municipal engineering tends to anticipate many decades, or even centuries, as the potential life of a given facility, manufacturing can change very rapidly as new products are developed or science and technology offer a novel approach to a product. Pollution control equipment must either be sufficiently flexible to accommodate such changes or must be expendable. Thus, while the majority of municipal wastewater treatment plants are built as massive structures of in-ground reinforced concrete, industrial plants tend to be of a less permanent nature and constructed, often above ground, of steel or glass-fibre.

Trends in industrial wastewater treatment
Pollution control is subject to short-term 'fashionable' changes, as are many other technologies. The installation of particular types of equipment — for example, surface or submerged aerators — or types of process — for example, biological or chemical — is influenced by a series of scientific, political and commercial factors. However, beyond these changes there are some perceptible trends, notably the recovery of byproducts, the integration of the wastewater treatment with manufacture, and the treatment and disposal of waste solids.

A major influence on many industrial activities has been the escalation of fuel costs and particularly of oil-based products. In several cases this has reduced the load to wastewater treatment plants. Many industrial processes produce 'waste oil' and when oil prices were low much of this was discharged into the drainage or sewer system. As the price of oil has increased, so has the incentive to recover or recycle oil, and the tendency to lose oil in wastewater has correspondingly decreased. Similarly, industries which rely heavily upon oil-based products — for example, the paint and solvent industries — have much greater incentives to recover losses of oil-based products; in this case the changes have been accelerated by more stringent air pollution legislation which has favoured water-based rather than solvent-based paints.

Higher energy costs have led to a re-examination of anaerobic processes for wastewater treatment. Anaerobic treatment can represent a double saving: the process does not need to be supplied with oxygen, hence the power and operating costs of aeration can be eliminated; and methane can be generated as a potential supplementary fuel. The value of methane is limited unless the supply can be continuous and can be utilized at the existing site. However, the mere conversion of carbon into a gas reduces the sludge mass for disposal, and this in itself can be a significant advantage. A disadvantage of anaerobic units has been that they were large, requiring many days retention and hence taking up an unacceptable proportion of potential manufacturing area. The development of high-rate contact anaerobic processes, such as the upflow blanket and the fluidized bed reactor, has overcome this major disadvantage and it can be assumed that anaerobic industrial wastewater treatment will continue to increase in importance.

The majority of wastewater treatment plants are for segregation rather than treatment of wastewaters. The raw wastewater is passed through a series of unit operations and is discharged as a less polluting, lower solids stream — the treated effluent. However, a high solids stream — waste sludge — remains and requires separate management. Some inorganic sludges, such as the mixed metal hydroxide sludges from plating works and the mercury-containing sludges from chlorine manufacture, contain toxic components and thus are unsuitable for direct disposal to landfill sites. For these intractable sludges chemical fixation methods have been devised to immobilize the species in a solid matrix. Such an option is relatively expensive but is a requirement for some waste solids. Organic sludges from biological processes are often a nuisance because of their large volume and tendency to become malodorous. In order to keep the size of a wastewater treatment plant to a minimum, aerobic processes with a high loading rate are installed and the resulting waste sludge is incompletely digested. The yield of sludge solids is often approximately 1 kg per kg BOD_5 removed and is wasted at a high moisture content (0.5–2% solids). The provision of sludge management for the wastewater treatment plant has to be considered as an integral part of the system rather than as a minor addition. The increasing costs of transporting waste sludges and more stringent regulation of disposal to landfill will increase the importance of sludge management.

Byproduct recovery, whether by converting high BOD effluents into methane gas or into a saleable product, or by recovering higher proportions of oils, solvents, metals or foodstuffs, must be advantageous. However, the potential for byproduct recovery is often limited by the mixed and erratic nature of the industrial wastewaters. Foodstuffs can be recovered from several food and beverage industries, but the discharge of chemicals such as detergents, alkalies, acids and disinfectants can devalue the product. Hence, while byproduct recovery is feasible in many new, well designed plants, it may be very restricted where the operation is long-established.

In fact, it is often found that major reductions in effluent loads can be achieved by simple flow inventories and improved 'housekeeping' rather than by very sophisticated recovery methods.

Conclusion

Industrial wastewaters are subject to wide variations of both flow and load. Moreover, rarely do any two factories produce similar effluents: no two tanneries, for instance, process the same mixture of hides by the same tanning methods, so their effluents cannot be similar. As manufacturing plants become older they tend to produce more polluted wastewaters as there is greater leakage from the manufacturing operations. The older manufacturing plants are likely to have been designed to use more water and power, to encourage the discharge of materials to the drainage system and to have been subject to modifications, all of which produce a wastewater which will be more costly to treat. These difficulties, combined with management attitudes which class pollution control equipment as 'non-productive', make it difficult to provide reliable effluent quality.

It is only by adopting an integrated approach to combine the manufacturing and pollution control operations that an industry can be fully optimized. In this way the requirements for effluent quality can be constructive, in that they encourage good housekeeping within the manufacturing area, provide an impetus to recycle and recover materials and minimize the environmental impact of the whole operation.

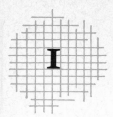

1 The treatment of wastes from the sugar industry

K E McNeil, *Sugar Research Institute, Mackay, Australia*

1.1 Introduction

This chapter reviews the origin and treatment or disposal of wastewaters from beet sugar factories, cane sugar factories, sugar refineries and distilleries which produce ethyl alcohol (ethanol) from molasses or cane juice.

Sugar beet and cane sugar factories have some similar problems in terms of water management and wastewater treatment. Firstly, much of the high water content of sugar beet and sugar cane is recovered during the evaporation of juice, resulting in a surplus of water even with extensive reuse. Secondly, large quantities of water are used for cooling of barometric condensers and the prevention of sugar loss to cooling-water during evaporation is imperative. In addition, a high standard of housekeeping is required because of potential losses of processing liquors of high BOD to effluent. Finally, the seasonal operation of both industries introduces problems, particularly with the start-up of intensive biological treatment plants.

The liquid effluent from beet sugar factories is dominated by an excess of water from the transportation and washing of sugar beets. Effluent from cane sugar factories can be surplus hot water, which may be polluted overflow from a spray pond or cooling tower where recirculation of condenser cooling-water is practised, and large volumes of cane wash-water in those countries where cane is washed. The effluents from beet and cane sugar factories contain mainly organic substances with only low levels of colouring and inorganic substances. The organic matter may be equivalent to that discharged from cities of up to 30 000 people.

Distilleries have an entirely different waste problem. The fermentation residue or stillage contains all the nonfermentable organic substances, colour and inorganic substances from the initial substrate. A large distillery has a pollution loading equivalent to that from cities of over 500 000 people. The volume of waste produced is over ten times the volume of alcohol produced. As waste disposal will be one of the major problems from large-scale production of ethanol from 'renewable' resources, considerable research is being undertaken on this matter at the present time.

1.2 Production and refining of raw sugar — sources of wastewater

1.2.1 Processing operations

Processing of sugar beet

The growing of sugar beet is confined mainly to the USA, Canada and Europe, including the Soviet Union which is by far the largest producer. The beet, *Beta vulgaris*, is a deep-rooted plant which is best suited to the northern latitudes (McGinnis, 1971). In 1978, world production of raw sugar from beet was 36.3 million tonnes.

In Western Europe, yields of beet in the early 1980s average 35.36 tonnes per hectare and contain 13.61% sugar, thus providing 4.84 tonnes of sugar per hectare (Reinefeld, 1979). Here, the average factory processes about 4000 tonnes per day of beet and there is an increasing trend towards factories processing beet to white or refined sugar (Reinefeld, 1979).

Sugar beets are mechanically harvested and processed during a short campaign, the length of which depends on geographical location. European campaigns, which occur in late autumn and early winter, last for 70 to 90 days, whereas those in the USA vary from 7 months in cooler states to 9 months in warmer areas (Train *et al.*, 1974a).

In cool climates, beets must be harvested before there is danger of frost damage and before they can be frozen into the ground. Thus stock-piling of beets for periods of up to 180 days is common. However, in California and other warm areas the time between harvest and processing must be less than 20 hours (McGinnis, 1971; Fischer and Hungerford, 1971). Losses of sugar during the piling operation are minimized by removal of mud and trash by mechanical means (McGinnis, 1971), by multijet washers or by ventilation systems (Reinefeld, 1979).

A simple flow diagram of a beet sugar factory is shown in Fig. 1.1. Beets are normally conveyed into the factory by means of a flume, although dry conveyors are sometimes used. Flumes, made of steel plate or concrete, are 1 m to 4 m in depth (McGinnis, 1971). The beets are then cleaned by being passed through beet feeders, sand separators and other mechanical apparatus before being washed. The surplus of water from the beet transport (fluming) and wash operations constitutes the major effluent from modern sugar beet factories.

Following the wash operation, the sugar beets are sliced into thin strips known as cossettes. The extraction of sugar is effected by a process called diffusion in which the cossettes are extracted by a countercurrent flow of water

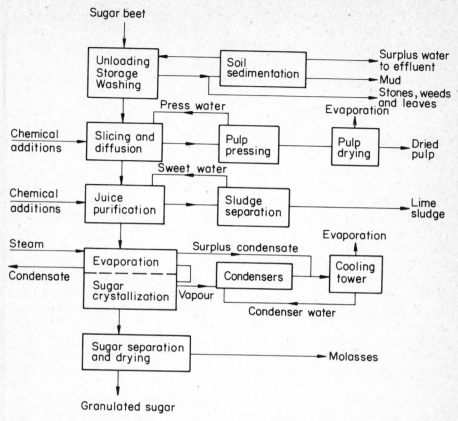

Fig. 1.1 Production of raw sugar from sugar beets (Smith, 1973)

at about 70°C. The cossettes are heated in hot juice before being conveyed to the extraction equipment.

There are four principal types of diffusion equipment (Reinefeld, 1979): the RT-diffuser, which is an extraction drum; the DDS-diffuser, which is an extraction trough; cross-flow extractors; and extraction towers. All modern diffusers are continuous in operation. The extracted cossettes (or pulp) are dewatered in pulp presses to 76–84% moisture and dried in a pulp drier. The press water is reheated and returned to the diffuser.

Following extraction, the juice is heated to 80°–85°C. The juice purification involves a first and a second carbonation. The first stage involves addition of lime and the bubbling of carbon dioxide to precipitate lime and calcium carbonate crystals within a sludge of precipitated impurities. The sludge is separated in a clarifier. The second carbonation removes the remaining lime in the juice as a further precipitate of calcium carbonate crystals. The sludge is dried on filters and disposed of.

The purified or thin juice is evaporated in a series of up to five evaporators — a multiple-effect system. In such a system, the steam evaporated from the juice is used to boil the succeeding evaporator. In this way, the juice is concentrated from 10–15% to 50–65% dissolved solids (McGinnis, 1971).

Modern beet sugar factories now purify thin and thick juice to a greater degree for the production of a high-grade or 'white' sugar.

The process of sulphitation or passage of sulphur dioxide into the thin juice stream for colour removal is used extensively (Silin, 1964; McGinnis, 1971). Filtration with kieselguhr, candle filters or centrifuges (Madsen *et al.*, 1978) is also used.

Ion-exchange resins are becoming a feature of many sugar factories. According to Landi and Mantovani (1975), common processes include methods for decalcification of second carbonation juices, for deionization of sugar juices (including the Quentin process) and, to some extent, for decolorization of juices and syrups. Disposal of wastewaters from the regeneration of ion-exchange resins has assumed significance in recent years (Bidan and Heitz, 1970).

Crystallization involves boiling the juice in vacuum pans to supersaturation and seeding with sugar. The combined sugar crystals and liquor are referred to as massecuite. The sugar crystals are centrifuged and the recovered liquid is reboiled for the recovery of an intermediate grade of sugar. The final low-grade boiling requires the use of crystallizers to extend the crystal growth time. These are cylindrical and horizontal tanks with slow-moving horizontal stirrers. The residual liquor, after recovery of the low-grade sugar, is final molasses.

A limited number of factories recover sugar from molasses using the Steffen or calcium saccharate process. The molasses is diluted with water to 6% (McGinnis, 1971) and calcium oxide is added within a cooled precipitator (below 24°C) to precipitate calcium saccharate. The calcium saccharate is filtered and washings are reheated to 85°C, to precipitate more calcium saccharate, and refiltered. The combined filter cakes are slurried and returned to the raw juice at the first carbonation stage. The hot filtrate is referred to as Steffen waste. The process recovers 97% of the sucrose in molasses. The waste contains sodium and potassium salts, betaine and amino acids. The effluent is difficult to dispose of and has been a contributing factor to a decline in the number of Steffen processes in the USA (Fischer and Hungerford, 1971). A detailed review of the process is presented by Hartmann (1974).

The refining process increases the sucrose content of sugar crystals to 99.9%. The basic process consists of the following stages (Silin, 1964):

(i)   dissolution of granulated sugar
(ii) purification by filtration (bone char)
(iii) boiling and crystallization
(iv) centrifugation and drying.

The basic refining processes are very similar to those described in more detail for cane sugar refining (see below). Unlike cane sugar refining, beet sugar refineries are invariably closely associated with the beet sugar factories and the same methods are employed for treatment of wastes.

Processing of raw sugar from sugar cane

Sugar cane, a grass of the genus *Saccharum*, has been developed into many varieties grown in tropical and subtropical countries. World production of raw sugar from sugar cane in 1978 was nearly 56 million tonnes.

The harvesting and processing season for sugar cane is generally in the cooler and drier months of the year. The length of the season varies greatly among cane-growing countries. The season lasts four months in Louisiana, five to six months in Australia and ten to twelve months in the Philippines and Hawaii.

The harvesting of cane is carried out by simple hand cutting, by machine cutting of whole stalks or by mechanical harvesters which remove the tops and cut the stalks into short lengths known as billets. In Hawaii and areas of mainland USA the cane is pushed and cut by bulldozers, a process which necessitates large washing facilities at the mills to remove extraneous matter which can be 70% of the total mass.

In some countries the cane is burnt before harvest. The method of transport of cane to mills varies from ox-cart to railroad systems and the length of time between cutting and processing varies greatly. In Australia, where the whole crop is chopper-harvested, this period must be maintained at less than 15 hours if cane deterioration and sugar loss are to be avoided.

The cane may be washed on arrival at the factory and then passes through either a milling or diffusion process for juice extraction. Various modifications of these processes are described by Meade and Chen (1977). The residual fibre (or bagasse) after juice extraction is used as fuel for steam generation. The overall factory process is illustrated in Fig. 1.2.

Milling involves chopping the cane by means of revolving knives and/or shredding it in a swing-hammer-type shredder. The cane then enters the first mill which expresses 40% to 75% of the juice. The mixed juice from the first and second mills passes to subsequent processes. Maceration or imbibition water is applied to the bagasse proceeding to the final mill. The final mill juice is applied to the bagasse feeding the preceding mill, and so on, back to the number two mill.

Diffusion requires the preparation of cane by shredding and/or by crushing in a mill. Either cane diffusion or bagasse diffusion may be used. In cane diffusion, all of the prepared cane is passed into the diffuser. With bagasse diffusion, up to 75% of the juice is extracted by one or two mills and the residual bagasse enters the diffusers. The bagasse is then dewatered by one or two mills.

Following extraction, the juice is limed to between pH 7.5 and pH 8.5 and is

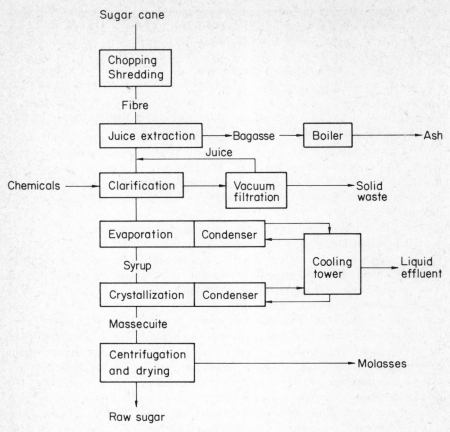

Fig. 1.2 Raw sugar manufacture from sugar cane

then heated to boiling point. This causes the precipitation of calcium phosphates, denatured organic complexes and insoluble matter such as soil and fine fibre particles which settle from the juice in a clarifier. Polyelectrolytes can be used to improve settling rates and juice clarity. In some countries carbonation or sulphitation processes are used for juice clarification. Juice associated with the insoluble matter (mud) is recovered by vacuum rotary filters and the solids are returned to farms as fertilizer. Some factories slurry the mud and pump it to special ponds or include it with effluent for irrigation.

Clarified juice is then concentrated by evaporation in multiple-effect evaporators. This increases the juice solids concentration from 15° Brix to 65°–70° Brix. The concentrated juice is then called syrup. Crystallization is accomplished batchwise in single-effect vacuum evaporators or vacuum pans. The syrup is converted to a mixture of crystalline sucrose and mother liquor (massecuite). The crystalline sugar is recovered by centrifugation. Three separate stages of boiling are required. After the third stage of boiling, crystallization is completed in large tanks or crystallizers, with slow stirring, until the mother liquor is exhausted of crystallizable sucrose. The recovered liquor is final molasses. The raw sugar crystals are dried in large rotating drums prior to shipment to refineries.

Refining of cane sugar

The refining of cane sugar involves processes designed to improve the quality of cane sugar as a foodstuff for direct consumption and for food and beverage manufacture and to improve its storage potential. The sucrose level is increased from over 97% to in excess of 99.9% by removal of solid particles, organic and inorganic substances and micro-organisms. Refineries are usually located in large cities remote from the growing areas where raw sugar is manufactured.

The refining process for production of crystalline sugar is shown by the simple flow diagram in Fig. 1.3. The affination process involves washing the raw sugar free of surface molasses and other impurities with syrup (recovered after the affination process). Raw sugar and hot syrup at about 72°C are mixed, or cold magma is heated in a coil mixer (Abram and Ramage, 1979). The magma is centrifuged to recover crystals which, in turn, are washed with hot water. The latter can consist of surplus condensate or low-density process liquors. The supernatant wash-water passes to vacuum pans for sugar recovery.

The crystals from the affination process and recovered sugar from the wash-liquid are then dissolved or 'melted' in water to about 65° Brix. After coarse screening to remove sand or scale, the melt liquor then undergoes defecation or clarification. This is accomplished by processes involving carbonation–filtration or phosphatation–flotation.

Phosphatation involves the addition of lime and phosphoric acid to form a calcium phosphate–impurity complex which can be air-floated to the surface of a clarifier. The scum (or mud) passes to a secondary clarifier or centrifuge for further sugar removal (desweetening). Carbonation involves mixing milk of lime and melter liquor followed by carbon dioxide gassing to precipitate calcium carbonate and impurities. The defecated liquor is then clarified by pressure filtration.

Decolorization is accomplished by either bone charcoal or activated carbon. Bone charcoal is capable of removing most impurities including divalent cations and anions and organic ions but does not remove monovalent ions. The bone char is stored in cisterns containing about $30 \, m^3$ of charcoal. The clarified liquor is fed down through the cisterns at about 75°C. When colour removal falls, the cistern is washed with hot water and the sugar is recovered by evaporation. When the solution has a Brix value of 0.5°–1.0° it is uneconomical to recover (Abram and Ramage, 1979) and becomes a wastewater. The char is regenerated in kilns.

Granular activated carbon adsorbs very little ash and its use does not produce as much waste wash-water as does the use of bone char (Train *et al.*, 1974b).

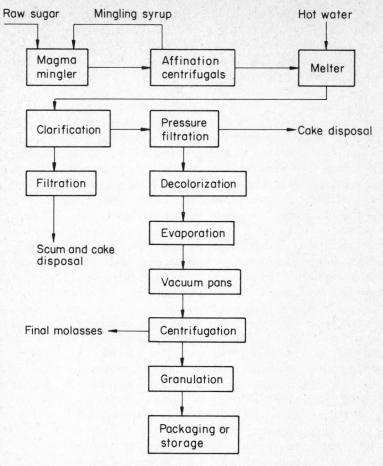

Fig. 1.3 Manufacture of crystalline refined sugar (Train *et al.*, 1974b)

Powdered activated carbon is used to some extent in small refineries and liquid sugar manufacture. Ion-exchange resins are sometimes used for removal of ash and additional colour.

The liquor is then evaporated in multiple-effect evaporators resulting in an increase from 65° Brix to 74°–75° Brix (Abram and Ramage, 1979). The condensation of the last effect vapours is achieved by contact with river or sea water. Losses of sugar due to entrainment in the vapours are a source of pollution. Sugar is crystallized in vacuum pans, recovered by centrifugation and dried.

A small number of refineries manufacture liquid sugar. Liquid sugar can refer to melted granulated sugar that has had some colour removed by powdered carbon or to high-grade liquor that has had full colour removal but

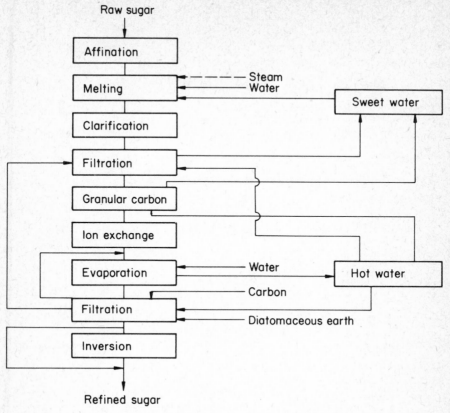

Fig. 1.4 Manufacture of liquid refined sugar (Train *et al.*, 1974b)

no crystallization (Meade and Chen, 1977). Liquid invert is produced by acid inversion of liquid sugar in strong acid cation-exchange resins. A flow scheme for a liquid sugar refinery is shown in Fig. 1.4.

1.2.2 Water use and sources of wastewater in a beet sugar factory
The principal water circuits in beet sugar factories are:
(i) the transport and wash-water circuit;
(ii) the barometric condenser cooling-water circuit;
(iii) water used or generated in processing including condensate, press-water for diffusion, ion-exchange circuits and Steffen filtrate;
(iv) cooling-water for turbogenerators, pump bearings and crystallizers; and
(v) water used in conveying beet pulp and lime sludges.

Figure 1.5 illustrates water flow for a factory employing maximum recirculation of water.

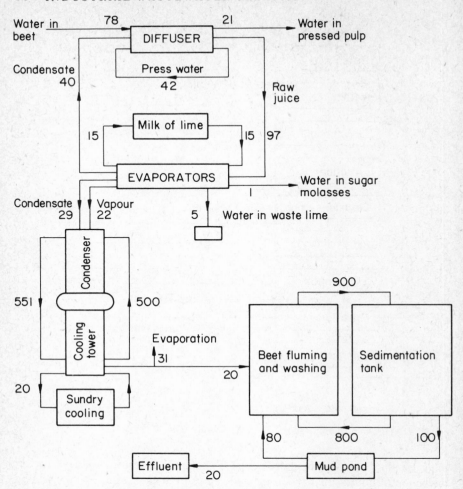

Fig. 1.5 Water flow in a beet sugar factory which employs maximum recirculation (Crane, 1968). Numbers denote tonnes per 100 tonnes beet

The total water requirement for processing a tonne of beet is $9-19\,m^3$ (Delvaux, 1974; Reinefeld, 1979). Until recently, most beet factories operated on a 'once-through' basis with very little recirculation of water. Antipollution legislation, modern technology and increased demands on diminishing water supplies have resulted in almost all factories adopting recirculation of process-water, transport-water and wash-water.

Recirculation and careful water management can reduce the surplus wash-water to $0.5-1\,m^3\,t^{-1}$ of beet (Heitz and Bidan, 1970; Delvaux, 1974; Reinefeld, 1979). However, these reductions in volume result in a concentrated effluent having a BOD as high as $4000-5000\,mg\,l^{-1}$ (Reinefeld, 1979).

A primary objective is the reduction or elimination of raw water intake. In the past, this has been estimated at between 1.25 and 25 m^3 t^{-1} of beet for factories in the USA (Train et al., 1974a). However, the high water content (78–80%) of the incoming beet makes impossible the complete elimination of water from a beet factory. Crane (1968) estimates that of this initial 78%, 31 parts would be lost by evaporation in cooling towers, 20 parts would be evaporated from pulp during drying, 5 parts are present in lime sludges and 2 parts are present in sugar, molasses and pulp. Thus, a minimum of 0.2 m^3 of water per tonne of beet must be disposed of. The maximum use of the water that is condensed after evaporation is necessary. At least 50% of the water in beets is condensed after evaporation and at least 20–30% of this is recoverable as surplus condensate (De Vletter, 1972).

Table 1.1 shows the volume and BOD of potential effluents from a beet factory.

Continuous diffusers allow complete recirculation of all pulp press-water to the diffuser. Pollution from the diffusion operation is now reported to be negligible (Reinefeld, 1979). The drying of pulp for animal feed has eliminated problems of drainage from wet pulp silos. Surplus condensate from the heating and evaporation of juice is high-quality hot water that can be used for boiler feed-water, diffuser supply water, washing of lime cake precipitate, centrifugal wash-water, factory cleaning and remelting of sugar (Fischer and Hungerford, 1971; Delvaux, 1974).

Lime cake from rotary–vacuum filtration of carbonation sludges is thixotropic. It contains 45–50% moisture and the solid matter comprises inorganic matter (43–48%), organic matter (6–7%), sugars (0.4–1.3%) and nitrogen (0.25–0.35%) (Heitz and Bidan, 1970). The established practice has been to slurry the cake and transport it to a separate storage basin for evaporation and seepage of the water. Technological advances have facilitated the transport of these sludges with minimum quantities of water (Schwieter, 1963; Lührs, 1963). Madsen et al. (1979) have recently developed a new filtration procedure for increasing dry substance to 67–69%. Processes for recalcification of dried lime cake in furnaces for lime recovery are being developed (Kominek et al., 1970; Schiweck et al., 1979). The organic matter is burnt off at temperatures over 900°C, and calcium carbonate is converted to calcium oxide and then slaked.

The most successful means of disposal of Steffen waste has been lime removal by carbonation and subsequent concentration to 60% solids by multiple-effect evaporation (Hartmann, 1974), then land disposal or use as an animal feed supplement. Ion-exchange resin eluants have proved to be a problem because of high salt content. A partial answer is irrigation and the development of improved procedures for resin regeneration (Heitz and Bidan, 1975). Many factories using the Quentin process simply allow the waste to mix with other effluent for discharge (Vellaud, 1978).

The cooling-water required for the barometric condensers on evaporators

Table 1.1 Volume and BOD of potential wastewaters from sugar beet processing

Source of water	Volume ($m^3 t^{-1}$ beet)	BOD ($mg\,l^{-1}$)	Country	Reference[a]
Transport-water and wash-water	9	150	UK	1
	5–10			2
	9.9	210	Canada	3
	5–8	300–5000	France	4
	8.8		USA	5
	5–17	6000–7000	USA	8
	10–12	200–1400	W. Germany	9
	8.5–10	[b]	W. Germany	10
Condenser cooling-water	5.4	45	UK	1
	4–8			2
	1.54–3.3	22–114	W. Germany	7
	4–6	[b]	W. Germany	10
	7.5	40	Canada	3
	4–10	5–30	France	6
	8		USA	5
	5.4–18.8	25	USA	8
Condensate	0.1–0.3		W. Germany	10
	1.63–1.97	48–65	W. Germany	7
	0.02–0.9	80	France	4
Cooling-water and sealing-water	0.2–1.0		W. Germany	10
Lime sludges	0.4–0.8	5000–10 000	France	6
	0.2		USA	5
	0.35	8600	Canada	3
	0.04–0.4	1060–278 000	USA	8
	0.05–0.1		W. Germany	10
Lime pond effluent	0.1	1750	UK	1
	0.29	1420	Canada	3
Decalcification	0.03–0.05	100	France	4
Ion-exchange (cationic)	0.3	200–2000	France	6
Ion-exchange (anionic)	0.2–0.3	200–4000	France	6
Steffen filtrate	0.34	13 250[c]	USA	5
	0.5	10 200	USA	8

[a] 1. Crane (1968); 2. Delvaux (1974); 3. Black and Teft (1965); 4. Heitz and Bidan (1970); 5. Fischer and Hungerford (1971); 6. Bidan and Heitz (1970); 7. Schneider *et al.* (1961); 8. Train *et al.* (1974a); 9. Beelitz (1971); 10. Hoffmann-Walbeck (1977).
[b] The BOD loading for combined transport-water and condenser cooling-water was $650\,g\,t^{-1}$.
[c] Calculated from data in reference 5.

and crystallizing pans is about $5-5.4\,m^3\,t^{-1}$ of beet (Crane, 1968; De Vletter, 1972). This water contains carbon dioxide, ammonia ($\sim 10\,mg\,l^{-1}$), volatile amino acids and sugars lost as a result of entrainment (Heitz and Bidan, 1970). Typical BOD values are from $15\,mg\,l^{-1}$ (Offhaus, 1965), ranging up to $20-50\,mg\,l^{-1}$ (Crane, 1968; De Vletter, 1972; Smith, 1973). The temperature of the returned water is $40°-55°C$.

Past practice was to discharge these waters directly to the river, i.e. factories operated on a single-pass basis. However, there is an increasing tendency to recirculate condenser water after cooling in spray ponds or cooling towers which also can reduce BOD_5 by about 50% (Slijkhius van der Haarst and van der Toorn, 1972). Sugar losses to cooling-water are minimized by the use of entrainment arresters. The mixing of surplus condensates with cooling-water adds approximately $0.2-0.25\,m^3\,t^{-1}$ of beet (De Vletter, 1972; Delvaux, 1974). Surplus water from the condenser circuit can be used as wash-water for filters (Heitz and Bidan, 1970), as spray-water on beet washers and for factory cleaning (Crane, 1968), as flume-water make-up (Brenton, 1972), or as supply-water for gas washers, carbon dioxide compressors, pump bearings and vacuum pumps (Åkermark, 1975).

The quantity of water required for fluming and washing of beet has been estimated as $4.5-15\,m^3\,t^{-1}$ in factories in the USA (Fischer and Hungerford, 1971) and as $7-10\,m^3\,t^{-1}$ in Europe (Crane, 1968; De Vletter, 1972). Delvaux (1974) estimated that the separate requirements were $4-8\,m^3\,t^{-1}$ for transport-water and $1-2\,m^3\,t^{-1}$ for wash-water.

One prime factor in the quantity of water used is the soil level (or dirt tare) associated with the beets. The quantity of soil introduced with sugar beet varies from area to area and is much higher in wet weather. In the USA the soil adhering to beets comprises 3–4% of the gross weight in dry seasons and in areas with light sandy soil but may exceed 10% in wet areas (Fischer and Hungerford, 1971). An average level of 12% has been reported in the UK (Crane, 1968). However, in wetter parts of Europe levels of 20–50% and even 80% are common (Heitz and Bidan, 1970; Delvaux, 1974). Crane (1968) reported that beet containing 8–25% dirt tare produces 10 to $30\,g\,l^{-1}$ solids in transport-water.

The maximum workable soil level in transport-water has been estimated at $300\,g\,l^{-1}$ (Heitz and Bidan, 1970). Thus, a continuous rejection of some transport-water is required to avoid excessive levels of solids. This wastage should be almost the total reject water from a beet sugar factory employing complete recirculation, i.e. $0.5-1\,m^3\,t^{-1}$ of beet.

As a first step in maintaining the recirculation system, attention must be given to minimizing damage to beets during mechanical harvesting and before processing, so as to leave most dirt in the field (Delvaux, 1974). Damage to beets results in considerable losses of sugar. The major losses occur during washing, with little loss during transport (Leclerc and Edeline, 1960; Oldfield *et al.*, 1972). Beets are present in transport-water for a maximum period of only three minutes whereas the wash period may be up to twelve minutes (De Vletter, 1972). Some workers therefore recommend separate circuits for transport-water and wash-water (Heitz and Bidan, 1970). Thus, transport-water would be composed primarily of mud solids and wash-water would contain the most sugar. Sugar losses by diffusion in a completely enclosed wash-water circuit could then be restricted as the sugar content rose (Blankenbach and Willison, 1969; Heitz and Bidan, 1970).

Sugar losses during transport and washing have been estimated at 0.05% (Leclerc and Edeline, 1960), 0.17–0.37% (Schneider, 1963) and 0.2% (De Vletter, 1972) on beet and as 100 to 300 mg l^{-1} in the water circuit (Oldfield et al., 1972). Oldfield et al. considered that the sugar losses from the cut face exposed by topping were minor and that greatest losses occurred as a result of abrasion and bruising of the beet. De Vletter (1972) estimated sugar loss of 2–3 mg cm^{-2} through an exposed beet slice washed for 4 min at 10°–20°C.

De Vletter and Wind (1975) propose the following formula for calculating sugar losses between two points (A and B) in the transport-water and wash-water circuit using the COD test:

$$F (COD_A - COD_B) = 1.35S$$

where F = flow (m^3 d^{-1})
S = sugar loss (g d^{-1}).

Numerous modified schemes for recycling flume-waters have been adopted (see, for example, Force, 1965; Nielsen, 1968; Crane, 1968; Blankenbach and Willison, 1969; Brenton, 1972; De Vletter, 1972; Slijkhuis van der Haarst and van der Toorn, 1972; Delvaux, 1974; Åkermark, 1975; Brunner, 1977).

After transportation of the beets, the water is screened of tops, leaves, roots and other particles. The screens usually have a minimum slit width of 2 mm (De Vletter, 1972). Crane (1968) describes the use of wedge-shaped bars of 3 mm aperture, whereas Blankenbach and Willison (1969) screened 19 000 m^3 d^{-1} of transport-water by means of a travelling water screen and four 1.2 × 2.4 m vibrating screens in series. Force (1965) describes the use of a liquid cyclone for grit removal followed by vibrating screens, 1.5 × 3.6 m, for the removal of organic matter and Brenton (1972) refers to the use of Dorr–Oliver parabolic screens.

Following screening, the water passes to a vessel for removal of soil. The vessel is commonly a sedimentation basin or clarifier; hydroclones are occasionally used but are sensitive to abrasion (Heitz and Bidan, 1970).

The choice of whether to use a basin or a clarifier depends very much on the available land near the factory and relative costs. Holding basins sited too close to a factory can cause odour problems. However, pipeline costs to more remote ponds, difficulties in finding land at reasonable cost, problems due to microbial activity and the potential pollution of groundwater are inherent disadvantages of sedimentation basins.

Sedimentation basins may provide a residence time of up to 24 h and include basic rectangular, canal or U-shaped types (Delvaux, 1974). Phipps (1960) reported that the most desirable shape was rectangular with a length-to-width ratio of 5:6. The desired circulation rate through the basin was in excess of 25 cm min^{-1}.

The system described by Brenton (1972) had two alternate primary basins for settling the greater proportion of mud followed by a secondary basin for settling finer particles. Primary and secondary settling basins should be provided if a litre of flume-water contains a settleable volume of particles in excess of 25 ml after two hours settling (Edeline and Leclerc, 1957).

The long retention times required in sedimentation basins can permit the development of anaerobic bacteria. As a result, odours develop and gases retard the settling of sludge. Organic matter becomes solubilized and causes heavier pollution of recycled water. The potential for anaerobic activity can be reduced by segregation of highly polluted wastewater from transport-waters, particularly those containing sugar and ammonia nitrogen (Heitz and Bidan, 1970; Delvaux, 1974). Dead spaces in the circuit can be avoided by circulating the flume-water at speeds of up to 0.1 m s^{-1}.

Mechanical clarifiers or thickeners usually provide a 1–4 h residence time (Smith, 1973). Modern installations in the UK are circular concrete tanks 30–60 m in diameter, having an inverted conical base (Crane, 1968). In France, diameters of 20–40 m are common (Heitz and Bidan, 1970). Slijkhuis van der Haarst and van der Toorn (1972) describe two types of clarifier in use in the Netherlands: clarifiers in which mud is scraped to a central cone and pumped out by an external pump usually achieve an underflow solids content of 10–12%; other tanks with a pump attached to the rotating scraping equipment are able to achieve solids levels of up to 20%.

The size of clarifier required depends on the flow, nature of the soil and dirt tare. Force (1965) used two clarifiers of 19 m diameter and 3.6 m high to handle an individual flow rate of 9.3 m^3 min^{-1}. In the UK, clarifiers are used to reduce the solids concentration from 10–30 g l^{-1} to 400 mg l^{-1} in the recirculated water (Crane, 1968). In France, water loaded with 50–60 g l^{-1} of soil is reduced to several grams per litre in the recirculated water, with the underflow solids concentration varying from 250–300 g l^{-1} (Laguerre, 1970). Fordyce and Cooley (1974) describe the operation of a clarifier of 35 m diameter and 2.4 m deep, handling 21 m^3 min^{-1} of transport-water. For an assumed mud tare of 5% from 4000 t d^{-1} of beet, the underflow solids concentration was 80 g l^{-1}. The operating parameters were daily hydraulic overflow 31.6 m^3 m^{-2}, solids loading 188 kg d^{-1} m^{-2}, daily hydraulic underflow 2.3 m^3 m^{-2}, daily weir loading 274 m^3 m^{-2} and detention time 1.9 hours. Zama *et al.* (1979) describe a 52 m clarifier used in Italy for a 6000 t d^{-1} factory.

Solids settling, control of microbial growth and control of corrosion by acid-forming bacteria are assisted by the addition of lime to maintain pH between 10 and 11. The dry lime requirements have been variously estimated at 1.0 (Nielsen, 1968), 1.1 (Brenton, 1972), 1.6 (Åkermark, 1975) and

0.3–2.5 kg t^{-1} beet (De Vletter, 1972). Other workers recommend adjustment of pH to 6.8–7.5 to minimize scale formation in pumps, in pipelines and on screens (Smith *et al.*, 1975). Saponins in beet promote frothing. This problem becomes more acute when lime is added to raise pH (Heitz and Bidan, 1975). Delvaux (1974) has suggested that liming to high pH is undesirable where subsequent biological treatment of excess flume-water is required. In his opinion, the encouragement of some microbial activity in flume-water by control of pH between 6 and 7 is a useful means of preventing excessive build-up of organic matter in flume-water.

Several factories use polyacrylamide flocculants to improve settling (Laguerre, 1970; De Vletter, 1972; Brunke and Voigt, 1974 and Åkermark, 1975). Devillers *et al.* (1969) noted that settling rates of sludge containing 100 g soil l^{-1} increased from 1.6 cm min^{-1} to between 4 and 30 cm min^{-1} for 1 mg l^{-1} of flocculant. At the same time the supernatant solids level decreased from 2.5 g l^{-1} to values ranging from 76 mg l^{-1} to 1 g l^{-1}. Roche (1969) observed that above flocculant addition of 1 mg l^{-1}, settling rates did not increase markedly. Evidence was presented which indicated a retarding effect of polymers on biological purification of treated waters.

Fordyce and Cooley (1974) studied the effect of polyelectrolytes on transport-water, pH 11.4, containing 8% solids. The water was fed to a pilot clarifier at 15 l min^{-1}, providing a retention time of 10 min. The required dose of anionic flocculants was 3–5 mg l^{-1} in order to achieve underflow solids levels of 166–290 g l^{-1}.

The mud in sedimentation ponds or from the underflow of clarifiers is pumped to mud storage ponds. The problems of disposal of this mud are becoming increasingly acute in densely populated countries. De Vletter (1972) estimated that 4 ha is required for 1000 t daily capacity. The supernatant liquor in storage ponds can be purified by biological treatment and returned to flume-water (Delvaux, 1974; Brunner, 1977). Fordyce and Cooley (1974) studied methods for improving the disposal of solids in the underflow from clarifiers. Vacuum filtration of the underflow was achieved at cake rates of 24–48 kg m^{-2} h^{-1}, provided that the feed solids level was 10% or preferably 20% by weight. Best results were obtained if the underflow was preheated to 80°–90°C and pH was increased to 11.

After decantation, recycled flume-water requires make-up with either fresh water or, preferably, water from elsewhere within the factory. In cold weather, the addition of warm barometric condenser-water has the advantage of preventing freezing of flume-water (Blankenbach and Willison, 1969) and thawing frozen beets (Fischer and Hungerford, 1971). Surplus steam condensates are also commonly used (De Vletter, 1972). The use of hot water make-up can result in warm flume-water with consequent higher microbial activity and increased sugar loss. The solution is to cool the water for use as bearing cooling-water and pump sealing-water and then use it as make-up for flume-water (Delvaux, 1974).

As mentioned earlier, the main effluent from a beet sugar factory is surplus transport-water and wash-water. The concentration of organic matter in this water depends to a great extent on the raw water intake. Factories having a high dirt tare tend to introduce more raw water and thus dilute the soluble organic matter which has been derived from sugar losses into the wash-water. In addition, the concentration of organic matter increases from the commencement of the campaign and reaches a plateau after one to two months. Heitz and Bidan (1975) quote a plateau of 2500–3000 mg l^{-1} of BOD in recycled transport-water in France, whereas a value of 6000–7000 mg l^{-1} was described as being typical for factories in the USA (Train et al., 1974a).

According to De Vletter (1972), factories with a sugar loss of 0.1% on beet would have a COD in flume-water of 6000–7000 mg l^{-1} or a BOD of 3500–4000 mg l^{-1}. Hoffmann-Walbeck and Pellegrini (1975) estimated the effluent volume from a factory with recirculation as 1.09 m^3 t^{-1} of beet with a BOD loading of 16.5 kg or a COD loading of 33 kg for each 1000 t of beet processed.

Reinefeld et al. (1975) studied the relationship of COD to BOD for beet factory wastes in varying stages of degradation. Above BOD$_5$ loadings of 700 mg l^{-1}, the COD was equal to about 1.5 times the BOD. However, the ratio of COD to BOD rose to 10 as the waste was progressively stabilized.

A detailed composition of beet factory effluent as described by Oswald et al. (1973) is given in Table 1.2.

1.2.3 Water use and sources of wastewater in a cane sugar factory

Water usage and the nature of the effluents disposed of vary greatly from country to country and also within individual countries.

The washing of cane is a major source of wastewater from factories in Hawaii, Louisiana, Brazil and some factories in Puerto Rico (Monteiro, 1975; Train et al., 1975) yet it is not practised at all in some other countries. Historically, factories located in dry areas or with limited supplies of fresh raw water have applied more stringent water economies. Thus, many factories have had to recirculate condenser cooling-water by means of cooling towers or spray ponds irrespective of requirements for reducing pollution. On the other hand, factories in dry areas have greater potential access to irrigation schemes, thus allowing less stringent control of wastewater discharges. Sugar factories in Hawaii which discharge wastewater to irrigation tend to have high volumes of discharge (Train et al., 1975).

Cane entering the factory contains 70% moisture. As a result, an excess of water has usually to be disposed of, even with the most stringent conditions of water reuse. The water cycle within a typical Australian sugar mill is described by Stuart (1976) and is shown in Fig. 1.6. This water circuit demonstrates that in the complete absence of any raw water intake about 0.08–0.1 m^3 of hot water per tonne of cane will be discharged together with a continuous flow of effluent

Table 1.2 Composition of effluent from a USA factory, after sedimentation[a] (after Oswald et al., 1973)

Parameter	Value[a]
Total nitrogen, N	16.4
Ammonia, N	6.3
Nitrate, N	2.6
Chlorides	400
Sulphate	210
Alkalinity, $CaCO_3$	538
Sulphide	0.68
Phosphate, P	3.4
Calcium	178
Magnesium	66
Sodium	222
Potassium	88
BOD, unfilt.	930
COD, unfilt.	1601
COD, filt.	1195
Suspended solids	1015
Suspended volatile solids	360
Suspended ash	655
Dissolved solids	2209
Dissolved volatile solids	1139
Dissolved ash	1070
Total solids	3224
Total volatile solids	1499
Total ash	1725
Sugar	1.25
Dissolved oxygen	0.0
Physical factors	
pH	7.06
Light penetration (cm)	45.6
Specific conductance (μmho)	300

[a] All values expressed in $mg\,l^{-1}$ unless indicated otherwise.

from the cooling tower (where recirculation is practised). The water circuit shows that almost all of the water used in a factory can be provided from the cane without the use of external raw water. However, in practice, virtually all factories have an intake of raw water. In factories in the USA, this intake varies from 3.96 to 13.4 $m^3\,t^{-1}$ of cane crushed but some is lost by evaporation and in filter cake and bagasse (Train et al., 1975). In these factories, the average overall discharge is less than the average raw water intake by an amount of 0.3–0.7 $m^3\,t^{-1}$ of cane crushed, indicating total use of water in cane together with some of the influent raw water.

The major requirements for water in a sugar factory are:

(i) cooling-water for barometric condensers;
(ii) maceration water, boiler feed-water and surplus condensates;
(iii) internal plant cooling-water;

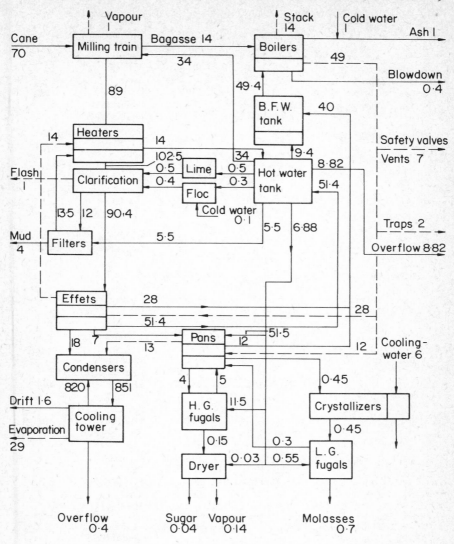

Fig. 1.6 Water flow in an Australian factory which processes sugar cane (Stuart, 1976)
——— Liquid, - - - - Vapour; Cane rate = 100 t h⁻¹ (units tonnes)

(iv) slurry-water for fly-ash and filter mud (in some instances);
(v) factory wash-water; and
(vi) cane wash-water (in some instances).

The quantities of those waters that may be wasted are shown in Table 1.3. This table is intended to show some of the variations in volume and BOD in

Table 1.3 Volume and BOD of potential wastewaters of cane sugar processing

Source of water	Volume (m^3 t^{-1} cane)	BOD (mg l^{-1})	Country/state	Reference[a]
Condenser cooling-water, spray pond overflow or cooling tower overflow	8.2–18.2[b]	10–40	Australia	1
	0.29–0.59[c]	800–2000	Australia	1
	8.7	36	Louisiana	2
	8.35–26	11–224	Louisiana	3
	13.8–21.4	6–2110	Florida/Texas	3
	5.49–15.7	6–71	Hawaii	3
	12–37	28–130	Puerto Rico	3
	—	51–312	South Africa	4
	0.66–1.49	150–350	India	5,6
	11.1	40–140	Brazil	10
Filter mud slurry	0–0.69	14 700	Louisiana	3
	0–0.282	12 600	Florida/Texas	3
	0–0.427	2140–7144	Hawaii	3
	0–1.18	15 800	Puerto Rico	3,7
	0.234	2090–3260	India	5
Surplus condensates	0.08	300–1000	Australia	1,8
Boiler blowdown	0–0.3	139	Louisiana	3
	0.007–0.035	?	Florida/Texas	3
	0.091	35	Hawaii	3
	0.004	250–300	Australia	1,8
Fly ash slurry	0.3	323	Louisiana	3
	0–0.31	?	Florida/Texas	3
	0.12–0.92	38–222	Hawaii	3
	0–0.3		Puerto Rico	3
Floor washings	0.0007–2.48	600	Louisiana	3
	0.001–0.91	?	Florida/Texas	3
Cane wash-water	0.98–20	81–562	Louisiana	3
	3.2–19.5	140–1190	Hawaii	3
	0–12.0	180–290	Puerto Rico	3
		260–700		9
	5	180–500	Brazil	10

[a] 1. McNeil (unpublished results); 2. Hendrickson and Grillot (1971); 3. Train et al. (1975); 4. Cox (1969); 5. Verma et al. (1978); 6. Aora et al. (1974); 7. Biaggi (1968); 8. Stuart (1976); 9. Meade and Chen (1977); 10. Monteiro (1975).
[b] Single-pass cooling factories.
[c] Factories with recirculated cooling-water.

some factories in various countries. However, in many instances these are not the final volumes ultimately discharged because of reuse, storage or irrigation.

The source of condenser cooling-water is either river water used in a single-pass system through the factory or recycled cooling-water from a cooling tower or spray pond. The quantity of water required by a factory depends on the temperature of the available water and also reflects overall water management within the factory. Stuart (1976) estimated the requirements for cooling-water at $8.2\,m^3\,t^{-1}$ of cane crushed. The actual usage for factories in the USA varies from 7.35 to $20.5\,m^3\,t^{-1}$ of cane (Train et al., 1975). Monteiro (1975) estimated the usage for Brazilian factories as $11.1\,m^3\,t^{-1}$ of cane.

Recirculation of condenser cooling-water became necessary for many mills as crushing rates increased relative to the capacity of adjacent rivers which provided their cooling-water. Spray ponds have been a universal feature of Indian sugar mills for many years. In Queensland there has been a gradual trend to the construction of spray ponds or cooling towers with the expansion of the industry. Here, experience has shown that the returned cooling-water from mills on large rivers has little detrimental effect if heavy wastes are separately treated. Recirculation of condenser cooling-water is practised in a different form in Louisiana and Hawaii, where it often serves as make-up for cane wash-water.

The principal pollutants of condenser cooling-water are heat, sugar and volatile substances such as ethanol. Heat is detrimental to small receiving streams. Sugar loss due to entrainment is the principal source of pollution of condenser cooling-water in most countries although in Queensland ethanol has assumed greater significance as a pollutant of condensate, and hence of cooling-water (see below). Single-pass factories in Queensland can be expected to discharge at least $3000\,m^3\,h^{-1}$ of water at 45°–50°C with a BOD_5 of $10-40\,mg\,l^{-1}$.

Control of entrainment losses requires careful operation of vacuum pans and evaporators, with provision of sufficient vapour height above the boiling liquid to allow liquid particles to fall back to the boiling liquor. Modern vessels are fitted with entrainment arresters, several types of which are in use (Meade and Chen, 1977). A louvre type (Fig. 1.7) designed by Frew (1971) has been particularly successful in Australia where sugar levels in single-pass cooling-water are now generally less than $5\,mg\,l^{-1}$.

The use of surface condensers in place of barometric condensers would prevent sugar losses to cooling-water as the vapours do not come into contact with the cooling-water. Two objections to their use are greater fouling potential and difficulties in vacuum control within the factory (Train et al., 1975).

Factories with closed-circuit cooling accumulate pollutants in the cooling tower or spray pond. The associated problems of micro-organism control, acidity in cooling-water and corrosion have to be contended with although an undetermined amount of BOD reduction does occur. The continuous overflow of polluted water with a BOD of $800-2000\,mg\,l^{-1}$ from the spray pond or cooling tower is a major source of pollution. Factories with excessive sugar losses or limited tower capacity often add raw water as a diluent. This, of course, displaces extra water to effluent.

Approximately $0.51\,m^3$ nett volume of hot condensates per tonne of cane is evaporated from juice and syrup during evaporation and crystallization and a further $0.14\,m^3\,t^{-1}$ is condensed from factory heaters (Stuart, 1976). The

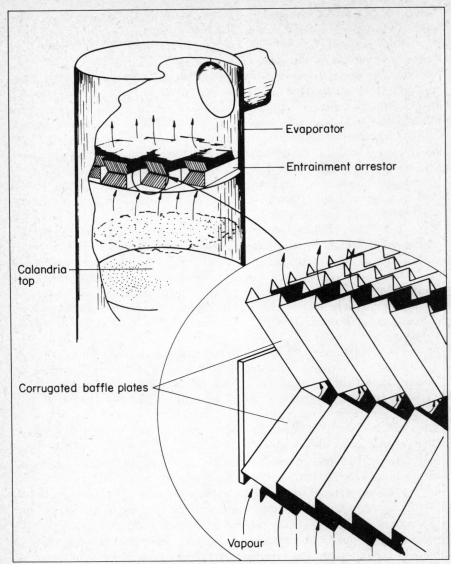

Fig. 1.7 Typical arrangement of the CSR Ltd corrugated louvre entrainment arrestor in final evaporator (Frew, 1971)

highest quality condensates are used directly for boiler feed-water. The condensed steam from the first evaporator does not come into contact with juice and is the preferred source of boiler feed-water. Condensate from vacuum pans is usually sent direct to the boiler and that from the heaters and other evaporators is pumped to a hot-water tank where it is available for maceration water, as a supplement to boiler feed-water, for wash-water for filters and centrifuges and for factory cleaning. Raw water intake can be minimized by using condensates.

Sugar in the boiler water can degrade into substances which can cause foaming and corrosion of boiler tubes. Thus, regular blowdown is necessary to maintain low sugar levels. This water contains boiler treatment chemicals and BOD due to the organic substances. However, in Australia substantial BOD levels are common, due to high ethanol concentrations (see below).

Surplus hot water has been generally considered to be pure and uncontaminated (Hendrickson and Grillot, 1971) with values of BOD of $10\,\text{mg}\,l^{-1}$ being reported (Train et al., 1975). However, under Queensland conditions, condensates and all sources of hot water derived from them are major sources of pollution (McNeil et al., 1974) with BOD values of $300-500\,\text{mg}\,l^{-1}$. The high BOD level has been shown to arise as a result of high ethanol levels (Table 1.4; Blake, 1976). Monteiro (1975) has reported BOD values of $500-1000\,\text{mg}\,l^{-1}$ in condensates from Brazilian factories although ethanol has not been specifically implicated.

Table 1.4 Composition of condensates from Queensland factories (after Blake, 1976)

Parameter[a]	Factory A		Factory B	
	No. 1 effect	No. 2 effect	No. 1 effect	No. 2 effect
pH	9.2	8.7	9.3	9.0
TOC		577	225	509
COD	1024	2487	672	2304
BOD	450	1936	813	2271
Kjeldahl, N	6.0	13.4	5.6	12.2
Amine, N	5.4	13.4	3.9	10.3
Sucrose		2.7	0.8	2.1
Ethanol	312	912		1164

[a] Concentrations in $\text{mg}\,l^{-1}$.

This phenomenon appears to be a direct result of synthesis of ethanol by micro-organisms in the period between cutting and crushing of cane. In Queensland, chopper harvesting of cane into short billets is believed to be a prime factor in promoting this synthesis. The BOD and alcohol levels are known to increase in the summer months and during periods of processing stale cane. Blake and McNeil (1978) further observed that the levels of ethanol were 60–70% lower in condensate and condenser cooling-water when processing green harvested cane in place of normal burnt cane.

Cold water is required for mill journals and bearings, oil coolers, crystallizers, sealing vacuum pumps and for pump bearings. In modern factories much of this water is in a closed circuit with internal towers for water conservation.

The processing liquors of a sugar factory have a very high BOD, e.g. mixed juice ($80-100\,\text{g}\,l^{-1}$), syrup ($400-500\,\text{g}\,l^{-1}$) and molasses ($400\,\text{g}\,l^{-1}$). Thus, all effort must be directed to preventing spillages and leaks from tanks, pumps and pipes. Floor washing should be replaced by dry-cleaning as far as is practicable. If necessary, surplus condensates should be used in preference to raw water for washing floors. Rainwater drains must be separate from process drains. In Queensland, mills return all spillages from around the milling train to process. There does not appear to be any deleterious effect if the drains are kept clean and the return to process is rapid (Bevan, 1973). If the juice is allowed to deteriorate the benefits are lost and increased production of molasses occurs. Some mills also return to process spillages from around other processing areas including the pan stage, juice tanks, centrifuges and crystallizers.

Spillages from around the filters are highly polluting and are unsuitable for recycle to process. The loss of mud from filter conveyor belts to the floor has to be avoided. In some countries, filter mud is slurried and disposed of as a waste stream and represents the most significant source of organic loading (Train *et al.*, 1975). Separate impoundment of these washings or preferably dry disposal on to farmland, as is the Australian practice, is essential.

Most cane sugar factories are periodically closed for cleaning. In Australia the factories close each weekend. In a short time up to $450\,m^3$ of wash-water can be discharged. These washings include milling train and floor washings, filter washings and occasionally the mud layer from a clarifier. Evaporators require cleaning of scale by caustic soda or other special products. The BOD of shutdown waste is between 30 000 and 50 000 mg l^{-1}. Many factories have special dams or holding ponds for these wastes. Others may cart them by truck for land disposal. These shutdown wastes are also discharged at the end of each season or during prolonged wet weather.

Methods for improved housekeeping are described by Bhaskaran *et al.* (1963), Ashe (1971), Hendrickson and Grillot (1971), Bickle (1972), Sestero and Logan (1972), Bevan (1973), McNeil *et al.* (1974), Train *et al.* (1975), and Verma *et al.* (1978).

Modern bagasse-fired boilers are required to be fitted with dry collectors or wet scrubbers to prevent release of fly ash to atmosphere. The fly ash contains 33% combustible material and 67% incombustible material, comprising 67% silicon dioxide and 23% aluminium oxide (Sawyer and Cullen, 1977). Dry collectors of 90–92% efficiency or wet scrubbers of 97–99% efficiency are required in order to meet an emission standard of 0.69 g per standard m^3.

Wet sluicing of ash is an integral part of wet scrubber systems and is used also for transport of ash from dry collectors. Closed-circuit systems for sluicing-water are essential for the prevention of water pollution. These in turn require very effective removal of ash from the water by screens and by settling tanks and filters. The designs of several different systems for fly ash removal have been described by Sawyer and Cullen (1975), McDougall *et al.* (1976), Sawyer and Cullen (1977), Milford (1977), Marie-Jeanne (1977), Watt and Morton (1979) and Crees *et al.* (1981). The use of large settling ponds has also been advocated (Jones and Dyne, 1977).

The use of condensates for supply-water to wet scrubbers has allowed some Queensland factories to evaporate large quantities of water which otherwise would pass to an effluent plant. Some $0.13\,m^3\,t^{-1}$ of cane can be evaporated in this way (Cullen, personal communication).

The washing of sugar cane is practised in Brazil, Hawaii and Louisiana and to some extent in Puerto Rico. Over 25% of the gross weight of cane delivered to the mill in Hawaii consists of rocks, soil and trash (Train et al., 1975). Here, the cane-washing plants have a system of rollers which remove rocks followed by a flotation bath which allows soil and rocks to settle from the cane. The cane passes along a conveyor for further washing and then through another set of rollers for trash extraction. The stalks are then finally prepared for milling. In Louisiana, the cane wash-water is sprayed on to the carrier.

The source of water for cane washing is fresh water, barometric condenser-water or recycled cane wash-water (Train et al., 1975). According to these authors the quantity of water used is $20\,m^3$ per tonne of cane in Louisiana and averages $11.5\,m^3$ per tonne in Hawaii. The BOD discharged in fresh water per tonne of cane is 1.46 kg in Louisiana, 5.0–5.1 kg in Hawaii and 1.87 kg in Puerto Rico. The suspended solids loadings per tonne of cane is 17 kg in Louisiana, 31.7–45.3 kg in Hawaii and 7.18 kg in Puerto Rico. Monteiro (1975) reports the discharge of $5\,m^3$ per tonne of cane from Brazilian factories, having a BOD of 180 to $500\,mg\,l^{-1}$ and containing 150 to $900\,mg\,l^{-1}$ suspended solids.

A report published by the US Environmental Protection Agency (EPA, 1971) estimated that for three mills in Hawaii the total suspended solids levels in wash-water were 4500, 10 700 and $5500\,mg\,l^{-1}$. Over 90% of the suspended solids were settleable solids and 70–90% were removable in clarifiers or sedimentation ponds.

In Hawaii, cane wash-water is mostly used for irrigation. Schantz and Kemmer (1969) describe a plant which removes 65% of solids by hydraulic cyclones before irrigation. Merle (1976) reports the operation of a combined system for handling boiler ash water and soil in cane wash-water. The soil was passed through a grit separator and combined with boiler ash water (also prestrained) and flocculant. The fine solids were precipitated in a hydroseparator with the aid of a flocculant and the underflow was dried to 55% solids with a vacuum filter. Experiments with dry cleaning methods for cane in place of wet cane cleaning (Middleton et al., 1971) offer a potential solution but, in the long term, modified harvesting practices may be more effective.

Excluding single-pass condenser cooling-water, all effluents from sugar mills are such as to require either discharge to irrigation or biological treatment. The composition of these effluents varies greatly from country to country because of the differences in water usage. In an Australian factory such effluent comprises polluted condensates and, in many cases, overflow from a cooling tower or spray pond. In Hawaii the effluent consists primarily of cane wash-water. Indian factories discharge overflow from spray ponds. Table 1.5 sets out

Table 1.5 Composition of cane sugar factory effluents[a]

Parameter	Australia[b]	Hawaii[c]	India[d]	South Africa[e]
pH	5.0–9.0	4.96–7.34 (6.04)	4.6–8.4	4.2–6.5
COD	500–2000		64–4380	358–1432
BOD_5	350–1400	264–1534 (976)	15–1916	
N, total	5–20	5.0–105 (27.2)	10–40	4.1–19.2
P, PO_4^{3-}	0–10	<0.1–<0.2	4.6–15.7	0–9.1
Total solids			478–3500	619–2231
Suspended solids	250–500	1320–25 375 (7908)	27–928	115–442
Settleable solids		4.0–98 (42)		
Dissolved solids			250–444	504–1719
Nitrate				0–8
Sugar				90–300
Chloride		10–950 (171)		
Oil and grease			16–106	9.6–68.6

[a] All concentrations are in $mg\,l^{-1}$; average values are given in parenthesis.
[b] McNeil, unpublished results.
[c] Hawaii Department of Health, cited by EPA (1971).
[d] Bhaskaran and Chakraborty (1966); Aora *et al.* (1974); Verma *et al.* (1978).
[e] Cox (1969).

the composition of effluent in certain countries. The feature common to all of the effluents is the low level of nitrogen and phosphorus relative to COD (or BOD). Aora *et al.* (1974) report COD/BOD ratios of 1.3–4.2. Ratios for fresh waste in Queensland vary from 1.3–1.5 (McNeil, unpublished results).

1.2.4 Sources of cane sugar refinery wastewater

The number of publications on wastewater from sugar refineries is limited. This may reflect the fact that most refineries are located in major cities where wastewaters can be discharged to municipal systems. The most comprehensive document is one published by the US Environmental Protection Agency (Train *et al.*, 1974b). Unless indicated otherwise, the information below is taken from this report.

Water balances for liquid sugar and crystalline sugar refineries are shown in Figs 1.8 and 1.9. The intake to the liquid sugar refinery is based on $1.67\,m^3\,t^{-1}$ of high quality fresh water and $20.9\,m^3\,t^{-1}$ of condenser cooling-water. Intake to a crystalline sugar refinery is based on $3.38\,m^3\,t^{-1}$ of fresh water and $41.7\,m^3\,t^{-1}$ of condenser cooling-water.

Barometric condenser cooling-water requirements for a liquid sugar refinery are $16.3\,m^3\,t^{-1}$ of melt while $36.5\,m^3\,t^{-1}$ are required for crystalline factories. Losses of sugar due to entrainment occur but are minimized by careful design and operation of evaporators and by the installation of entrainment arresters. The BOD of cooling water from crystalline sugar refineries is $4.39\,mg\,l^{-1}$ or

THE SUGAR INDUSTRY 27

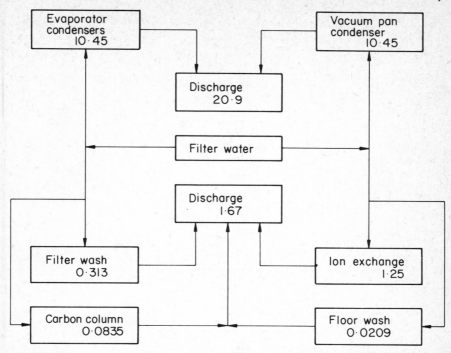

Fig. 1.8 Water balance for a liquid sugar refinery (Train *et al.*, 1974b), values in $m^3 t^{-1}$ melt

$0.07-1.8$ kg t^{-1} of melt, whereas the corresponding figures for liquid sugar factories are $6-31$ mg l^{-1} or $0.16-0.44$ kg t^{-1} of melt. The location of refineries in urban areas has generally restricted the use of cooling towers or spray ponds for recirculation. Condensates from evaporation are used as boiler feed-water.

Filter cake is transported dry for land disposal or may be slurried for impounding or discharge to a municipal plant. Diatomaceous earth, which is used as a filter aid, can be recovered by heating in a kiln. The BOD of the slurry is about 735 mg l^{-1}. The design of a system for desweetening and disposal of carbonation muds is described by Hohnerlein (1973).

Trucks and rail cars that are used for liquid sugar transport require washing. The washings have a BOD of approximately $17\,250$ mg l^{-1}. Washings can be returned to process if recovered before large-scale dilution occurs.

Most of the large crystalline sugar refineries use bone char for decolorization. Char wash-water is a major source of pollution with a volume of $0.22-0.84$ $m^3 t^{-1}$ of melt and BOD_5 of $500-2000$ mg l^{-1} or $0.15-1.7$ $m^3 t^{-1}$ of melt. Granular activated carbon is used in a minority of crystalline sugar refineries and in most liquid sugar refineries for colour removal. The columns are washed to remove sugar and heated to remove organic substances. Approximately 0.08 $m^3 t^{-1}$ of wash-water containing 0.1 kg BOD t^{-1} of melt is

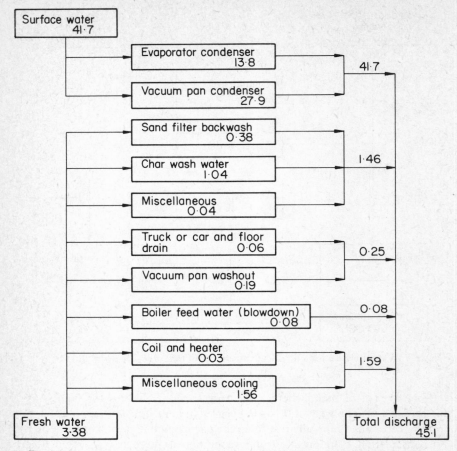

Fig. 1.9 Water balance for a crystalline sugar refinery (Train *et al.*, 1974b), values in m^3 t^{-1} melt

wasted. Wash-water from the regeneration of bone char and carbon columns is reduced by recovery of the initial higher purity wash-waters.

Ion-exchange columns are used for colour and ash removal, particularly in liquid sugar refineries. Often these columns follow granular carbon columns and thus serve mainly to remove inorganic substances. Ash removal requires both cationic and anionic exchange columns whereas colour removal requires an anionic exchange resin.

Langley and Bohlig (1973) describe a process for reducing levels of solid materials in refinery effluents. The solid substances included bone char fines, diatomaceous earth from filters, calcium phosphate, grit from the grit eliminator for melt and back-washings for sand filters. Liquid wastes from boiler blowdown and acid and soda washings from equipment cleaning were also included. The wastewaters were neutralized and sedimented in a vessel with the aid of phosphoric acid and polyelectrolyte. The underflow was dewatered on a rotary belt filter and solids were disposed of on land.

1.3 Manufacture of sugar — treatment and disposal of wastewater

1.3.1 Irrigation

Irrigation with wastewater is common in both the beet and the cane sugar industries. In the case of beet sugar, it has been practised for many years (Schulz-Falkenhain, 1964). In France, some 12% of the factories dispose of settled mud on to the land and 3% dispose of both settled mud and clarified waters (Vellaud, 1978). On the other hand, land disposal has been discontinued in the Braunschweig region of West Germany (Hoffmann-Walbeck, 1977). One half of the cane sugar factories in Queensland irrigate sugar cane or cattle pasture with untreated wastewater and a number of others use irrigation in association with other methods of treatment. Irrigation with cane wash-water and factory wastewater is widely practised in Hawaii (Train et al., 1975). However, published reports of irrigation with cane factory wastewater are rare.

Irrigation has some clear advantages. These include conservation of water, the relatively simple technology involved and the fact that all the components of the wastewater are returned to the land.

Irrigation with beet factory wastewater

Beet factories may irrigate with either clarified or nonclarified transport-water, sometimes including carbonation sludge. In France the land disposal of $50-300 \, m^3 \, h^{-1}$ of water between October and December is equivalent to $100-500 \, mm$ of rainfall (Catroux et al., 1974b).

For totally successful irrigation the soil must allow filtration and retention of water, while acting as a support for the decomposition of organic matter by microflora. Careful monitoring of the reduction in organic levels by soil micro-organisms is required, to avoid pollution of the aquifer and overloading of the irrigated fields (Mühlpforte, 1962).

A mathematical expression for BOD removal in soil is proposed by Kramer (1960):

$$A_T = A_0 / 10^{T/C}$$

where A_0 is the BOD_5 ($mg \, l^{-1}$) of irrigated water
A_T is the BOD_5 ($mg \, l^{-1}$) at depth T
T is the depth (cm)
C is a coefficient which varies with soil type and temperature.

The value of C was determined as 98 for an irrigation of 50 mm, and as 168 for an irrigation of 450 mm.

Irrigation with water containing mud has the principal advantage of returning soil to farmland. Devillers (1968) estimates the land requirements for a 3000 t d^{-1} factory as 50 to 200 ha. Catroux et al. (1974b) report that 60 m^3 h^{-1} of water containing 300 g l^{-1} of soil can be applied by spray irrigation at a rate of 1 ha d^{-1} or 80–100 ha per campaign. This results in a water application of 150–200 mm and a soil application of 2–3 cm. The same land can be irrigated for three successive years. The major disadvantages of spray irrigation are the wear on pipes, blockage of nozzles, loss of soil permeability and ground subsidence.

Flood irrigation with water that contains mud requires considerable land preparation; the land has to have a slight slope and must be contour-ploughed with earth ridges 100–150 cm apart. The wastewater is released at a high point and circulates by overflow of the earth ridges (Catroux et al., 1974b), the channel between the earth ridges being progressively filled with soil.

The deposited soil and water form a relatively impermeable mud layer at the surface. The percolation rate is slow enough to allow the adsorption of dissolved substances by the subsoil (Heitz and Bidan, 1970). The greater degree of purification occurs in the upper layer of the subsoil. The rapidity with which the soil becomes impermeable to water restricts applications of water to 500 to 1000 mm.

The soil deposits remain saturated with water for 15 to 30 days and odours may arise. Following flood irrigation the soil is mixed by ploughing perpendicular to the ridges. Catroux et al. (1974b) estimate that three men are required in order to irrigate with 100 m^3 h^{-1} of water containing 300 g l^{-1} soil at a loading of 4200 m^3 ha^{-1}.

The application of 420 mm of water can introduce 100 t ha^{-1} of soil containing 13 to 16 t of organic matter, 150 to 200 kg of nitrogen, 50–100 kg P_2O_5 and approximately 1 tonne each of potassium and sodium (Catroux et al., 1974b).

Irrigation with clarified or decanted water permits more intensive irrigation and the use of sprays. However, the storage of large quantities of dirt is then unavoidable. The benefits of this method are lower maintenance costs on pipes and sprays and lower land requirements. Further, the water can be held in the clarification ponds for irrigation in the warmer months (Catroux et al., 1974b).

Kramer (1961) recommends a limit for spray irrigation of 500 mm water per campaign. The land requirement is estimated as 30–80 ha of grassland or 60–160 ha of arable land for waste volumes of 1–4 m^3 t^{-1}. Parkhomenko (1964) reports the irrigation of 527 ha at the rate of 7900 m^3 d^{-1} over a 150 day campaign. Permeable soil has been irrigated at rates of up to 100 m^3 h^{-1} on 10 ha (Devillers, 1968). Catroux et al. (1974b) report irrigation of rye grass and festuca at applications of up to 12 500 m^3 ha^{-1} over a three week period. Devillers (1968) found that with irrigation water of 600 mg l^{-1} BOD, the BOD of subsurface water increased to 20 mg l^{-1} during irrigation and then returned to zero after two or three months. The retained levels of potassium and nitrogen were 5 t ha^{-1} and 450 kg ha^{-1} respectively.

Soviet authors have studied improvements in crop yield from irrigation with wastewaters. The application of 1500 m^3 ha^{-1} of unclarified wastewater resulted in improved production from sugar beet (50.9%), wheat (22.2%), oats (87.37%) and corn (35.3%) (Bereznikov and Novikov, 1976). Parkhomenko (1964) estimated that the irrigation of 7900 m^3 d^{-1} over 527 ha during a 150 day campaign increased production of beets by 65–70% and replaced 40–70% of fertilizer requirements.

1.3.2 Treatment in ponds or aerated lagoons

Ponds or lagoons have achieved great popularity for wastewater treatment, particularly in the beet sugar industry. Vellaud (1978) estimated that 85% of beet factories in France used ponding systems. A third of the cane sugar factories in Queensland have ponding systems, often in association with irrigation systems.

The overall benefits of such systems include adaptability to seasonal operations of very short duration, potential use as reservoirs for irrigation and low operation and maintenance costs. The level of technology required makes them especially suitable for sugar producers in Third World countries. Capital costs for construction may also be lower. This depends on whether relatively cheap land can be obtained at reasonable proximity to the factory. For example, in Queensland land close to the mills is invariably under cultivation and is expensive. The cost of pipeline to land as far away as 8 km can result in construction costs comparable to those of activated sludge plants. The major disadvantages are land requirement, odour production and potential infiltration of polluted water into groundwater.

Ponding of beet factory wastewater

Extensive use has been made of lagoons, ponds or holding dams for complete or partial treatment of beet factory wastes. The very brief beet campaign makes ponding particularly attractive as a method of treatment. Almost invariably, the process is simple batch ponding in which the wastewater is held from the end of the campaign, through the ensuing winter and then discharged prior to the next campaign.

The mode of treatment involves initial fermentation to organic acids followed by breakdown of nitrogenous substances (Delvaux, 1974). These first anaerobic phases usually coincide with the winter months. When the BOD is sufficiently reduced, aerobic conditions become evident, a development which occurs in the warmer months. Red and green algae proliferate as temperatures increase to 15°–18°C.

The size of the ponds is normally decided by the capacity required to hold the wastes of the whole campaign. Some factories have used mechanical aeration to minimize odours and to increase the rate of treatment.

The rate and extent of treatment has been examined by workers in several countries. Carruthers *et al.* (1960) report a BOD reduction from 1239 mg l^{-1} to 38 mg l^{-1} between March and October. In the USA, Barr (1962) describes the operation of several ponds, two of 18 ha and 4 m in depth, one 20 ha and 3 m in depth and another 17 ha and 1.8 m in depth. Nine tonnes of BOD per day of average concentration of 455 mg l^{-1} were released to the ponds during a campaign and released after winter storage. The BOD was reduced by 53% over this period. Brenton and Fischer (1970), Brenton (1972) and Fischer (1974), report the treatment obtained in a 0.7 ha, 4.6 m deep pond batch-operated after a campaign. With the assistance of two 6.6 kW mechanical aerators, the BOD was reduced from 67 to 0.45 t and the COD from 104 to 1.2 t over a 32 week period. The final effluent quality was 31 mg l^{-1} BOD and 87 mg l^{-1} COD.

Chekurda and Parkhomets (1968) studied the treatment of wastewaters from a Ukrainian factory in 3 ha ponds of 0.8–1.2 m depth. At an average ambient temperature of 7°C, twelve months treatment was required to reduce the BOD from 2000–3000 mg l^{-1} to 12 mg l^{-1}. The pond area required for a plant discharging daily about 2500 m^3 of wastewater over a 4–5 month campaign was 30 ha.

Continuous recirculation ponds are described by Viehl *et al.* (1974). Waste was reduced in BOD from 1800 mg l^{-1} to less than 25 mg l^{-1} in four ponds in series, 2 m deep and total area 20 ha. The contents of the final pond were recirculated to the first.

French workers have made a number of studies of ponding of beet factory wastes. Devillers *et al.* (1970) report BOD$_5$ reduction ranging from 65% to 98.7% over an 8 month period, in lagoons for five factories. Details are given of the performance of ponds ranging from 0.9 to 1.5 m in depth, treating wastewater of various strengths. Devillers and Lescure (1971) studied the operation of a system of ponds of total area 30 ha and of 1 m depth treating refinery wastes. Over a 6 month period from January to June, BOD reduction in three ponds was from 1090, 1140, and 1450 mg l^{-1} to 12, 13 and 25 mg l^{-1} respectively. Curis (1972) reports that the mean rate of purification for ponds measured over periods of 4–7 months was 5.9–17.5 mg BOD l^{-1} d^{-1}. As a result of studies of 9 ponds of varying depth, Lescure (1973) suggests that the optimum depth of ponds is 2–3 m.

Schneider and Hoffmann-Walbeck (1968) demonstrated that the decomposition of organic matter in ponds containing both water and mud was slower than would occur in a pond where mud had been separately disposed of. Their results show rapid reduction in organic loading of seepage water at distances of 15–150 m from the storage ponds.

The time for treatment in lagoons has been reduced from 2–3 months to 10 to 20 days with mechanical aeration (Laguerre, 1970). Power requirements were estimated at $2-2.5\,\text{kWh}\,\text{kg}^{-1}$ of BOD_5 removed. Similarly, Klapper (1970) describes the performance of two large ponds, 24 000 and 20 000 m² in area and 1.3 m deep, one of which was aerated. The aerated pond achieved a BOD_5 reduction from $1100\,\text{mg}\,\text{l}^{-1}$ to $30\,\text{mg}\,\text{l}^{-1}$ in two months compared with a drop from $1550\,\text{mg}\,\text{l}^{-1}$ to $240\,\text{mg}\,\text{l}^{-1}$ in the other pond.

Teichmann and Leswal (1976) studied the performance of two aerated ponds. One, 28 000 m³ in volume and 2 m deep, was aerated to provide a total oxygen transfer of $96\,\text{kg}\,\text{h}^{-1}$. The second, with a volume of 85 000 m³ and 4 m depth, was aerated at an oxygen transfer rate of $164\,\text{kg}\,\text{h}^{-1}$. The first pond reduced the waste from a BOD_5 of $4399\,\text{mg}\,\text{l}^{-1}$ and COD of $8600\,\text{mg}\,\text{l}^{-1}$ to $55\,\text{mg}\,\text{l}^{-1}$ BOD_5 and $500\,\text{mg}\,\text{l}^{-1}$ COD over 5 months. The second pond achieved a reduction from $3367\,\text{mg}\,\text{l}^{-1}$ BOD_5 and $6014\,\text{mg}\,\text{l}^{-1}$ COD to $12\,\text{mg}\,\text{l}^{-1}$ and $646\,\text{mg}\,\text{l}^{-1}$, respectively, over a similar period. All aerators were about 4000 h in operation. Oxygen transfer in the first pond was $2.0\,\text{kg}\,(\text{kWh})^{-1}$ with BOD_5 removal of $1.75\,\text{kg}\,(\text{kWh})^{-1}$ and in the second was $1.65\,\text{kg}\,(\text{kWh})^{-1}$ with BOD_5 removal of $1.36\,\text{kg}\,(\text{kWh})^{-1}$.

Tsugita *et al.* (1969) and Oswald *et al.* (1973) describe studies of three pilot ponds in series designed as anaerobic, facultative and aerobic ponds respectively. The anaerobic pond was 0.4 ha in area and 4.6 m deep, the facultative pond 0.8 ha and 2 m deep and the aerobic pond 1.2 ha and 1 m deep. These ponds were loaded at rates varying from 505 to $2530\,\text{kg}\,\text{ha}^{-1}\,\text{d}^{-1}$ with removal of 94–99% of the incoming BOD. At the maximum loading, the retention time was 28 days and flow rate was $1.2\,\text{m}^3\,\text{min}^{-1}$. The corresponding maximum COD loading was $4700\,\text{kg}\,\text{ha}^{-1}\,\text{d}^{-1}$. The so-called facultative pond operated anaerobically when loaded in excess of $54\,\text{kg}\,\text{ha}^{-1}\,\text{d}^{-1}$.

Oswald *et al.* established that the anaerobic pond was able to remove 80% of the incoming load up to a maximum daily BOD loading of $2240\,\text{kg}\,\text{ha}^{-1}$. However, at the maximum loading, odours were intense. Aeration of the anaerobic pond at a rate of $136\,\text{kg}\,\text{d}^{-1}$ of oxygen permitted daily loads of $1120\,\text{kg}\,\text{BOD}\,\text{ha}^{-1}$ without odour problems. Following anaerobic treatment, 224 kg of BOD $\text{ha}^{-1}\,\text{d}^{-1}$ were treated in the aerobic pond, operated as a well-mixed algae pond, and $112\,\text{kg}\,\text{ha}^{-1}\,\text{d}^{-1}$ were treated in a facultative pond, without odour problems. Recirculation of aerobic effluent at a rate of 50% to 100% into the anaerobic pond improved performance. In order to comply with effluent standards, it was found necessary to filter algae and other microbial solids from the effluent to remove large quantities of insoluble BOD. The pond area required for a $4500\,\text{t}\,\text{d}^{-1}$ beet factory was estimated as at least 29 ha.

Ponding of cane factory wastewater

Treatment by ponding has been reported in India (Bhaskaran *et al.*, 1961, 1963; Bhaskaran and Chakraborty, 1966; Parashar, 1965, 1969; Gupta, 1965), Puerto Rico (Guzman, 1962), Australia (Anon., 1968; Bond and McNeil, 1976), South Africa (Lewis and Ravnoe, 1976) and Louisiana (Chen *et al.*, 1971).

The type of ponding system most commonly referred to in the literature is the 'deep anaerobic pond'. However, given the high BOD of sugar mill wastewater, there is little possibility that any mill ponding system operates other than as an anaerobic pond for the initial phase of treatment, irrespective of depth. Aerobic or facultative operation can only be ensured at a surface loading less than 60 kg BOD ha^{-1} d^{-1}. Thus, a factory discharging 2160 kg d^{-1} would require 36 ha of pond area to ensure aerobic conditions in at least the upper layer of the pond. Therefore, initial anaerobic treatment is necessary to reduce the area of land required.

The high levels of carbohydrate and other organic matter and the relative absence of protein or nitrogenous material in mill effluents cause most anaerobic ponds to register a fall in pH values to as low as pH 4.0–4.5. In these circumstances, anaerobic removal of organic matter is very slow. The continuous addition of lime (Bruijn, 1977) or sewage effluent (Lewis and Ravnoe, 1976) to anaerobic lagoons results in more stable operation. Lewis and Ravnoe reported on a 4455 m^3 lagoon which initially treated 1500 m^3 d^{-1} of waste, reducing the COD of 2512 to 1539 mg l^{-1}. During the course of treatment, the pH fell from 8.2 to 5.5. However, after the addition of up to 260 m^3 d^{-1} of treated sewage and reduced loading from the factory, the COD was reduced from 1919 to 266 mg l^{-1} and from 1527 to 212 mg l^{-1} in two separate trials. Final pH was 6.8–6.9.

Indian workers have favoured anaerobic lagoons followed by oxidation ponds seeded with algae and water hyacinth. Bhaskaran and Chakraborty (1966), in pilot-scale studies in an anaerobic system, reduced the BOD of waste from 1600 to 550 mg l^{-1} in 7 days, at a loading of 0.23 kg BOD m^3 d^{-1}. Oxidation ponds, loaded at 316 kg ha^{-1} d^{-1}, reduced the BOD from 307 to 34–180 mg l^{-1} in 13 days. With 7 days retention, the BOD was reduced from 272 to 41–118 mg l^{-1} at a loading of 325 kg ha^{-1} d^{-1}. Gupta (1965) reports that 40–59% of BOD was removed in anaerobic ponds operated at a loading of 0.08–0.56 kg BOD m^{-3} d^{-1}. The subsequent aerobic stage achieved 25–64% BOD reduction at a loading of 635 to 1684 kg BOD m^{-3} d^{-1}. The aerobic plant was seeded with algae and aquatic plants. Sinha and Sinha (1969) report that anaerobically digested wastewater seeded with water hyacinth was treated from a BOD of 258 to 16 mg l^{-1} in 7 days compared with a reduction to 116 mg l^{-1} in a control.

The most successful Australian experience has been with shallow ponds of 1 m depth operated as a batch system (Bond and McNeil, 1976). These ponds undergo an initial anaerobic phase of treatment for a period of three weeks during which 70–90% of the BOD is removed. The ponds then become aerobic and the BOD is reduced to less than 30 mg l^{-1} in a further period of 3–5 weeks retention. The treatment cycle is slower in the winter months and occasionally an irrigation spray has to be used as an 'aerator' to assist treatment. The combined factors of wind action and shallow pond depth allow anaerobic treatment to take effect without a rapid fall in pH. The results of the operation of this scheme over a season are shown in Table 1.6.

Table 1.6 Performance of a shallow ponding system treating cane factory wastes

Pond No.	Area (ha)	Months of operation	Initial BOD (mgl^{-1})	Final BOD (mgl^{-1})	Suspended solids (mgl^{-1})	Treatment time (d)	Removal rate (kgha^{-1}d^{-1})
1	2.0	Jul–Aug–Sep	1635	26	20	63	255
		Oct–Nov	1480	18	6	28	522
2	2.83	Jul–Aug–Sep	520	40	92	42	114
		Oct–Nov	900	24	7	41	208[a]
3	5.5	Aug–Sep	620	25	30	42	142
4	2.43	Jun–Jul–Aug	200	11	47	56	34
		Sep–Oct	570	19	26	35	157[b]
5	2.55	Jun–Jul–Aug	1100	12	5	56	194
		Sep–Oct	970	10	27	42	229[c]

[a] Assisted by use of an irrigation spray for 100 hours.
[b] Assisted by use of an irrigation spray for 80 hours.
[c] Assisted by use of an irrigation spray for 40 hours.

1.3.3 Activated sludge plants

Activated sludge treatment is not so commonly used in beet and cane sugar manufacture as are ponding systems or irrigation. However, activated sludge systems have been preferred to other intensive treatment methods such as biofiltration or anaerobic treatment.

The seasonal nature of the sugar industry and, in particular, the very short duration of the beet campaign are disadvantageous to the introduction of activated sludge treatment. This method of treatment is usually adopted where land is unavailable for alternative methods. Costs of supervision, construction, nutrients and electricity and potential variations in loading to the plant are factors that predispose against the introduction of these plants. An additional problem is the requirement to treat large volumes of high-strength shutdown wastes at the close of the season for up to three or four weeks. Nevertheless, factories which have installed these plants have achieved excellent performance after overcoming these difficulties.

The seasonal nature of the sugar industry introduces the immediate disadvantage of annual start-up of plant on sewage and molasses. Most full-scale plants operate at a sludge solids level of 2000–4000 mg l^{-1} in plants having a capacity of over 5000 m^3. Thus, initially, some 9–18 t dry weight of sludge solids has to be synthesized in time for the commencement of crushing. Bathgate et al. (1977) estimated that 27 t of molasses and 5 weeks preparation were required to achieve a suspended solids level of 2000 mg l^{-1} in two ponds of a total 8800 m^3 capacity. Those requirements were reduced by about 40% when the sludge from the previous season was stored and 'reconditioned' in the following season.

For beet factories, the disadvantage of a short campaign does not apply to the same extent to factories which have a refinery operation or which process stored juices outside the normal campaign (Zama *et al.*, 1979). A more stable operation can also be expected where the activated sludge treatment is preceded by an extensive ponding system (Langen and Hoeppner, 1964; Zama *et al.*, 1979).

Probably the major problem that has arisen in activated sludge plants in sugar factories has been the formation of bulking sludge, a condition in which sludges do not settle readily in a clarifier, resulting in rapid loss of sludge from the system. Commonly, this condition is promoted by or associated with the development of filamentous micro-organisms in the sludge flocs. Conditions known to induce 'bulking' with different types of wastewater are:

(i)   inadequate nitrogen and phosphorus in relation to BOD;
(ii) acid pH;
(iii) low dissolved oxygen in the range $0.5-1.0\,\mathrm{mg\,l^{-1}}$;
(iv) a high sludge loading, in excess of $0.7\,\mathrm{kg\,BOD\,d^{-1}\,kg^{-1}\,MLSS}$.

Other factors such as flow dynamics, concentration of macro- and micro-elements and hydraulic retention time may have an influence (Beccari *et al.*, 1980).

There is a possibility that wastes containing carbohydrate have an enhanced potential to promote 'bulking' of sludge, in particular by promoting the growth of organisms producing a polysaccharide sheath (Delvaux, 1974). However, in practice it is usually difficult to determine whether the observed 'bulking' of sludge is a direct effect of carbohydrate composition of the waste or an indirect result of a situation where sudden losses of sugar have led to any or all of the conditions referred to above.

The supply of adequate nutrients, particularly nitrogen and phosphorus, has been shown many times to be essential for adequate treatment and the prevention of bulking. The requirement depends basically on whether the plant is operated as an intensive high-rate plant with maximum conversion to excess sludge or as a low-rate plant where sludge production is minimal. For sugar cane wastes, inputs varying from BOD : N : P of 100 : 2.5 : 0.5 (Simpson *et al.*, 1972) to 100 : 5 : 1 (Miller, 1971; McNeil *et al.*, 1974) have been recommended. When reviewing the operation of five plants in German beet factories, Reinefeld *et al.* (1979) found a ratio of 100 : 2 : 0.3 was ample for plants operated at a loading of $0.25\,\mathrm{kg\,BOD\,kg^{-1}\,MLSS\,d^{-1}}$. For a high rate RT–Lefrancois process (see below) the requirements are 100 : 2.6 : 0.7 (Brunner, 1977). Catroux *et al.* (1974a) used a working ratio of 100 : 5 : 1.

Simpson *et al.* (1972) carried out extensive studies of the sludge loading to achieve optimum sludge settling rates in plants treating cane mill waste. The optimum sludge settleability was achieved at a loading of $0.6\,\mathrm{kg\,COD\,kg^{-1}}$ sludge solids $\mathrm{d^{-1}}$. The observation by those authors that effluent COD (filtered) did not increase significantly over the range of $0.4-1.4\,\mathrm{kg\,COD\,kg^{-1}}$

sludge solids d^{-1} indicated that the retention time required to maintain an optimum sludge loading exceeded that required to achieve adequate treatment.

Reinefeld *et al.* (1979), after several years of surveying the operation of activated sludge plants in the beet industry, confirmed that operation at less than $0.25\,\text{kg}\,\text{BOD}\,\text{kg}^{-1}$ sludge solids d^{-1} was less likely to result in the development of filamentous organisms. *Sphaerotilus natans, Thiothrix* sp. and *Haliscomenobacter hydrossis* were able to develop at higher sludge loadings, even in the presence of adequate nutrients.

The anaerobic pretreatment of sugar-containing wastes has resulted in a substantially reduced development of filamentous micro-organisms (Schneider *et al.*, 1964; Kollatsch, 1969). This has been attributed to a lower propensity for fermentation end-products to promote bulking of sludge. Bathgate *et al.* (1977) report a similar observation with wastewaters from a sugar mill.

The operation of an activated sludge plant by a sugar factory demands stringent control both in the factory and at the plant site. The sensitivity of activated sludge plants to shock loads demands the prevention of sudden large losses of process liquors to the plant. Relatively small volumes of molasses or syrup can easily promote difficulties. Factories must be particularly alert to froth-over of evaporators and spillages of sugar and clarifier mud. The establishment of reclaim areas to return spillages to process is desirable, provided that the drains are efficient and clean. The ever-present potential of shock loads justifies the use of low-rate high-volume plants such as extended aeration systems and oxidation ditches where maximum dilution can be obtained.

Shutdown and weekend wastes are released over a very short period of time. The established practice for sugar mills is to collect these wastes in a special pond and aerate them for several days until they can be released gradually into the main system (Bevan, 1971; Miller, 1971; McNeil *et al.*, 1974; Bathgate *et al.*, 1977). The ponds must be sized to cope with the prospect of a midweek shutdown caused by rain and a substantial quantity of end-of-season wastes. The filter washings provide sufficient phosphorus but the addition of nitrogen is required to assist treatment. A pond with a $22.4\,\text{kW}$ aerator is capable of reducing the COD of over $400\,\text{m}^3$ of waste from $6000-12\,000\,\text{mg}\,\text{l}^{-1}$ to $200\,\text{mg}\,\text{l}^{-1}$ in about five days (McNeil *et al.*, 1974).

The sizing of the main aerobic treatment stage of sugar mill plants is straightforward. The desirable upper limit for sludge solids concentration is $4000\,\text{mg}\,\text{l}^{-1}$. Above this level, settling rates of the sludge are slow. An overriding objective is to maintain a sludge which does not 'bulk'. Simpson *et al.* (1972) recommend an optimum daily sludge loading of $0.6\,\text{kg}\,\text{COD}\,\text{kg}^{-1}\,\text{MLSS}$ and plants in Queensland have been observed to operate with greatest stability at a daily loading of $0.3-0.4\,\text{kg}\,\text{BOD}\,\text{kg}^{-1}\,\text{MLSS}$. Thus, the tank volume can be calculated for a waste of known BOD and for a desired level of mixed liquor suspended solids. For a waste of $1000\,\text{mg}\,\text{l}^{-1}$ BOD, mixed liquor solids concentration of $4000\,\text{mg}\,\text{l}^{-1}$ and sludge loading of

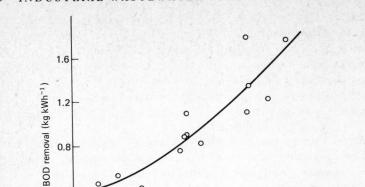

Fig. 1.10 Relationship of BOD removal to sludge loading for activated sludge wastewater treatment

0.3, the residence time provided for treatment is 20 hours. In pilot plant experiments (McNeil, unpublished results), a BOD reduction from 1000 mg l^{-1} to 20 mg l^{-1} has been achieved in less than 10 hours.

The oxygen requirement of an activated sludge plant is the sum of that required for oxidation of organic matter and that required for 'maintenance' of the sludge. Plants operating at a high sludge loading convert a high proportion of organic matter to surplus sludge, so they require less oxygen than those operating at a low sludge loading. The variation of BOD removal in terms of aerator power versus sludge loading was determined in a pilot-scale activated

Table 1.7 Experimentally determined loading rates for secondary clarifiers treating sugar mill wastewaters

Sludge sample	Mixed liquor solids (mg l^{-1})	Clarifier loading rate (m^3 m^{-2} h^{-1})
A	3060	2.90
	3500	1.92
	3930	1.77
	4370	0.74
	4800	0.66
B	2430	4.16
	3650	1.42
	4870	0.99
	6080	0.21

sludge plant operating at a dissolved oxygen concentration of $1.0-2.0\,\mathrm{mg\,l^{-1}}$ (McNeil, unpublished results). The relationship is shown in Fig. 1.10. At a sludge loading of 0.3, the BOD removed was $0.6\,\mathrm{kg\,(kWh)^{-1}}$. The requirement for operating at a low sludge loading in order to avoid sludge bulking causes a high demand for oxygen.

Clarification requirements are determined by batch settling tests. The clarifier loading rates for various mixed liquor solids concentrations and a thickening factor of 3 have been determined (McNeil, unpublished results — Table 1.7). The loading rate at a mixed liquor suspended solids concentration of $4000\,\mathrm{mg\,l^{-1}}$ was $1\,\mathrm{m^3\,h^{-1}\,m^{-2}}$. This loading rate in practice is the sum of the influent flow rate and sludge return rate.

Activated sludge plants in beet factories

Carruthers et al. (1960) demonstrated on the laboratory scale that activated sludge treatment could reduce BOD from $410\,\mathrm{mg\,l^{-1}}$ to $20\,\mathrm{mg\,l^{-1}}$ in 24 hours. Langen and Hoeppner (1964) describe a plant capable of treating $18\,\mathrm{m^3\,h^{-1}}$ of effluent overflow from a lagoon system from $1000\,\mathrm{mg\,l^{-1}}$ to $73\,\mathrm{mg\,l^{-1}}$ BOD. The plant had two 25 kW aerators in a pond providing a retention time of 24 hours. In a subsequent paper (Meyer, 1968), the same plant was reported to treat $18-32\,\mathrm{m^3\,h^{-1}}$ of lagoon overflow to $15-35\,\mathrm{mg\,l^{-1}}$. Although the lagoon overflow BOD decreased from $2000\,\mathrm{mg\,l^{-1}}$ in January to $95\,\mathrm{mg\,l^{-1}}$ in August, the effluent quality of the activated sludge plant remained relatively constant.

Schneider et al. (1964) and Kollatsch (1969) treated wastewater to achieve a BOD of $15-20\,\mathrm{mg\,l^{-1}}$ by anaerobic pretreatment followed by conventional activated sludge treatment. The anaerobic stage resulted in 30% BOD removal in 1.5 days or 50% in 5 days at a pH controlled between 6.7 and 7.3. Exceptional sludge settleability in the anaerobic stage and a sludge volume index of 50 were reported. Anaerobic pretreatment plants have been loaded at $13-14\,\mathrm{kg\,COD\,m^{-3}\,d^{-1}}$ with a reduction of COD from $5000-5500\,\mathrm{mg\,l^{-1}}$ to $1000\,\mathrm{mg\,l^{-1}}$ (Hoffmann-Walbeck and Pellegrini, 1978). The stable operation of aerobic plants to achieve a better than $30\,\mathrm{mg\,l^{-1}}$ BOD in final effluent requires a load on the sludge of less than $0.25\,\mathrm{kg\,BOD\,d^{-1}\,kg^{-1}}$ sludge solids (Reinefeld et al., 1979). Huss (1979) has extended the anaerobic pretreatment principle to methane gas generation in an anaerobic contact process, followed by activated sludge treatment to achieve $25\,\mathrm{mg\,l^{-1}}$ BOD.

Demidov and Demidov (1973) have developed design criteria for the operation of two-stage activated sludge plants. High-rate adsorption of organic matter was achieved in the first stage followed by oxidation of organic matter in the second stage. Although the combined volumes of two-stage systems were smaller, the requirement for air was greater than for a single-stage plant. The low level of colloidal or insoluble matter in cane sugar factory wastewaters would not justify the use of contact adsorption processes.

An installation treating $1000\,\mathrm{m^3\,d^{-1}}$ of effluent in a $3400\,\mathrm{m^3}$ basin was described by Zama et al. (1979). An 18 m diameter clarifier was used with sludge recycle at 100% of the influent waste flow. The influent waste had a BOD of 2700 mg l^{-1} and a COD of 3800 mg l^{-1}, 96.8% of which was removed in the plant. Oxygen requirements were 1.64 kg kg^{-1} BOD removed (1.17 kg kg^{-1} COD removed) and power requirements were 1.04 kW kg^{-1} BOD removed (0.74 kW kg^{-1} COD removed).

Catroux et al. (1974a) carried out extensive laboratory studies in a 1.5 litre vessel with air supply sufficient to maintain a dissolved oxygen concentration of 1 to 6 mg l^{-1}. At a daily loading equivalent to 4.8 kg COD m^{-3}, the influent COD was reduced from 4000 mg l^{-1} to 560 mg l^{-1} (BOD$_5$ 336 mg l^{-1}). This loading represented a residence time of 1.2 days and a daily sludge loading of 0.6 kg COD kg^{-1} sludge solids. The surplus sludge generated was 0.54 g g^{-1} of COD removed and oxygen requirements were 0.82 g of oxygen g^{-1} of COD removed. When the loading was increased to 16.5 kg COD m^{-3} d^{-1}, the effluent COD was 700 mg l^{-1}. Sludge accumulation was 0.47 g g^{-1} of COD removed and the oxygen requirement was 0.73 g g^{-1} COD removed.

Catroux et al. (1974a) also determined the values of the coefficients a and b in the equation:

$$O_2 = (a \times \text{BOD removed d}^{-1}) + (b \times \text{mass of sludge})$$

Coefficient a had an average value of 0.45 kg O$_2$ kg^{-1} COD removed and b had a value of 0.28 kg O$_2$ kg^{-1} sludge d^{-1}.

These studies were extended to a 300 litre vessel operated at a loading of 6–17.2 kg COD m^{-3} d^{-1} and a sludge loading of 0.6–3.6 kg COD d^{-1} kg^{-1} sludge solids. The sludge solids were maintained at 4.5 to 14.4 g l^{-1}. The effluent COD varied from 340 to 600 mg l^{-1}.

A number of workers have used an intensive aeration system known as the RT–Lefrancois fermenter as a modified activated sludge system. Revuz (1971) operated an 8 m^3 vessel and 4 m^3 clarifier with an influent flow rate of 2 m^3 h^{-1} and a high sludge solids content of 20 to 60 g l^{-1}. The plant loadings were 8–10 kg l^{-1} of BOD, 1.2–1.5 kg BOD d^{-1} kg^{-1} sludge or 25–30 kg BOD m^{-3} d^{-1}. The COD was reduced from 3500–6000 mg l^{-1} to 250–500 mg l^{-1} (180–300 mg l^{-1} BOD). The power requirements were 0.5–0.65 kWh kg^{-1} of BOD removed.

Pieck (1974), Simonart et al. (1975) and Dubois (1976) report the treatment of waters in a 240 m^3 Lefrancois fermenter with a 500 m^3 clarifier. The plant was operated at a flow rate of 60–70 m^3 h^{-1} with 75–90 m^3 h^{-1} of sludge recycle, thus providing a retention time of 3–4 hours in the fermenter. At a COD loading of 9000–11 000 kg d^{-1} and BOD$_5$ of 7000–9000 kg d^{-1}, 80% of the COD and 90% of the BOD were removed. The sludge solids concentration was 10–20 g l^{-1} resulting in a loading of 1.5–2.5 kg BOD d^{-1} kg^{-1} sludge solids. The settling qualities of sludge were kept at a sludge volume index of 50 to 85. The energy consumption was 0.4–0.6 kWh kg^{-1} COD removed. Excess sludge production was 0.75 kg kg^{-1} COD removed.

Teichmann and Leswal (1976) describe a plant with a $300\,m^3$ aeration tank and a $436\,m^3$ secondary clarifier that was designed for the treatment of $55-60\,m^3\,h^{-1}$ of waste. The aeration tank provided a retention time of $5-5.5\,h$, and a total air input of $10\,000\,m^3\,h^{-1}$. The BOD_5 loading was $25\,kg\,m^{-3}\,d^{-1}$. The refinery wastewater was reduced from $7000-9000\,mg\,l^{-1}$ COD and $4000-6500\,mg\,l^{-1}$ BOD_5 to $1700-3000\,mg\,l^{-1}$ COD and $500-900\,mg\,l^{-1}$ BOD_5. The sludge retained good settling properties of $20-30\,ml\,g^{-1}$ when operated at an overall solids level of $30-50\,g\,l^{-1}$. Brunner (1977) treated effluent of $10\,000\,mg\,l^{-1}$ to $1000\,mg\,l^{-1}$ at a solids level of $20-30\,g\,l^{-1}$. Energy consumption was $1.2-1.7\,kWh\,kg^{-1}$ COD removed.

High-rate processes such as the RT–Lefrancois process appear suitable for pretreatment only and would not achieve in a single stage an effluent suitable for discharge to a watercourse. However, the effluent can be recycled to the transport water circuit.

The Pasveer oxidation ditch has been installed and studied in some beet factories (Devillers, 1968; Smith et al., 1975). Such installations in the UK are loaded at $0.2-0.25\,kg\,BOD\,m^{-3}\,d^{-1}$ or $0.03-0.04\,kg\,BOD\,d^{-1}\,kg^{-1}$ sludge solids. The necessary retention time varies from 20 days during the campaign, when influent BOD is high, to 2 days in the post-campaign period.

The Pasveer ditch described by Smith et al. (1975) received the overflow from an extensive lagoon system. The ditch had a total capacity of $2700\,m^3$ and was aerated by two $7.3\,m$ paddles driven by $11\,kW$ motors. The maximum design loadings were $550\,kg\,BOD\,d^{-1}$ or $0.2\,kg\,BOD\,m^{-3}$ of ditch capacity. When the influent BOD dropped in the post-campaign period, problems were encountered with the very fine flocs produced.

Some beet sugar refineries are having to comply with restrictions of nitrogen level in final effluent. Faup et al. (1978) and Heitz (1979) report the development and application of an 'anoxic' or low oxygen process designed to reduce nitrogen levels in a combined effluent of refinery wastewater and wastewater from a local community. The influent waste passed into the anoxic reactor where it was mixed with four parts of recycled effluent from a conventional activated sludge plant in series with the anoxic reactor. In the anoxic reactor micro-organisms converted to nitrogen gas the nitrates produced by nitrification in the activated sludge plant. Over 85% removal of ammonia nitrogen has been achieved.

The full-scale installation in a French refinery treated $1440\,m^3\,d^{-1}$ of refinery wastewater and $300\,m^3\,d^{-1}$ of communal waste. The combined BOD, COD and total nitrogen levels per day were 2450, 4415 and $500\,kg$, respectively. The anoxic reactor was a $700\,m^3$ rectangular basin mixed by two $11\,kW$ aerators. The main aerator basin had a capacity of $7000\,m^3$ and was equipped with four $70\,kW$ aerators designed to maintain an oxygen level of $2\,mg\,l^{-1}$. The oxygen transfer was $1.62\,kg$ per kW absorbed. A $154\,m^3$ clarifier recycled sludge to the main aeration basin at 100% of the influent flow. The removal of COD was 95%.

Activated sludge plants in cane sugar factories

Several full-scale activated sludge plants have been built adjacent to cane sugar factories following earlier pilot-scale trials by Shukla and Kapoor (1962), Shukla and Varma (1964), Bevan (1971), Miller (1971), and Simpson et al. (1972). Although these early studies established the potential of the activated sludge process for sugar mill waste, the transition to routine operation of full-scale plants was difficult.

McNeil et al. (1974) describe a treatment plant which combined an activated sludge plant, for primary treatment, in series with an oxidation ditch for final treatment. The activated sludge plant had a capacity of 910 m^3 with a 56 kW aerator for oxygen transfer, and the oxidation ditch had a capacity of 2280 m^3 with two 22.4 kW aerators. The plant has since been equipped with a 15 m diameter clarifier returning sludge from the oxidation ditch into the activated sludge system. The treatment of 40 m^3 h^{-1} of waste of COD 1600–4000 mg l^{-1} to 50–150 mg l^{-1} of COD or 6–25 mg l^{-1} BOD$_5$ was reported at a mixed liquor suspended solids concentration of 3500–4500 mg l^{-1}. Surplus sludge was disposed with mill mud to the field.

Ashe (1976) and Bruijn (1977) report the development of an oxidation ditch currently treating 2250 m^3 d^{-1} of overflow from an anaerobic pond from a COD of 700 mg l^{-1} BOD (350 mg l^{-1}) to a COD of 108 mg l^{-1}. The sludge solids level was 1000–3000 mg l^{-1}. Clarification was achieved in a section of the ditch which had been duplicated and in which the sludge was allowed to settle while aerators were not working. The ditch had a capacity of 1600 m^3 and was aerated at any time by eight 5.5 kW aerators. The oxygen requirements were 3 kg oxygen kg^{-1} BOD applied.

Bathgate et al. (1977) describe a two-stage activated sludge plant in which the first stage was operated at a zero level of dissolved oxygen, by providing insufficient aeration to satisfy the oxygen demand of the waste, and the second stage at an oxygen level maintained at 30% of saturation. Each stage of the system had separate clarification and return of sludge solids. Flow from the 'anaerobic' stage into the second stage was controlled by an oxygen electrode. The anaerobic pond was maintained at pH 5.8–6.0 by the addition of lime. The anaerobic stage was considered to act as a buffer tank and to eliminate problems of bulking sludges. It also provided a means of obtaining maximum oxygen transfer by the aerators for the purpose of BOD removal. The anaerobic pond of volume 3800 m^3 was aerated with two 16.8 kW aerators, with clarification effected by a 9.8 m diameter clarifier. The main pond had a volume of 5000 m^3 with aeration by two 50.4 kW aerators. The final effluent clarifier had a diameter of 16 m. The plant was reported to treat 3598 m^3 d^{-1} of waste of BOD$_5$ 949 mg l^{-1} to an average BOD$_5$ of 26 mg l^{-1}.

Simpson and Hemens (1978) describe an activated sludge plant which treated sugar mill waste together with bark effluent at various times. The plant had a capacity of 1840 m^3 and was equipped with two 40 kW aerators and an 89 m^3 settling tank. The sugar mill effluent at a flow of 340 m^3 d^{-1} was treated from a COD of $7864\,mg\,l^{-1}$ and BOD_5 of $4636\,mg\,l^{-1}$ to a COD of $788\,mg\,l^{-1}$ and a BOD of $85\,mg\,l^{-1}$. The sludge volume index was 155. Phosphate deficiency and the introduction of bark mill effluent tended to promote bulking of sludge.

Activated sludge plants in refineries
Tippens *et al.* (1978) describe the operation of a pilot and full-scale 'Activox' activated sludge system for treatment of char wastewater. The full-scale plant treated 738 to $1832\,m^3\,d^{-1}$ of char wastewater with an average BOD of $800\,mg\,l^{-1}$. Process efficiency has been maintained at over 98%. The process was a modified oxidation ditch system and incorporated two tanks of combined volume of $5300\,m^3$. Each tank was aerated by a 44.7 kW aerator which directed flow in the tanks at a velocity of $30.5\,cm\,s^{-1}$. The 15.2 m diameter clarifier was designed for $10.2\,m^3\,d^{-1}\,m^{-2}$ surface flow and sludge recycle of 100% of influent flow. Once operation started, the plant mixed liquor solids concentration was maintained at $7500\,mg\,l^{-1}$. Oxygen requirements were determined at $1.1\,kg\,O_2\,kg^{-1}\,BOD$ removed. Nitrogen and phosphorus were added in the ratio BOD:N:P of 100:2.7:0.36 to increase the existing nutrient levels in char wastewater to a total level of 100:5:1. Surplus sludge was thickened and discharged to irrigate a pecan orchard.

1.3.4 Biofiltration or trickling filters

The application of trickling filtration on a large scale has been inhibited by the long induction period required, the high recirculation rates required for high-strength wastes and the risk of blockages.

Biofiltration in beet factories
Carruthers *et al.* (1960) examined the performance of plants treating lagoon effluent from two factories. A reduction in BOD ranging from 55 to 96% was obtained over a 12 month period at loadings of 0.04 to $0.45\,kg\,BOD\,m^{-3}$ of filter. An experimental unit consisting of two filters in series treated daily $1.42\,m^3$ of wastewater m^{-3} of filter from $105-180\,mg\,l^{-1}$ to less than $20\,mg\,l^{-1}$. The maximum possible daily loading for a single-stage unit was $0.4\,m^3\,m^{-3}$. Crane (1968) reports the successful operation of a full-scale two-stage plant treating a waste lagoon overflow from as high as $2800\,mg\,l^{-1}$ to $7.5\,mg\,l^{-1}$ of BOD_5. High-strength influent waste was diluted with up to 7 volumes of river water and 8–9 volumes of recirculated effluent to effect purification. The average daily loading to the first- and second-stage filters was 3 kg and $0.07\,kg\,BOD\,m^{-3}$, respectively.

Biofiltration in cane sugar factories
Early experiments by Shukla *et al.* (1960) and Shukla and Kapoor (1961) demonstrated that 70% reductions in BOD could be obtained by biofiltration

of sugar factory effluent mixed 20:1 with sewage. Bevan (1971) reports the operation of a 'Flocor' plastic-medium biofiltration plant. The plant was 4.9 m deep and of 0.37 m cross-sectional area with an effective surface area of 160 m^2. The extent of recirculation of treated effluent varied from 1.8 to 4. The plant was reported to reduce the BOD of 4 m^3 h^{-1} of waste from 500 mg l^{-1} to 100 mg l^{-1}.

The treatment of a synthetic waste based on molasses and sucrose was studied by Bruijn (1975). The plant consisted of two filters in series, each of 2 m in height, with a sedimentation tank after each filter. The first filter was a high-rate plant, 0.22 m in diameter, composed of plastic medium ('Cloisonyle') of specific surface 220 m^2 m^{-3}. The low-rate filter was 0.2 m in diameter and was packed with gravel (specific surface 40 m^2 m^{-3}). The addition of nitrogen and phosphorus in the ratio COD:N:P 100:4:0.8 was essential for adequate BOD removal. The optimum performance was achieved (Bruijn, 1975; 1977) when 2 kg m^{-3} d^{-1} of waste of 1400 mg l^{-1} BOD was applied to the high-rate filter and a loading of 0.26 kg m^{-3} d^{-1} was applied to the low-rate filter, each with a recirculation ratio of 3 parts of treated effluent to one part of the incoming waste. The COD reduction was from 1400 mg l^{-1} to 440 mg l^{-1} in the high-rate filter and from 440 mg l^{-1} to 147 mg l^{-1} in the low-rate filter. The corresponding BOD removal was from 650 mg l^{-1} to 110 mg l^{-1} to 6 mg l^{-1}.

A full-scale plant consisting of a primary and secondary filter in series, followed by individual settling tanks, is described by Lewis and Ravnoe (1976). The plant received the overflow from a large anaerobic pond and the surplus sludge produced was discharged back to the pond. Each filter was 23 m in diameter and contained 920 m^3 of stone of 100–150 mm size-grading and a specific surface area of 40 m^2 m^{-3}. Each filter was followed by two sedimentation tanks, 7.62 m by 3948 m with a peripheral weir length of 18.3 m. The design flow rate was 30 l s^{-1} (including 1:1 recirculation) and this constituted a surface loading to the tanks of 1.16 m^3 m^{-2} h^{-1}. The design load on the primary filter was 1 kg COD m^{-3} d^{-1}. The plant was reported to treat the anaerobic pond overflow to less than 120 mg l^{-1} COD. A constant removal of 77% of the applied COD was observed. These results indicated that the COD applied to the filters could not exceed 520 mg l^{-1} if the effluent was to be discharged at less than 120 mg l^{-1} COD.

1.3.5 Anaerobic methane-generating processes
The increasing interest in anaerobic treatment processes has resulted in the development of pilot-scale and full-scale plants, particularly in France and the Netherlands.

Anaerobic treatment of beet factory wastewater
A laboratory-scale mesophilic process (35°C) was operated by Devillers *et al.* (1977) at daily loadings ranging from 0.34–1.33 kg TOC m^{-3} with 84 to 97%

removal of organic matter. (The BOD:TOC and COD:TOC ratios were 2:1 and 3:1 respectively.) The methane gas content varied from 61% to 90%. Lescure and Bourlet (1977; 1978) report 90% removal of organic carbon in a 90 m^3 vessel at a daily loading of 3 kg COD m^{-3}. The required retention time for 90% removal of organic matter was 18 h.

Larger installations are currently being developed in France (Lescure and Bourlet, 1979a, b; Wagner, 1980) which have been operated at daily loadings of up to 6 kg COD m^{-3}. The vessel was divided into an 800 m^3 reaction vessel and a settling basin having a surface area of 400 m^2. A reduction in COD from 5896 mg l^{-1} to 1107 mg l^{-1} was reported (Lescure and Bourlet, 1979b) at an average flow of 28 m^3 h^{-1}. The gas contained 82.4% methane.

A major advance in the anaerobic treatment of sugar factory wastewaters appears to have come from the adaptation of the Upflow Anaerobic Sludge Blanket (UASB) process (Lettinga *et al.*, 1980) to beet factory wastes (De Vletter, 1979; Pette, 1979). The UASB process relies on the development of a well flocculated anaerobic sludge in a bottom-fed reactor. The reactor incorporates a gas-separation and sludge-settling device which ensures minimum sludge loss from the system.

Pette *et al.* (1980) describe in detail the performance of the UASB process in handling liquid sugar waste and beet factory wastes. Liquid sugar waste (COD 17 000 mg l^{-1}) has been treated in a 30 m^3 vessel at daily organic loadings of 13.3 kg COD m^{-3} in a retention time of 30 h. Over 94% removal of COD was reported with gas production of 0.56 m^3 kg^{-1} COD removed and a methane content of 68%. Overflow from mud ponds in a beet factory has been treated in a full-scale 800 m^3 vessel. The influent waste (COD 3000 mg l^{-1}) was reduced in organic content by 88% at daily loadings of 16.25 kg COD m^{-3} in a retention time of 4.5 h. The gas production was 0.38 m^3 kg^{-1} COD removed, containing 90% methane. High sludge solids levels of 30 to 40 kg m^{-3} were retained in the reaction vessels.

Significantly, Pette *et al.* have demonstrated that the process can be operated over a short beet campaign and can then be restarted in a matter of days at the commencement of the next campaign.

Effluent from the UASB process operating on beet factory wastewater may be used as make-up for transport-water or can be treated by aerobic methods for discharge to a river. Huss (1979) also describes an anaerobic methane-generating process which has been applied in Swedish factories.

Anaerobic treatment of cane factory wastewater

Bhaskaran and Chakraborty (1966) operated a pilot plant at a daily loading of 0.13–0.21 kg BOD m^{-3} and a retention time of two days. The BOD was reduced from 886–1416 mg l^{-1} to 136–685 mg l^{-1}, a reduction of 65%. Sinha and Thakur (1967) achieved an 88% reduction in BOD of a wastewater in a pilot-scale digester at a daily loading of 1.77 kg BOD m^{-3}. The BOD was reduced from 3612 mg l^{-1} to 396 mg l^{-1}.

High-rate processes such as the UASB process could no doubt be applied with equal success in cane sugar factories. However, the methane gas is of less value to cane sugar factories as they are generally more than self-sufficient in bagasse fuel.

1.4 Fermentation of molasses or sugar cane juice — sources of distillery wastewater

In recent years, the major source of industrial ethanol has been the petrochemicals industry. Molasses from sugar beet or sugar cane manufacture has been an established substrate for rum manufacture and for limited production of industrial alcohol. In addition, molasses is used as an animal feed supplement and to a limited extent as a substrate for production of citric acid, glutamic acid and other fermentation byproducts. There is increasing world interest in the production of ethanol as a fuel extender by fermentation of carbohydrates extracted from crops. Molasses supplies are limited and, in the long term, large-scale production of sugar cane or other crops such as wheat or cassava would be required for fermentation ethanol to have any significance as a fuel substitute.

As far as the sugar industry is concerned, virtually all published information on treatment of distillery wastes has related to stillage from the fermentation of molasses.

The molasses is pasteurized and diluted to 9–11% sugar for rum manufacture or to 15–20% for industrial alcohol manufacture. Molasses contains about 30–40% sucrose and 12–18% invert sugar. Following acidification with sulphuric acid, and the addition of nutrients, it is fermented by a yeast for 24–48 hours at between 25° and 35°C. The fermentation generates heat and cooling-water is required. The final yield of ethanol is approximately half the initial fermentable sugar concentration. Most yeasts become sensitive to alcohol levels above 9–10% and this limits the initial sugar concentration and the fermentation rate. Yeast recycle is extensively practised.

Following fermentation the alcohol passes to a stripping column and then to rectifying columns for concentration up to 95% ethanol. The principal waste is the stillage emanating from the first stripping column, otherwise known as vinasse, vinhoto, slops or dunder. Other wastewaters arise from the washing of fermenters. Thus about 8–12 m^3 of stillage are discharged for each kilolitre of alcohol produced.

The composition of stillages from cane juice and molasses differ in that the former contains lower levels of ash and organic matter (Table 1.8). Cane juice stillage is light brown in colour whereas molasses stillage is dark brown to black. Molasses stillage varies according to the source of the molasses. Australian molasses stillage is generally high in ash and organic matter (Jackson, 1966; Bieske, 1979). The composition of a sample of Australian molasses stillage is

Table 1.8 Comparison of stillage composition from cane juice and molasses (after Jackman, 1977)

Parameter	Cane molasses (g l^{-1})	Cane juice (g l^{-1})
Organic matter	63.4	19.5
Total nitrogen	1.2	0.3
Sulphate (SO$_4$=)	8.4	0.6
Calcium (CaO)	3.6	0.7
Phosphorus (P$_2$O$_5$)	0.2	0.2
Magnesium (MgO)	1.0	0.2
Potassium (K$_2$O)	7.8	1.2

shown in Table 1.9. A detailed description of stillage from the fermentation of beet molasses was published by Basu (1975).

All of the studies described below refer principally to molasses stillage treatment. Stillage from cane juice fermentation has not been as widely examined but the disposal problem is not as great as it is for molasses stillage.

Table 1.9 Composition of molasses stillage (Australian sample)

Parameter	Concentration
pH	4.8
COD, total (mg l^{-1})	110 000
COD, soluble (mg l^{-1})	102 000
BOD, total (mg l^{-1})	73 000
BOD, soluble (mg l^{-1})	70 000
Total solids (%)	11.9
Organic matter (%)	8.8
Ash (%)	3.0
Sulphate (mg l^{-1})	3 000
Total nitrogen (mg l^{-1})	2 300
Phosphate (mg l^{-1})	40
Potassium (mg l^{-1})	12 000
Sodium (mg l^{-1})	330
Calcium (mg l^{-1})	2 800
Magnesium (mg l^{-1})	1 700
Iron (mg l^{-1})	40
Manganese (mg l^{-1})	13

1.5 Distillery wastewater — treatment, disposal and utilization

1.5.1 Reduction of effluent levels by process modification

A significant reduction in effluent volume is obtainable by fermentation at the highest possible initial sugar concentrations. Such advances may come from the development of alcohol-tolerant yeasts. Clarified juice should be concentrated to at least 20–22% sugar level before fermentation. Kujula (1979) reports the achievement of final alcohol concentrations of up to 12%.

Distilleries are able to recycle or 'slop back' some of the stillage for molasses dilution. Recycle volumes of up to 20% are possible (Kravets et al., 1970) but clear limitations are imposed by the quality of the initial molasses. Kujala (1979) describes a process whereby molasses pretreatment can allow slop-back volumes of up to 60%. The pretreatment includes pasteurization, acid hydrolysis of non-fermentables, elimination of volatile organics and removal of gums and calcium sulphate.

1.5.2 Land disposal of stillage

Distilleries located in rural areas can opt for irrigation under suitable conditions of climate, soil type and rainfall. The concept of disposal back to land of the nutrient substances in stillage is attractive. However, central distilleries would find it possible to irrigate only a fraction of the original growing area if massive transportation and pipeline costs were to be avoided. Careful monitoring of the soil and plant yields are required to avoid excessive salinity, deterioration in cane quality and rising levels of ash in juice. In addition, nitrogen and phosphorus supplements are required to avoid fertilizer imbalance.

Studies in Brazil have established that irrigation of stillage can result in improvements in soil and cane quality. Early reports quoted application rates of $419 \, m^3 \, ha^{-1}$ (Guimares et al., 1968) and even up to $1000 \, m^3 \, ha^{-1}$ (Monteiro, 1975). However, more recent work in Brazil used dosage levels of only $35 \, m^3 \, ha^{-1}$ (Brieger, 1977; 1979; Magro and da Gloria, 1977).

Cooper (1975) reports the application of stillage in Trinidad at rates of 61 to $185 \, m^3 \, ha^{-1}$, and notes that although cane sugar yields improved, the results were less favourable at the upper level. In Australia, Bieske (1979) has determined that the maximum application rate to sugar cane appears to be about $12 \, m^3 \, ha^{-1}$. This is due to the higher level of inorganic substances in Australian molasses. Bieske observed that at this rate stillage was as effective a source of potassium as muriate of potash in supplying the cane plant's requirements. Usher and Willington (1979) have reported the commencement of a trial in Australia with application rates of $100 \, m^3 \, ha^{-1}$.

1.5.3 Evaporation, incineration and potassium recovery

Some distilleries concentrate stillage for application as a fertilizer or as an animal feed supplement. The degree of concentration need only be to 30°–35° Brix but for purposes of transportation it should be concentrated to 60°–65° Brix. Spray-dried material is hygroscopic and is therefore difficult to handle. In France, Lewicki (1978) reports application of 60% stillage at a rate of 2.5 to $3 \, t \, ha^{-1}$. Dubey et al. (1977) concentrated stillage to 75°–80° Brix for mixing with cane factory filter mud for sale as a fertilizer.

A patented process described by Bass (1974) heated concentrated dunder (60°–80° Brix) to near boiling to induce thickening and then added soluble phosphate at 80°C. Further heating at 105°–120°C induced coagulation after which the material was dried to form a friable non-hygroscopic solid which was suitable as a fertilizer or fodder. The phosphate dose for making fertilizer was 4% triple superphosphate; 5% ammonium dihydrogen phosphate was added for a fodder product.

The application of stillage as a fodder additive appears to have limitations due to vagaries of market conditions for animal sales and the prices of other fodders. The value of the stillage as a food supplement is limited by the laxative effects of the inorganic constituents (Jackman, 1977). In addition, the cost of evaporation is high for distilleries that do not generate their own fuel (e.g. bagasse). Autonomous distilleries would find it preferable to concentrate stillage to 60°–65° Brix for incineration. The combustion at these concentrations is autothermal and the released energy is said to be sufficient to maintain the evaporating station (Jackman, 1977). Further, the ash is rich in potassium which may have market value as a fertilizer.

The combustion temperature must not exceed 700°C, above which ash fusion results in a worthless product. However, there are difficulties in obtaining complete combustion at this temperature.

Reich (1945) developed a process for ultimate recovery of a high-grade carbon and potash. Removal of calcium sulphate from the molasses feedstock helped reduce ultimate processing costs for stillage treatment. The fermented liquor was centrifuged and the yeast was recovered for animal feed. The liquor was then passed direct to a quadruple-effect evaporator. The condensates from the first two vessels were distilled for ethanol and the final vessels concentrated the stillage to 70 to 80% solids. The stillage was neutralized above pH 8.5, passed through low temperature (343°C) carbonizing retorts and was then activated at 870°C. The material was extracted with water to recover carbon and a liquor solution composed of potassium chloride and potassium sulphate.

Chakraborty (1964) in India neutralized and filtered stillage and then passed it into a quadruple-effect evaporator for concentration to 75% solids. The waste passed through the incinerator in inclined baffle plates countercurrent to the hot gases. The ash was leached with water, neutralized with sulphuric acid and concentrated, and crystalline potassium sulphate (72.25%) and potassium chloride (20.75%) were recovered. Forced circulation evaporation was claimed to reduce foaming and scale formation.

A fluidized bed combustion unit at 700°C was employed for incineration by Gupta et al. (1968). Potash recovery of 70% was obtained. Kujala et al. (1976) recommended using a fluidized bed as a gasifier, operating below the fusion temperature of the ash. They suggested that the ash could be separated from volatiles and gases in a cyclone separator and that an after-combustion chamber could be used for oxidation of volatiles.

Evaporation and incineration appear to represent a means of obtaining ultimate disposal of stillage. However, hot stillage is corrosive and neutralization with lime gives rise to enormous problems of scaling. The economics of the process depends on the value of the potassium recovered. The low potassium levels in Brazilian cane make the process uneconomic in that country (Jackman, 1977). The economics of potassium recovery in India is also doubtful (Bhandari *et al.*, 1979). Kujala (1979), however, considered that distilleries processing 200 t day^{-1} of molasses could generate US$0.5M in potassium sales per year.

1.5.4 Biological treatment

Biological treatment of distillery wastes has received extensive study in recent years although few distilleries appear yet to have constructed biological treatment plants. Distillery wastes, particularly stillage from molasses, have a high level of inorganic salts that are not removable by biological treatment alone. Thus, large rivers or estuaries would be required if discharge to a watercourse were envisaged. Alternatively, expensive tertiary treatment might be required. However, these inorganic substances would be better returned as a fertilizer to the agricultural land from which they were received.

The application of biological treatment can be envisaged in a number of situations. Firstly, the waste could be assimilated directly by a municipal plant if substantial dilution by sewage was available. Sawyer and Anderson (1949) diluted to 1% with sewage a rum distillery waste of BOD$_5$ 24 500–50 000 mg l^{-1} and applied this to a two-stage trickling filter system with a 3:1 recycle ratio. The waste was reduced from 485 mg l^{-1} BOD$_5$ to 20 mg l^{-1} at a daily loading rate of 0.95 kg BOD m^{-3} or 17.1 m^3 m^{-3}. Burnett (1973) diluted rum stillage of COD 90 000 mg l^{-1} and BOD 25 000–30 000 mg l^{-1} 10% in domestic sewage. An average reduction in loading of 39.9% was achieved by preneutralization and application to a trickling filter at a rate of 4.8–6.08 kg COD d^{-1} m^{-3} with a recirculation of 0.5:1. A daily loading of 10.24–11.68 kg COD m^{-3} and a recirculation of 0.3:1 resulted in only 25.6% reduction although an increase in recirculation rate to 1.5:1 resulted in an increased removal of 39.6% at a daily loading of 12.32–13.44 kg COD m^{-3}. Apart from the unattractively low removal, problems of odour, fly-breeding and fungal overgrowth occurred. Thus, treatment to a low BOD effluent could be expected to be achieved only by even greater dilution and a higher recirculation rate. A similarly low performance of 22.8% removal was obtained by activated sludge treatment for 4.5 h at a daily loading of 62.9 kg COD m^{-3}.

Secondly, biological treatment by anaerobic methods to remove the major proportion (~90%) of organic matter would be a suitable method of pretreatment before disposal to a municipal system, to land disposal or even an ocean outfall. Ultimate disposal would be facilitated and the anaerobic treatment of the distillery waste would generate methane gas as a fuel for steam generation.

Jackman (1977) estimated that digestion of distillery waste would generate 580–720 litres of gas per kg of BOD removed consisting of 65% (v/v) methane and having a calorific value of 6000 kcal m^{-3}. In the case of an autonomous molasses distillery where fuel has to be imported, this gas would largely replace other fuels. The picture is not as favourable for a distillery situated as an adjunct to a cane sugar factory. Wright and McNeil (1980) estimate that a factory processing 50 000 kilolitres of ethanol per annum from 559 000 t of cane would generate daily 42 500 m^3 of methane. This gas would be of no economic value to a sugar mill unless the bagasse displaced could be sold. The heat content of the methane gas would be about 30% of the heating value of the bagasse produced by the sugar factory. The high temperature of distillery waste would ensure that very little of the gas would be required to maintain the temperature of the anaerobic reaction vessel.

There is some concern that distilleries operating on a seasonal basis may be confronted with difficulties in bringing plants to full operation before factory operation commences. Most of the studies cited later pay particular attention to the long acclimatization periods and careful start-up required. Periods of three months or more are commonly mentioned. However, studies by Lettinga *et al.* (1980) suggest that, once established, anaerobic processes can be shut down for many months and brought back to full activity in a few weeks.

The high volume of discharge of distillery wastes means that any practical anaerobic process must be a high-rate process in order to minimize the tank volumes required. Considerable research has been undertaken in order to develop suitable processes for distillery wastes.

Batch processes have been demonstrated as being completely impractical because of long lag times of the order of 85 days (Sen and Bhaskaran, 1962; Basu and Leclerc, 1973). Early studies (Stander and Snyders, 1950) established solids recycle as an integral part of maintaining process stability and high performance. However, conventional clarification techniques have limitations because settling rates of sludges can be slow due to the presence of gas bubbles. The generated gases can have the effect of lifting large quantities of settled sludge to the clarifier surface, resulting in loss of micro-organisms from the system. The use of vacuum degasification and flocculants has been offered as a solution by Shea *et al.* (1974), who degasified the clarifier feed at a vacuum of 15 mm Hg. A polymer requirement of 15 g d^{-1} m^{-3} stillage and a clarifier loading rate (overflow) of 9.7 m^3 d^{-1} m^{-2} were recommended. The Upflow Anaerobic Sludge Blanket (UASB) process developed by Lettinga *et al.* (1980), or related anaerobic processes, may overcome these difficulties.

Inorganic components, particularly sulphate and potassium, are present in molasses stillage at levels which approach or exceed reported levels of toxicity to anaerobic bacteria. The determination of the precise levels at which various cations are inhibitory is complicated by synergistic and antagonistic effects between the ions in solution (Kugelman and Chin, 1971). However, dilution of stillage has been shown to result in increased performance with reduced retention times up to the point where throughput is hydraulically limited (Radhakrishnan *et al.*, 1969).

Sulphate is converted by sulphate-reducing bacteria to sulphide which is then released as hydrogen sulphide gas or precipitated by metals (Roth and Lentz, 1977). High sulphate levels in molasses dunder were shown by Stander and Elsworth (1950) to be inhibitory to anaerobic digestion. These workers demonstrated that mesophilic and thermophilic digesters performed far better treating desulphated stillage containing 500 mg l^{-1} sulphate than with neutralized stillage containing 6750 mg l^{-1} sulphate. This was evidenced by higher throughput, lower levels of volatile acids and higher percentage conversion to gas of organic carbon in feed. Further experiments with deionized stillage to which a low level of required nutrients and determined quantities of sulphate were added confirmed these results. The level of sulphide in digester liquor which was considered inhibitory was 150 mg l^{-1} in the mesophilic digester and 125 mg l^{-1} in the thermophilic digester.

A number of explanations of the toxicity of high sulphate levels have been proposed. Competition by sulphate-reducing bacteria and methanogens for available hydrogen and acetate (Bryant *et al.*, 1977; Winfrey and Zeikus, 1977) appears the most likely cause, although the raising of the redox potential of the environment by sulphate (Macgregor and Keeney, 1973) and direct inhibition by sulphide (Cappenberg, 1974) are also possible.

Among the proposed means of overcoming toxicity, dilution of the waste appears to be the most popular. However, in practice large quantities of dilution water of either low or high quality are probably unlikely to be available for many distilleries. Many would question the use of good-quality dilution water to help dispose of an already highly polluting wastewater.

A more practical method may be to scrub digester gas of hydrogen sulphide and recirculate this gas back into the digester liquor (Hiatt *et al.*, 1973; Shea *et al.*, 1974). This has the dual effect of assisting the removal of hydrogen sulphide from the digester liquor and circulation of the contents.

Studies of anaerobic digestion of distillery wastes include separate research in the thermophilic range 52°–55°C and in the mesophilic range 33°–37°C. The high temperature of distillery wastes makes thermophilic treatment a realistic proposition. Boruff and Buswell (1932) treated at 53°C a distillery waste of 17 000 mg l^{-1} BOD to 3000 mg l^{-1} BOD at a daily loading of 2.8 kg BOD m^{-3}. The level of volatile acids was 500 mg l^{-1} and gas production was 0.3 l d^{-1} l^{-1} of digester volume. The performance of another reactor at a daily loading rate of 8.5 kg BOD m^{-3} was less stable, with volatile acid levels of 2000 mg l^{-1}. Gas production here was 7 l d^{-1} l^{-1} of digester volume and the final effluent was 5700 mg l^{-1} BOD. The gas produced contained 55–58% methane. Buswell and Le Bosquet (1936) used a two-stage thermophilic digestion process to treat a maximum loading of 2.4 kg BOD d^{-1} m^{-3} from a BOD of 15 500 mg l^{-1} to 1500–2000 mg l^{-1} with a retention time of 13 days. The gas production from the waste (2.4–3.2% volatile solids) was 0.7 m^3 kg^{-1} of volatile solids. More recently, Sonoda and Tanaka (1968), with stillage diluted 33% in water, obtained 68–70% removal of BOD at maximum daily loading of $14.4\,kg\,BOD\,m^{-3}$. Gas production of $0.62\,m^3\,kg^{-1}$ BOD removed was obtained.

Stander and Snyders (1950) examined the mesophilic treatment of molasses stillage using a technique of periodic reinoculation with an artificially prepared inoculum and centrifugation and recycle of effluent solids. Using a stillage diluted 1 : 1, a daily loading of $8.8\,kg$ volatile solids m^{-3} was applied, equivalent to a retention time of 3.75 days. Sen and Bhaskaran (1962) studied the mesophilic treatment of molasses stillage in two digesters in series, the first providing a retention time of 40 days and the second 20 days. At a daily loading of $0.74\,kg\,BOD\,m^{-3}$, the BOD was reduced from $30\,000\,mg\,l^{-1}$ to $775\,mg\,l^{-1}$ in the first stage and to $320\,mg\,l^{-1}$ in the second. Further experiments demonstrated that the maximum safe daily loading rate was $3\,kg\,BOD\,m^{-3}$ for a retention time of 8 days.

Radhakrishnan et al. (1969) examined the mesophilic treatment of molasses stillage (BOD_5 $30\,000\,mg\,l^{-1}$) over a range of dilutions and varying retention times but at a relatively uniform organic loading rate of 1.6–$2.0\,kg\,BOD\,d^{-1}\,m^{-3}$. The percentage removal of BOD increased slightly from 87.5% to 89% as the retention time was decreased from 20 days to 5 days. However, the process was unstable below a retention time of 4 days, indicating wash-out of solids in the absence of solids recycle. By mixing 1.5 parts of stillage with 1 part of water, a daily loading of $3.6\,kg\,BOD\,m^{-3}$ was achieved at a retention time of 5 days.

Excellent results were also reported by Hiatt et al. (1973) for mesophilic treatment of molasses stillage. Undiluted stillage ($100\,g\,l^{-1}$ COD) was treated at a daily loading of $5.9\,kg\,m^{-3}$ and retention time of 16.7 days, resulting in a reduction in COD of 71.9%. The level of volatile acids was $650\,mg\,l^{-1}$. An impressive daily loading rate of $11.5\,kg\,COD\,m^{-3}$ (5.6 days retention time) was achieved when the stillage was diluted to $65\,g\,l^{-1}$. Gas production was 0.25–$0.32\,l\,g^{-1}$ of COD removed and the gas contained 62–66% methane. Typical solids levels in the reactor were 35 to $65\,g\,l^{-1}$.

Shea et al. (1974) studied the mesophilic treatment of rum stillage of $33\,000$–$55\,000\,mg\,l^{-1}$ BOD in an anaerobic contact process with solids recycle. The process had a daily loading rate of 0.086–$1.19\,kg\,BOD\,m^{-3}$ or 0.4–$3.44\,kg\,COD\,m^{-3}$, with hydraulic retention times varying between 140 days and 19 days, respectively. The COD removal varied between 60% and 80%. The total solids level in the reactor was 1500 to $6000\,mg\,l^{-1}$. Gas production was 0.28–$0.33\,l\,g^{-1}$ COD removed, 53 to 60% of which was methane. A daily loading rate of $3.9\,kg\,COD\,m^{-3}$ for a stillage of $54.6\,g\,l^{-1}$ COD was applied by Roth and Lentz (1977), achieving a 77–83% reduction in a hydraulic retention time of 14 days. The addition of 0.15% yeast extract allowed the daily loading rate to be increased to $9.9\,kg\,COD\,m^{-3}$, thus reducing the retention time to 5.5 days. Solids recycle was achieved by conventional settling.

Some authors have studied the relative efficiencies of thermophilic and mesophilic treatment on the same waste. Jackson (1956) reports improved performance by thermophilic processes relative to mesophilic processes, whereas Sen and Bhaskaran (1962) were not able to detect any great improvement. Ono (1965) refers to full-scale processes operating in Japan. Thermophilic processes were reported to be operating at 6 kg volatile solids $d^{-1} m^{-3}$ and mesophilic processes at 2.5 kg volatile solids $d^{-1} m^{-3}$. A BOD reduction of 80% to 90% from 30 000–40 000 mg l^{-1} was reported. Gas production of 500 to 700 $m^3 t^{-1}$ volatile solids was recorded.

An 80.5% reduction at 35°C and an 84.2% reduction at 55°C were reported by Basu (1970) for digesters loaded at 3 kg BOD $d^{-1} m^{-3}$ (retention time 10 days) treating stillage of 30 000 mg l^{-1} BOD. Stirred reactors operated at a daily loading of 3.5 kg m^{-3} achieved a 73% reduction at 35°C and 70% reduction at 55°C. Performance was better at a daily loading of 3.2 kg m^{-3} where 96% reduction was obtained at 35°C and 97.7% reduction at 55°C. Further comparisons by Basu and Leclerc (1972; 1973) with stirred digesters equipped with solids recycle confirmed a maximum daily loading rate of 3.0–3.2 kg BOD m^{-3} for a 10 day retention time at each temperature. These workers concluded that only minor benefits would be obtained with thermophilic digestion.

Thus, overall, no conclusive statement can be made as to whether thermophilic operation is more desirable than mesophilic. Perhaps varying levels of toxic substances in the different types of stillage account for the discrepancies. Stander and Elsworth (1950) observed that, for molasses stillage with high sulphate levels, treatment by mesophilic digestion appeared preferable to thermophilic digestion. The reverse situation applied in the case of molasses stillage in which the sulphate level was reduced. At this stage, pilot studies of individual wastes must be recommended.

At a practical level, most distilleries would be able to get larger volumes of fresh mesophilic sludges from sewage plants in the event of some operational catastrophe or for initial plant start-up. Further, reported sensitivities of thermophilic processes to load variations and temperature changes suggest that caution should be exercised before opting for a thermophilic process.

Anaerobic treatment must be followed by aerobic treatment to reach low BOD levels. Ono (1965) reports that distilleries in Japan achieved levels of 40–120 mg l^{-1} BOD in activated sludge plants at daily loadings of 3.5 kg BOD m^{-3}. Skogman (1979) describes a full-scale anaerobic–aerobic process capable of achieving 125 mg l^{-1} BOD. Surplus sludge produced in the aerobic stage was wasted through the anaerobic phase. Overall power requirements were 0.15 kWh kg^{-1} of treated organic matter.

The direct aerobic treatment of distillery waste is almost certainly uneconomical. Cillie *et al.* (1969) considered that aerobic treatment should not be attempted on distillery wastes above a COD level of 4000 mg l^{-1}. The enormous quantities of oxygen required and high conversion of BOD to sludge solids would almost certainly make direct aerobic treatment unattractive.

The satisfactory treatment of stillage in lagoons has been reported although the high land requirement would preclude it for many distilleries. Rao (1972) studied treatment in two pilot lagoons in series. The first lagoon, 1.8 m deep, was loaded at 0.604 kg BOD d^{-1} m^{-3} and a BOD reduction from 40 000 mg l^{-1} to 2000–3000 mg l^{-1} was obtained for a retention time of 66 days. The second lagoon, 0.9 m deep, reduced the BOD to 300–600 mg l^{-1} at a daily loading of 0.07 kg BOD m^{-3} for a retention time of 43 days. No significant difference in performance was observed if the pond depth varied between 0.9 and 3.0 m. The effluent from lagoons in India is not suitable for discharge and is used for irrigation (Rao, 1972; Bhandari et al., 1979).

1.5.5 Production of yeast or fungal biomass

A limited amount of stillage is used for the production of biomass. Certain yeasts and fungi are able to grow in stillage by utilizing some of the residual sugar, organic acids and glycerol. The potential yield of biomass will depend on the organic content of the stillage. Kujala et al. (1976) consider an average yield from molasses stillage would be 1 tonne of yeast from 100 tonnes of stillage.

The large-scale production of biomass from stillage seems very doubtful. Firstly, protein prices greatly affect the economics of the process. Secondly, the production costs are relatively high. Nutrient additions of nitrogen and phosphorus are required, contributing up to 50% of the cost (Jackman, 1977). Heat is generated at the rate of about 3870 kcal kg^{-1} of yeast (Kujala et al., 1976) and expensive cooling equipment is required in hot climates. Aeration costs are also high. Finally, the reduction in organic matter is usually only 50–80%. Thus, there would still be a highly polluting effluent from the biomass production.

Il'ina et al. (1969) obtained biomass yields of 12–16 kg m^{-3} from species of *Candida*, *Oidium*, *Hansenula* and *Rhodotorula*. Yields from a *Trichosporon* sp. were 28 to 50% higher. Lin (1970) cultured *Pichia tainania* and *Candida utilis* with yields of 42.1% and 46.4% protein after 42 h culture. The stillage was supplemented with 0.5% urea and 0.1% P$_2$O$_5$. Chang and Yang (1973), with *C. utilis*, obtained a yield of 16 kg m^{-3} of stillage with nutrient additions of 0.15% N and 0.1% P from a stillage of 9–11° Brix. The protein content was 46%.

Zabrodskii et al. (1973) obtained yields of 280 kg from the production of 1 kilolitre of alcohol in a two-stage system treating undiluted beet stillage of 8°–9° Brix. The retention times of the two stages were 6 h and 8–10 h. Paz and Lopez Hernandez (1973) obtained yields of *C. utilis* of 18.75 kg m^{-3} in a retention time of 9–12 h. Rolz et al. (1975) obtained COD reductions of 56–85% in batch culture tests of five fungi with mycelial yields of 11–19 kg m^{-3}. Wang et al. (1977) obtained increased utilization by supplementing the stillage with an equal volume of molasses containing 5% sugar. The residual effluent still contained 13–21% of the original BOD.

1.6 Conclusions

The sources and treatment of wastewater from the sugar industry have been reviewed.

Beet sugar factories have limited their principal wastewater discharge to surplus transport-water by means of extensive recirculation. The wastewater discharge from cane sugar factories varies greatly and includes surplus condensates, cane wash-water and overflow from spray ponds or cooling towers. The elimination of large volumes of cane wash-water from factories in some countries may in the long term be better effected by improved harvesting techniques. Technology is available for recovery of fly ash and filter mud and preventing their disposal to watercourses.

Many factories and refineries use river water in a single-pass system for cooling of barometric condensers. Careful design and operation of evaporators and vacuum pans and the installation of entrainment arresters can virtually eliminate sugar losses to cooling-water.

Ponding or irrigation or both of these methods have been widely adopted in both the beet and cane sugar industries. In practical terms, they appear preferable where adequate land is available and climate is suitable. These methods are especially suited to the seasonal nature of the raw sugar industry. Overall, batch ponding seems to have been most successfully applied.

Treatment by the activated sludge process has been successful. Operation at low sludge loadings ($0.2-0.3$ kg BOD d^{-1} kg^{-1} sludge solids) appears essential to avoid bulking of sludge and to achieve low effluent BOD. High-rate processes such as the RT–Lefrancois process have been successful for pretreatment. Anaerobic pretreatment of the waste has also been successful in avoiding problems associated with poorly settling sludges.

Anaerobic methane-generating processes have been successfully developed for the first-stage treatment of beet factory wastewaters. There is no doubt that they could also be applied to high-strength cane factory effluents, although the methane gas would be less valuable in that cane factories are usually self-supporting with bagasse fuel.

Distilleries are faced with an enormous effluent problem. Land disposal and evaporation followed by incineration appear to offer a means of ultimate disposal. The former method has been applied extensively in Brazil, but climate, soil type and other factors related to crop fertilization may limit its application in other cases. Evaporation of stillage has the disadvantage of severe problems of scaling.

The anaerobic contact process is under increasing investigation as an alternative. As with incineration, an independent distillery is able to generate much of its power from waste treatment. However, biological treatment alone does not result in removal of colour or salts and ultimate land disposal or tertiary treatment may still be required.

The production of fodder yeast from stillage and its use as a cattle food appear to have limited potential and are unlikely to offer a universal solution, particularly for a large-scale ethanol industry.

Acknowledgements

The author is indebted to Professor R. Beelitz, Dr B. Åkermark, Dr A. Canuti, Dr R. Pieck, Dr J. Fischer, Dr R. Munroe, Dr P. Bidan, Dr J. J. Sprink, Dr J. F. T. Oldfield, Dr R. De Vletter and Dr K. Dostal for supplying copies of papers and reports which would otherwise have been difficult to obtain. The generous assistance of Denis Foster, Deputy Director of Sugar Research Institute, in reading this manuscript is also acknowledged.

References

Abram, J.C. and Ramage, J.T. (1979) Sugar refining: present technology and future developments, in *Sugar Science and Technology*, Birch, G.G. and Parker, K.J. (Eds), Applied Science, London, 49–95.
Åkermark, B. (1975) The system for effluent treatment and measures taken to reduce the water requirement in beet sugar factories in Sweden, *Socker Handl.*, **27**, 1–22.
Anon. (1968) Racecourse mill effluent disposal, *Aust. Sugar J.*, **60**, 434–435.
Aora, H.C., Routh, T., Chattopadhya, S.N. and Sharma, V.P. (1974) Survey of sugar mill effluent disposal. Part II. A comparative study of sugar mill effluents characteristics, *Indian J. Environ. Health*, **16**, 233–246.
Ashe, G.G. (1971) Treatment of mill effluent and sewerage with aerators at Umfolozi mill, *S. Afr. Sugar J.*, **55**, 523–529.
Ashe, G.G. (1976) Water pollution control, *Rev. Agrico. Sucr. Ile Maurice*, **55**, 224–226.
Barr, W.W. (1962) Lagooning and treatment of wastewater, *J. Am. Soc. Sugar Beet Technol.*, **12**, 181–191.
Bass, H.H. (1974) Preparation of fertilizer and animal feed from molasses fermentation residue, US Patent 3 983 255, cited by *Sugar Ind. Abstr.*, **39**, 77–457P.
Basu, A.K. (1970) Contribution a l'étude du traitement des eaux résiduaires de distilleries, *Trib. CEBEDEAU*, **23**, 127–136.
Basu, A.K. (1975) Characteristics of distillery wastewater, *J. Water Pollut. Control Fed.*, **47**, 2184–2190.
Basu, A.K. and Leclerc, E. (1972) Mesophilic digestion of beet molasses distillery wastewater, *Proc. 6th Int. Conf. Int. Assoc. Water Pollut. Res.*, 581–590.
Basu, A.K. and Leclerc, E. (1973) Comparative studies on treatment of beet molasses distillery waste by thermophilic and mesophilic digestion, *Water Res.*, **9**, 103–109.
Bathgate, R.R., Keniry, J.S. and Strong, A.W. (1977) Treatment of sugar mill wastewaters, *Proc. 16th Congr. Int. Soc. Sugar Cane Technol.*, 2509–2517.

Beccari, M., Mappelli, P. and Tandoi, V. (1980) Relationship between bulking and physicochemical–biological properties of activated sludges, *Biotechnol. Bioeng.*, **22**, 969–979.

Beelitz, R. (1971) Moderne Abwasserreinigungsanlagen für die Zuckerindustrie, *Lebensm. Ind.*, **18**, 453–457.

Bereznikov, G.A. and Novikov, A.D. (1976) Use and purification of sugar refinery wastewaters in irrigation fields, *Sakh. Prom-st.*, **50**, 12–16.

Bevan, D. (1971) The disposal of sugar mill effluents in Queensland, *Proc. 14th Congr. Int. Soc. Sugar Cane Technol.*, 1504–1516.

Bevan, D. (1973) Recycling — or reclamation?, *Proc. 40th Conf. Qd. Soc. Sugar Cane Technol.*, 133–139.

Bhandari, H.C., Mitra, A.K. and Malik, V.K. (1979) Treatment of distillery effluent, *Proc. 43rd Annu. Conv. Sugar Technol. Assoc. India*, G65–G72.

Bhaskaran, T.R. and Chakraborty, R.N. (1966) Pilot plant for treatment of cane-sugar waste, *J. Water Pollut. Control Fed.*, **38**, 1160–1169.

Bhaskaran, T.R., Chakraborty, R.N., Das, N. and Sinha, S.N. (1961) Treatment and disposal of sugar factory effluents, *Indian Counc. Med. Res. Tech. Rep. Ser.*, **39**, 1–19.

Bhaskaran, T.R., Chakraborty, R.N., Das, N. and Sinha, S.N. (1963) Sugar factory effluents, *Effluent Water Treat. J.*, **3**, 323–325.

Biaggi, N. (1968). The sugar industry in Puerto Rico and its relation to the industrial waste problem, *J. Water Pollut. Control Fed.*, **40**, 1423–1433.

Bickle, R.E. (1972) Reduction of sugar mill effluent by recycling, *Proc. 39th Conf. Qd. Soc. Sugar Cane Technol.*, 153–156.

Bidan, P. and Heitz, F. (1970) Beet factory waste waters, in *Les Eaux Résiduaires des Industries Agricoles et Alimentaires*, Junk, W. (Ed.), 12th Symp. Int. Budapest, 339–374.

Bieske, G.C. (1979) Agricultural use of dunder, *Proc. 1st Conf. Aust. Soc. Sugar Cane Technol.*, 139–141.

Black, H.H. and Teft, R.A. (1965) Treatment of beet sugar wastes, *Proc. 33rd Annu. Conv. Sugar Technol. Assoc. India*, 45–46.

Blake, J.D. (1976) An investigation of condensates and their contribution to effluent disposal from sugar mills. Part I. Studies on composition of condensates, *Int. Sugar J.*, **78**, 131–137.

Blake, J.D. and McNeil, K.E. (1978) A comparative study of alcohol concentrations in green and burnt cane and the changes occurring during milling, *Proc. 45th Conf. Qd. Soc. Sugar Cane Technol.*, 127–132.

Blankenbach, W.W. and Willison, W.A. (1969) Waste water recirculation as a means of river pollution abatement, *J. Am. Soc. Sugar Beet Technol.*, **15**, 396–402.

Bond, J.F. and McNeil, K.E. (1976) Treatment of sugar mill waste by shallow ponding, *Proc. 43rd Conf. Qd. Soc. Sugar Cane Technol.*, 317–318.

Boruff, C.S. and Buswell, A.M. (1932) Power and fuel gas from distillery wastes, *Ind. Eng. Chem.*, **24**, 33–36.

Brenton, R.W. (1972) Treatment of sugar beet wastes by recycling, *Eng. Bull. Purdue Univ., Eng. Ext. Ser.*, **140**, 119–131.

Brenton, R.W. and Fisher, J.H. (1970) Concentration of sugar beet wastes for economic treatment with biological systems, *Proc. 1st Nat. Symp. Food Process. Wastes*, 261–280.

Brieger, F.O. (1977) Observations on the distribution of vinasse or distillery juice in São Paulo State, *Bras. Acucareiro*, **90**, 23–30.

Brieger, F.O. (1979) Distribution of distillery slops in São Paulo, Brazil, *Sugar Azucar*, **74**, 42–43; 46–47; 49.

Bruijn, J. (1975) Treatment of sugar factory effluent in biological trickling filters, *Proc. Annu. Congr. S. Afr. Sugar Technol. Assoc.*, **49**, 22–28.

Bruijn, J. (1977) Effluent treatment in the South African sugar industry, *Proc. 16th Congr. Int. Soc. Sugar Cane Technol.*, 2313–2327.

Brunke, H. and Voigt, D. (1974) Flume water treatment units in the sugar industry, *Zucker*, **27**, 129–135.

Brunner, H.R. (1977) Wastewater treatment and related problems, *Int. Sugar J.*, **79**, 213–216; 247–250.

Bryant, M.P., Campbell, L., Reddy, C.A. and Crabill, M.R. (1977) Growth of *Desulfovibrio* in lactate or ethanol media low in sulfate in association with H_2-utilizing methanogenic bacteria, *Appl. Environ. Microbiol.*, **33**, 1162–1169.

Burnett, W.E. (1973) Rum distillery wastes: laboratory studies on aerobic treatment, *Water Sewage Works*, **120**, 107–111.

Buswell, A.M. and Le Bosquet, M. (1936) Complete treatment of distillery wastes, *Ind. Eng. Chem.*, **28**, 795–797.

Cappenberg, T.E. (1974) Interrelations between sulfate-reducing and methane-producing bacteria in bottom deposits of a fresh water lake. I. Field observations, *Antonie van Leeuwenhoek J. Microbiol. Serol.*, **40**, 285–295.

Carruthers, A., Gallagher, P.J. and Oldfield, J.F.T. (1960) Biological treatment of sugar beet factory waste streams, *Int. Sugar J.*, **62**, 277–282.

Catroux, G., Bidan, P., Iwema, A. and Heitz, F. (1974a) Possibilité d'épuration biologique à forte charge des eaux résiduaires de sucrerie, *Ind. Aliment. Agric.*, **91**, 939–950.

Catroux, G., Germon, J.-C., Heitz, F. and Bidan, P. (1974b) L'épandage des eaux résiduaires de sucreries, *Ann. Agron.*, **25**, 307–337.

Chakraborty, R.N. (1964) Potash recovery — a method of disposal of distillery wastes and saving foreign exchange, *Symp. Ethyl Alcohol Production Techniques, New Delhi, India*, Noyes Development Corp., New York, 93–97.

Chang, C.T. and Yang, W.L. (1973) Study on feed yeast production from molasses distillery stillage, *Taiwan Sugar*, **20**, 200–203; 205.

Chekurda, A.F. and Parkhomets, A.P. (1968) Purification of wastewater from sugar refineries in nonflow-through basins, *Sakh. Prom-st.*, **42**, 21–25.

Chen, J.C.P., Walters, C.O., Blanchard, F.J., Caballero, M.P. and Picou, R. (1971) Handling of sugar factory waste streams, *Proc. 14th Congr. Int. Soc. Sugar Cane Technol.*, 1537–1543.

Cillie, G.G., Henzen, M.R., Stander, G.J. and Baillie, R.D. (1969) Anaerobic digestion—IV. The application of the process in waste purification, *Water Res.*, **3**, 623–643.

Cooper, B.R. (1975) Distillery waste as a fertilizer, *Annu. Rep. Res., Caroni Research Station*, cited in *Sugar Ind. Abstr.*, **38**, 76–1271.

Cox, S.M.H. (1969) Investigation of sugar mill effluents, *Proc. S. Afr. Sugar Technol., Assoc., 43rd Congr.*, 219–226.

Crane, G.W. (1968) The conservation of water and final treatment of effluent, *Br. Sugar Corp., 19th Technical Conf., Folkestone.*

Crees, O.L., Jacklin, J.D., Topfer, M.G. and Whayman, E. (1981) The application of polyelectrolytes to flyash settling, *Proc. 3rd Conf. Aust. Soc. Sugar Cane Technol.*, 195–201.

Curis, M. (1972) Epuration naturelle des eaux résiduaires de sucreries conservées en bassins, *C.R. Séances Acad. Agric., Fr.*, **58**, 814–828.

Delvaux, L. (1974) Wastewater treatment in the beet sugar industry, *Sugar Technol. Rev.*, **2**, 95–136.

Demidov, O.V. and Demidov, L.G. (1973) Biological purification of sugar factory wastewaters in two-stage aerobic mixing tanks, *Sakh. Prom-st.*, **47**, 32–34.

Devillers, P. (1968) L'eau en sucrerie, *Sucr. Fr.*, **109**, 95–99; 147–150.

Devillers, P., Cornet, P. and Lescure, J.P. (1969) Emploi des adjuvants de floculation en vue de la décantation des eaux boueuses, *Sucr. Fr.*, **110**, 280–283.

Devillers, P., Curis, M. and Lescure, J-P. (1970) Epuration des eaux de sucrerie par lagunage naturel, *Sucr. Fr.*, **111**, 81–88.

Devillers, P. and Lescure, J-P. (1971) Epuration des eaux résiduaires par lagunage à la sucrerie — raffinerie de Bresles, *Sucr. Fr.*, **112**, 189–193.

Devillers, P., Lescure, J-P. and Bourlet, P. (1977) Nouvelles perspectives pour le traitement des eaux en sucrerie: la fermentation méthanique mésophile, *Sucr. Fr.*, **118**, 173–183.

De Vletter, R. (1972) Measures against water pollution in beet sugar processing industries, *Pure Appl. Chem.*, **29**, 113–128.

De Vletter, R. (1979) Twenty years of experience and research concerning wastewater control at CSM, *C.R. 16ᵉ Assem. Gén. Comm. Int. Tech. Sucr., Amsterdam*, 1–16.

De Vletter, R. and Wind, E. (1975) COD balances as a tool in the wastewater management of sugar factories, *C.R. 15ᵉ Assem. Gén. Comm. Int. Tech. Sucr., Vienna*, 515–524.

Dubey, R.S., Varma, N.C., Patil, M.K. and Rao, S.N.G. (1977) Scheme for the disposal of distillery spent wash, cited by *Sugar Ind. Abstr.*, **40**, 78–571.

Dubois, J.P. (1976) Le génie chimique appliqué au traitement des eaux résiduaires industrielles. Traitement d'eaux résiduaires à forte charge organique par le fermenteur R.T.-Lefrancois. — Application en sucrerie, *Trib. CEBEDEAU*, **29**, 211–220.

Edeline, F. and Leclerc, E. (1957) cited by Delvaux (1974).

EPA (1971) Industry waste study. *The Hawaiian Sugar Industry Waste Study*, Report PB-238931, US Environmental Protection Agency, San Francisco, California.

Faup, G.M., Picard, M.A. and Del Zappo, C. (1978) Nitrification–denitrification of wastewaters with a high organic and ammonia nitrogen content (waste effluents from sugar refineries), *Prog. Water Technol.*, **10**, 493–501.

Fischer, J.H. (1974) *Biological Treatment of Concentrated Sugar Beet Wastes*, Environ. Prot. Technol. Ser. EPA-660/2-74-028, US Environmental Protection Agency, Washington, DC.

Fischer, J.H. and Hungerford, E.H. (1971) *State-of-art, Sugar Beet Processing Waste Treatment*, Water Pollut. Control Res. Ser. 12060 DSI 07/71, US Environmental Protection Agency, Washington, DC.

Force, S.L. (1965) The Findlay flume and condenser water system, *J. Am. Soc. Sugar Beet Technol.*, **13**, 478–491.

Fordyce, I.V. and Cooley, A.M. (1974) *Separation, Dewatering and Disposal of Sugar Beet Transport Water Solids*, Environ. Prot. Technol. Ser. EPA-660/2-74-093, US Environmental Protection Agency, Corvallis, Oregon.

Frew, R. (1971) Entrainment prevention in sugar mill evaporation plant, *Proc. 14th Congr. Int. Soc. Sugar Cane Technol.*, 1499–1503.

Guimares, E., Betke, E.G. and Bassinello, J.L. (1968) Determination of economic dose of vinasse as sugar cane fertilizer, *O Solo*, **60**, 87–91, cited by *Sugar Ind. Abstr.*, **32**, 70–412.

Gupta, S.C. (1965) Problem of industrial waste treatment with particular reference to the treatment of wastes from sugar factories, *Proc. 33rd Conv. Sugar Technol. Assoc. India*, 59–67.

Gupta, S.C., Shukla, J.P. and Shukla, N.P. (1968) Recovery of crude potassium salts from spent wash of molasses distilleries by fluidised incineration, *Proc. 36th Annu. Conv. Sugar Technol. Assoc. India*, XXXXIII–1–XXXXIII–7.

Guzman, R.M. (1962) Control of cane sugar wastes in Puerto Rico, *J. Water Pollut. Control Fed.*, **34**, 1213–1218.

Hartmann, E.M. (1974) The calcium saccharate process, *Sugar Technol. Rev.*, **2**, 213–252.

Heitz, F. (1979) Station d'épuration de la pollution glucidique et azotée en sucrerie raffinerie de betteraves, *C.R. 16ᵉ Assem. Gén. Comm. Int. Tech. Sucr., Amsterdam*, 79–117.

Heitz, F. and Bidan, P. (1970) Quelques problèmes concernant les eaux résiduaries de sucreries de betteraves, *Ind. Aliment. Agric.*, **87**, 879–891.

Heitz, F. and Bidan, P. (1975) Réutilisation de l'eau et recyclage dans la sucrerie de betterave, in *Maitrise de la Pollution et Valorisation des Effluents et Résidus dans l'Industrie Alimentaire*, Int. Conf. Grenoble, 2–3 October, 79–149, Assoc. Promotion Industrie Agriculture, Paris.

Hendrickson, E.P. and Grillot, F.A. (1971) Raw sugar factory wastes and their control, *Proc. 14th Congr. Int. Soc. Sugar Cane Technol.*, 1544–1551.

Hiatt, W.C., Carr, A.D. and Andrews, J.F. (1973) Anaerobic digestion of rum distillery wastes, *Proc. 28th Ind. Waste Conf.*, Purdue University, Indiana, 966–976.

Hoffmann-Walbeck, H.P. (1977) Environmental protection and the sugar industry, *Zucker*, **30**, 61–64.

Hoffmann-Walbeck, H.P. and Pellegrini, A. (1975) Level of waste water technology in the sugar industry and waste water legislation, *Zucker*, **28**, 527–534.

Hoffmann-Walbeck, H.P. and Pellegrini, A. (1978) Progress in the treatment of sugar factory waste waters, *Z. Zuckerind.*, **103**, 841–847.

Hohnerlein, O.G. (1973) Desweetening and disposal of carbonatation muds using vacuum filters at Savannah sugar refinery, *Proc. 32nd Annu. Meet. Sugar Ind. Technol. Inc.*, 19–23.

Huss, L. (1979) Treatment of sugar factory waste waters, *Sugar J.*, **41**(8), 9–12.

Il'ina, L.D., Osovik, A.N., Zabrodskii, A.G. et al. (1969) Selection of the most productive yeast-like fungi for culture on molasses distillery residue, *Fermentn. Spirt. Prom.*, **35**, 18–21, cited by *Sugar Ind. Abstr.*, **33**, 71–868.

Jackman, E.A. (1977) Distillery effluent treatment in the Brazilian national alcohol programme, *Chem. Eng. (London)*, **319**, 239–242.

Jackson, C.J. (1956) Whisky and industrial alcohol distillery wastes, *Inst. Sewage Purif. J. Proc.*, Pt 2, 206–214.

Jackson, C.J. (1966) Fermentation waste disposal in Great Britain, *Proc. 21st Ind. Waste Conf.*, Purdue University, Indiana, 19–32.

Jones, R.N. and Dyne, R.A. (1977) Boiler ash separation by ponding, *Proc. 44th Conf. Qd. Soc. Sugar Cane Technol.*, 335–340.

Klapper, H. (1970) Die Stapelteichbelüftung, eine wirtschaftliche Methode zur vollständigen biologischen Reinigung von Zuckerfabrikabwässern, *Wasserwirtsch. Wassertech.*, **20**, 174–178.

Kollatsch, D. (1969) Discussion of paper: Succession of microbial processes in the anaerobic decomposition of organic compounds, *Proc. 4th Int. Conf. Int. Assoc. Water Pollut. Res.*, 425.

Kominek, E.G., Lash, L.D. and Thompson, R.B. (1970) Treatment of wastes from sugar manufacturing, paper presented to 3rd Jt Mtg Inst. Ing. Quim. Puerto Rico and Am. Inst. Chem. Eng., San Juan, Puerto Rico, 17–20 May.

Kramer, D. (1960) Purification of sugar factory wastes by the soil, *Zucker*, **13**, 436–441; 489–493.

Kramer, D. (1961) Design, operation and success of plants for treating sewage from sugar factories by the soil, *Zucker*, **14**, 36–41; 61–66.

Kravets, Y.M., Kats, V.M. and Karanov, Y.A. (1970) Partial recirculation of spent mash for molasses dilution, *Ferment. Spirt. Prom.*, **36**, 20–21, cited in *Sugar Ind. Abstr.*, **33**, 71–1029.

Kugelman, I.J. and Chin, K.K. (1971) Toxicity, synergism, and antagonism in anaerobic waste treatment processes, in *Anaerobic Biological Treatment Processes*, Gould, R.F. (Ed.), American Chemical Society, Washington, DC, 55–90.

Kujala, P. (1979) Distillery fuel savings by efficient molasses processing and stillage utilization, *Sugar Azucar*, **74**(10), 13–16.

Kujala, P., Hull, R., Engström, F. and Jackman, E. (1976) Alcohol from molasses as a possible fuel, *Sugar Azucar*, **71**(3), 28–39.

Laguerre, H. (1970) Le traitement des eaux en sucrerie, *Sucr. Fr.*, **111**, 245–248.

Landi, S. and Mantovani, G. (1975) Ion exchange in the beet-sugar industry, *Sugar Technol. Rev.*, **3**, 1–67.

Langen, A. and Hoeppner, J. (1964) Distribution of waters of the Ameln raw sugar factory after applying the Vortair process, *Zucker*, **17**, 546–552.

Langley, P.J. and Bohlig, C.J. (1973) Waste treatment at Crockett refinery, *Sugar J.*, **36**(1), 20–24.

Leclerc, E. and Edeline, F. (1960) Étude de la mise en solution des matières organiques dans le circuit des eaux de transport et de lavage des betteraves d'une sucrerie, *Trib. CEBEDEAU*, **47**, 58–61.

Lescure, J-P. (1973) Étude du traitement par lagunage des eaux résiduaires de sucrerie, *Sucr. Fr.*, **114**, 241–247.

Lescure, J-P. and Bourlet, P. (1977) La dépollution des eaux, *Sucr. Fr.*, **118**, 103–109.

Lescure, J-P. and Bourlet, P. (1978) Traitement des eaux résiduaires par fermentation méthanique mésophile, *Sucr. Fr.*, **119**, 107–114.

Lescure, J-P. and Bourlet, P. (1979a) Traitement des eaux résiduaires, *Sucr. Fr.*, **120**, 99–105.

Lescure, J-P. and Bourlet, P. (1979b) Epuration des eaux de sucrerie par fermentation méthanique mésophile, *C.R. 16ᵉ Assem. Gén. Comm. Int. Tech. Sucr., Amsterdam*, 29–78.

Lettinga, G., van Velsen, A.F.M., Hobma, S.W., de Zeeuw, W. and Klapwijk, A. (1980) Use of the upflow sludge blanket (USB) reactor concept for biological wastewater treatment, especially for anaerobic treatment, *Biotechnol. Bioeng.*, **22**, 699–734.

Lewicki, W. (1978) Production, application and marketing of concentrated molasses – fermentation – effluent (vinasses), *Process Biochem.*, **13**(6), 12–13.

Lewis, J.W.V. and Ravnoe, A.B. (1976) Effluent treatment at Felixton Mill, *Proc. 50th Annu. Congr. S. Afr. Sugar Technol. Assoc.*, 242–245.

Lin, S. (1970) Research on the cultivation of fodder yeast in vinasse, *Report Taiwan Sugar Experiment Station*, cited by *Sugar Ind. Abstr.*, **36**, 74–625.

Lührs, H. (1963) Transportation of carbonatation sludges through pipe lines without any water addition, *Zucker*, **16**, 256–258.

Macgregor, A.N. and Keeney, D.R. (1973) Methane formation by lake sediments during *in vitro* incubation, *Water Resour. Bull.*, **9**, 1153–1158.

Madsen, R.F., Kofod Nielsen, W. and Winstrom-Olsen, B. (1978) Juice purification system, sugar house scheme, and sugar quality, *Sugar J.*, **41**(4), 15–19.

Madsen, R.F., Kofod Nielsen, W. and Johnsen, A.F. (1979) A new method for increasing dry substance in lime sludge, *C.R. 16ᵉ Assem. Gén. Comm. Int. Tech. Sucr., Amsterdam*, 231–252.

Magro, J.A. and da Gloria, N.A. (1977) Fertilization of sugar cane ratoons with vinasse complemented with nitrogen and phosphorous, *Bras. Acucareiro*, **90**, 31–34.

Marie-Jeanne, S. (1977) Separation of fly ash and other solid wastes from factory effluents by combined decanting and sieving over a precoated screen, *Proc. 16th Congr. Int. Soc. Sugar Cane Technol.*, 2459–2467.

McDougall, E.E., Messiter, G.M. and Sawyer, G.M. (1976) Racecourse boiler ash handling system, *Proc. 43rd Conf. Qd. Soc. Sugar Cane Technol.*, 157–163.

McGinnis, R.A. (1971) *Beet-sugar Technology*, 2nd edn, Beet Sugar Development Foundation, Fort Collins.

McNeil, K.E., Bond, J.F. and Gampe, W.C. (1974) Effluent treatment at Farleigh mill, *Proc. 41st Conf. Qd. Soc. Sugar Cane Technol.*, 235–241.

Meade, G.P. and Chen, J.C.P. (1977) *Cane Sugar Handbook*, Wiley-Interscience, New York.

Merle, J.P. (1976) The water treatment system at Honokaa Sugar Company, *Hawaiian Sugar Technol. Rep., 35th Conf.*, 78–84.

Meyer, F.W. (1968) Purification of waste water by means of activated sludge, *Zucker*, **21**, 338–341.

Middleton, F.H., Rhodes, L.J., Sloane, G.E. and Gibson, W.O. (1971) Dry vs wet cane cleaning at Laupahoehoe sugar company, *Proc. 14th Congr. Int. Soc. Sugar Cane Technol.*, 1393–1404.

Milford, B.J. (1977) Mossman mill ash separation system, *Proc. 44th Conf. Qd. Soc. Sugar Cane Technol.*, 341–344.

Miller, J.R. (1971) Treatment of effluent from raw sugar factories, *Proc. 14th Congr. Int. Soc. Sugar Cane Technol.*, 1529–1536.

Monteiro, C.E. (1975) Brazilian experience with the disposal of waste water from the cane sugar and alcohol industry, *Process Biochem.*, **10**(11), 33–41.

Mühlpforte, H. (1962) The present position of waste disposal in sugar factories, *Wasserwirtsch. Wassertech.*, **12**, 65–70.

Nielsen, F.S. (1968) Amalgamated Sugar Company combats water pollution, *Sugar Azucar*, **63**(6), 28–31.

Offhaus, K. (1965) Elimination of wastes in sugar factories with special regard to the situation in Bavaria, *Zucker*, **18**, 539–545.

Oldfield, J.F.T., Dutton, J.V., Morgan, N.D. and Teague, H.J. (1972) Determination of sugar losses in beet fluming and washing, *Sucr. Belge*, **91**, 433–441.

Ono, H. (1965) Discussion of paper by Bhaskaran, T.R., in *Adv. Water Pollut. Res.*, **2**, 100–104.

Oswald, W.J., Tsugita, R.A., Golueke, C.G. and Cooper, R.C. (1973) *Anaerobic–Aerobic Ponds for Beet Sugar Waste Treatment*, Env. Prot. Technol. Ser. EPA-R2-73-025, US Environmental Protection Agency, Washington, DC.

Parashar, D.R. (1965) Treatment of sugar factory effluents — self oxidation and purification, *Indian Sugar*, **15**, 325–333.

Parashar, D.R. (1969) Treatment of sugar factory effluents in relation to the tolerance limit of Biochemical Oxygen Demand, *Indian Sugar*, **18**, 879–880; 883–885.

Parkhomenko, A.N. (1964) Irrigation of fields with sugar factory waste waters, *Sakh. Prom-st.*, **38**, 748–751.

Paz, H.A. and Lopez Hernandez, J.A. (1973) Aprovechamiento de materia prima, subproductos y desechos de la industria azucarera en la obtencion de levadura-alimento, *Rev. Agron. Noroeste Argent.*, **10**, 205–214.

Pette, K.C. (1979) Anaerobic waste water treatment at CSM sugar factories, *C.R. 16ᵉ Assem. Gén. Comm. Int. Tech. Sucr.*, Amsterdam, 17–28.

Pette, K.C., de Vletter, R., Wind, E. and van Gils, W. (1980) Full scale anaerobic treatment of beet-sugar waste water, *Proc. 35th Purdue Ind. Waste Conf.*, Ann Arbor Science, Ann Arbor, Michigan, in press.

Phipps, O.H. (1960) Methods of water treatment. Legal aspects and physical treatment, *Br. Sugar Corp., 13th Tech. Conf.*, 1–18.
Pieck, R. (1974) Le problème de la pollution des eaux de sucrerie et les solutions adoptées aujourd'hui, *Tech. Eau. Assainissement*, **334**, 31–36.
Radhakrishnan, I., De, S.B. and Nath, B. (1969) Evaluation of the loading parameters for anaerobic digestion of cane molasses distillery wastes, *J. Water Pollut. Control Fed.*, **41**, R431–R440.
Rao, B.S. (1972) A low cost waste treatment method for the disposal of distillery waste, *Water Res.*, **6**, 1275–1282.
Reich, G.T. (1945) Production of carbon and potash from molasses distiller's stillage, *Trans. Am. Inst. Chem. Eng.*, **41**, 233–252.
Reinefeld, E. (1979) Progress in the technology of beet-sugar, in *Sugar Science and Technology*, Birch, G.G. and Parker, K.J. (Eds), Applied Science, London, 131–149.
Reinefeld, E., Hoffmann-Walbeck, H.P. and Wittek, J. (1975) Analytical investigations of sugar factory waste waters, particularly the COD of storage pond water, *Zucker*, **28**, 165–173.
Reinefeld, E., Hoffmann-Walbeck, H.P., Pellegrini, A. and Wittek, J. (1979) Investigations on activated sludge degradation in the treatment of sugar factory waste waters as well as on the constituents difficult to degrade, *Z. Zuckerind.*, **104**, 931–939.
Revuz, B. (1971) Epuration par oxygénation rapide dans un fermenteur. Application aux eaux résiduaires de sucrerie, *Ind. Aliment. Agric.*, **88**, 1031–1037.
Roche, M. (1969) La floculation des argiles et la décantation des eaux boueuses de sucrerie, *Ind. Aliment. Agric.*, **86**, 919–923.
Rolz, C., de Cabrera, S., Espinosa, R., Maldonado, O. and Menchu, J.F. (1975) The growth of filamentous fungi on rum distilling slops, *Ann. Technol. Agric.*, **24**, 445–451.
Roth, L.A. and Lentz, C.P. (1977) Anaerobic digestion of rum stillage, *Can. Inst. Food Sci. Technol. J.*, **10**, 105–108.
Sawyer, C.N. and Anderson, E.J. (1949) Anaerobic treatment of rum waste, *Water Sewage Works*, **14**, 112–114.
Sawyer, G.M. and Cullen, R.N. (1975) Dewatering of flyash from bagasse-fired boilers, *Proc. 43rd Conf. Qd. Soc. Sugar Cane Technol.*, 311–317.
Sawyer, G.M. and Cullen, R.N. (1977) A review of ash handling systems in Queensland sugar mills, *Proc. 16th Congr. Int. Soc. Sugar Cane Technol.*, 2263–2275.
Schantz, J.C. and Kemmer, F.N. (1969) Aspects of water pollution by the Hawaiian sugar industry, *Hawaiian Sugar Technol. Rep., 28th Conf.*, 11–18.
Schiweck, H., Cronewitz, T. and Schoppe, F. (1979) Das Rückbrennen von Carbonatationsschlamm in einer Hochgeschwindigkeitsreaktionskammer, *C.R. 16e Assem. Gén. Comm. Int. Tech. Sucr.*, Amsterdam, 189–230.
Schneider, F., Hoffmann-Walbeck, H.P. and Kollatsch, D. (1961) Water balances in sugar factories with regard to the waste water problem, *Zucker*, **14**, 619–626.
Schneider, F. (1963) Some aspects of the campaign 1962, *Zucker*, **16**, 435–443.
Schneider, F., Hoffmann-Walbeck, H.P. and Kollatsch, D. (1964) On the aerobic degradation of waste water, *Zucker*, **17**, 393–399.

Schneider, F. and Hoffmann-Walbeck, H.P. (1968) Waste water of sugar factories, *Zucker*, **21**, 396–402.

Schulz-Falkenhain, H. (1964) Measures of the sugar industry for maintaining waters in a pure state, *Zucker*, **17**, 518–523.

Schwieter, A. (1963) Pneumatic transportation of carbonatation sludge, *Zucker*, **16**, 253–256.

Sen, B.P. and Bhaskaran, T.R. (1962) Anaerobic digestion of liquid molasses distillery wastes, *J. Water Pollut. Control Fed.*, **34**, 1015–1025.

Sestero, J.R. and Logan, J.T. (1972) Mill housekeeping, *Proc. 39th Conf. Qd. Soc. Sugar Cane Technol.*, 165–168.

Shea, T.G., Ramos, E., Rodriguez, J. and Dorion, G.H. (1974) *Investigation of Rum Distillery Slops Treatment by Anaerobic Contact Process*, Environ. Prot. Technol. Ser. EPA-660/2-74-058, US Environmental Protection Agency, Washington, DC.

Shukla, J.P. and Kapoor, B.D. (1961) Stabilization of treated and untreated sugar factory effluents, *Proc. 29th Annu. Conv. Sugar Technol. Assoc. India*, 181–184.

Shukla, J.P. and Kapoor, B.D. (1962) The examination of activated sludge process for the treatment of sugar factory effluents, *Proc. 30th Annu. Conv. Sugar Technol. Assoc. India*, 122–127.

Shukla, J.P., Kapoor, B.D. and Gupta, S.K. (1960) Treatment of sugar factory effluent by high rate trickling filter, *Proc. 28th Annu. Conv. Sugar Technol. Assoc. India*, 231–236.

Shukla, J.P. and Varma, N.K. (1964) The examination of activated sludge process for the treatment of sugar factory effluents, *Proc. 32nd Annu. Conv. Sugar Technol. Assoc. India*, 63–72.

Silin, P.M. (1964) *Technology of Beet-sugar Production and Refining*, Israel Program for Scientific Translations, Jerusalem.

Simonart, A., Dubois, J.P. and Pieck, R. (1975) Possibilité de valorisation des eaux condensées ammoniacales en tant que complement nutritif azote dans l'épuration biologique RT-Lefrancois, *C.R. 15ᵉ Assem. Gén. Comm. Int. Tech. Sucr.*, Vienna, 525–542.

Simpson, D.E. and Hemens, J. (1978) Activated sludge treatment of sugar mill/wattle bark mill effluents at Dalton, Natal, *Proc. Annu. Congr. S. Afr. Sugar Technol. Assoc.*, **52**, 20–25.

Simpson, D.E., Hemens, J. and Cox, S.M.H. (1972) Aerobic treatment of sugar mill effluent with the addition of nutrients, *Proc. Annu. Congr. S. Afr. Sugar Technol. Assoc.*, **46**, 40–53.

Sinha, S.N. and Sinha, L.P. (1969) Studies on use of water hyacinth culture in oxidation ponds treating digested sugar wastes and effluent of septic tank, *Environ. Health*, **11**, 197–207.

Sinha, S.N. and Thakur, B. (1967) Anaerobic digestion of cane sugar wastes, *Environ. Health*, **9**, 118–125.

Skogman, H. (1979) Effluent treatment of molasses based fermentation wastes, *Process Biochem.*, **14**, 5–6.

Slijkhuis van der Haarst, M.E. and van der Toorn, G.J.J. (1972) *Pollution Abatement of Beet-sugar-factory Waste Water. Methods and Experience in the Netherlands*, Technisch Adviesbureau van der Unie van Waterschadden, Haarlem, Netherlands.

Smith, J.N. (1973) Water conservation in the beet sugar industry, *Chem. Ind. (London)*, **12**, 546–548.

Smith, J.N., Branch, M.F. and Rogers, R.H. (1975) Role of surface aeration in effluent treatment and its application at Wissington factory, *Sucr. Belge*, **94**, 41–56.

Sonoda, Y. and Tanaka, M. (1968) Anaerobic digestion of low concentration wastes. I. Continuous digestion tests of some industrial wastes, *J. Ferment. Technol.*, **46**, 789–795.

Stander, G.J. and Elsworth, J.F. (1950) Effluent treatment from the fermentation industries. III. The influence of high concentrations of sulphates on the anaerobic digestion method of treatment at thermophilic (55°C) and mesophilic (33°C) temperatures, *Inst. Sewage Purif. J. Proc.*, 303–312.

Stander, G.J. and Snyders, R. (1950) Effluent treatment from the fermentation industries. V. Re-inoculation as an integral part of the anaerobic digestion method of purification of fermentation effluents, *Inst. Sewage Purif. J. Proc.*, 447–458.

Stuart, K.A. (1976) The water cycle in a sugar mill, *Proc. 43rd Conf. Qd. Soc. Sugar Cane Technol.*, 319–321.

Teichmann, H. and Leswal, H.D. (1976) The biological treatment of waste-water from sugar refineries, *Wasserwirtsch.*, **66**, 275–280.

Tippens, D.E., Chou, C.C. and Hayes, L.C. (1978) New biological wastewater treatment process at Domino Refinery, Chalmette, Louisiana, *Proc. 37th Annu. Meet. Sugar Ind. Technol.*, 410–472.

Train, R.E., Sanson, R.L., Cywin, A. and Watkins, R.V. (1974a) *Development Document for Effluent Limitation Guidelines and Standards of Performance for New Sources. Beet Sugar Processing Subcategory of the Sugar Processing Point Source Category*, EPA-440/1-74-002-b, US Environmental Protection Agency, Washington, DC.

Train, R.E., Strelow, P., Cywin, A. and Dellinger, R.W. (1974b) *Development Document for Effluent Limitations Guidelines and New Source Performance Standards. Cane Sugar Refining Segment of the Sugar Processing Industry*, EPA-440/1-74-002-C, US Environmental Protection Agency, Washington, DC.

Train, R.E., Agee, T.L., Cywin, A. and Dellinger, R.W. (1975) *Development Document for Interim Final Effluent Limitations Guidelines and Proposed New Source Performance Standards for the Raw Cane Sugar Processing Segment of the Sugar Processing Point Source Category*, EPA-440/1-75-044, US Environmental Protection Agency, Washington, DC.

Tsugita, R.A., Oswald, W.J., Cooper, R.C. and Golueke, C.G. (1969) Treatment of sugar beet flume waste water by lagooning: a pilot study, *J. Am. Soc. Sugar Beet Technol.*, **15**, 282–297.

Usher, J.F. and Willington, I.P. (1979) The potential of distillery waste as sugar cane fertiliser, *Proc. 1st Conf. Aust. Soc. Sugar Cane Technol.*, 143–146.

Vellaud, J.P. (1978) Bilan du programme de lutte contre la pollution dans les sucreries de betteraves, *Sucr. Fr.*, **119**, 281–286.

Verma, S.R., Bahel, D.K., Pal, N. and Dalela, R.C. (1978) Studies on the sugar factories and their wastes in Western Uttar Pradesh, *Indian J. Environ. Health*, **20**, 205–218.

Viehl, K., Teichmann, H. and Leswal, M-D. (1974) Purification of sugar factory waste water with the aid of the pumping over method, *Z. Zuckerind.*, **24**, 536–541.

Wagner, P. (1980) Methanization of polluted water — the sugar factories set an example, cited by *Sugar Ind. Abstr.*, **42**, 80–1054.

Wang, L., Kuo, Y. and Chang, C. (1977) Distribution of organic matters during molasses alcohol fermentation and feed yeast cultivation on the slop, *Report Taiwan Sugar Research Institute*, cited by *Sugar Ind. Abstr.*, **40**, 78–1177.

Watt, J.W. and Morton, A.L. (1979) Combined mill waste disposal, *Proc. 1st Conf. Aust. Soc. Sugar Cane Technol.*, 167–171.

Winfrey, M.R. and Zeikus, J.G. (1977) Effect of sulfate on carbon and electron flow during microbial methanogenesis in fresh water sediments, *Appl. Environ. Microbiol.*, **33**, 275–281.

Wright, P.G. and McNeil, K.E. (1980) Costs associated with large scale ethanol production by the sugar industry, *Fuel Ethanol Research and Development Workshop, 4th–5th February*, Dept. National Development and Energy, Canberra.

Zabrodskii, A.G., Osovik, A.N., Polyanskaya, E.A. and Strizhenyuk, E.V. (1973) Possibility of getting higher yields of fodder yeast from vinasse, *Fermentn. Spirt. Prom.*, **39**, 40–43, cited by *Sugar Ind. Abstr.*, **36**, 74–772.

Zama, F., Accorsi, C.A. and Mantovani, G. (1979) Eaux résiduaires de sucrerie — paramètres physiques, chimiques et biologiques pour obtenir la meilleure épuration avec une consommation minimale d'énergie, *C.R. 16ᵉ Assem. Gén. Comm. Int. Tech. Sucr., Amsterdam*, 119–150.

2 The treatment of wastewater from the beverage industry

H M Rüffer and K-H Rosenwinkel, *Institut für Siedlungswasserwirtschaft, Universität Hannover*

2.1 Introduction

The consumption of beverages per head of the population has greatly increased since the early 1970s although, as a result of rationalization within the industry, there has during the same period been a marked decline in the total number of manufacturers. The enlargement of individual plants has placed additional loads on some localities, especially where existing wastewater purification facilities are already used by the industry. The choice of the effluent treatment methods that these plants are to adopt is influenced by economic considerations; however, most firms are able to use the facilities provided by the local authority.

The amount of wastewater and pollution arising from the production of drinks is illustrated by the calculated wastewater load from drink production in the Federal Republic of Germany (West Germany) in 1979 (Table 2.1). Regional differences in consumption are illustrated by Table 2.2, which gives per capita consumption of drinks in the countries of the European Economic Community (EEC) for 1977.

The West German beverage industry in 1979 generated an annual volume of wastewater amounting to 88 million m^3 and a corresponding BOD load of 90 626 t. These values are equivalent to those for populations of 1.6 million and 4.1 million respectively (based on a per capita daily production of 150 litres and 60 g BOD). In terms of BOD load, this is about 7% of the total wastewater load arising in the whole of West Germany.

The increase in the consumption of drinks per head of the West German population in the years 1970–79 is shown in Fig. 2.1 and Table 2.3, according to the individual beverages consumed. This shows that, with the exception of spirits, there was a continuous per capita increase in all types of drink consumed over the period indicated. Over these nine years, this per capita increase amounted to 22% in the total consumption of drinks in the entire population. In considering this, the considerable economic importance of the

Table 2.1 Total volume of wastewater and pollutional load arising from the production of beverages in West Germany in 1979[a]

Beverage	Output (1000 m³)	Wastewater volume		BOD$_5$ load	
		Specific (m³ m^{-3})	Total (1000 m³)	Specific (kg m^{-3})	Total (t)
Milk	3 418	2.0[b]	6 836	2.5[b]	8 545
Soft drinks and mineral water	5 839	3.0[c]	17 517	2.4[c]	14 013
Juice	991	6.0[d]	5 946	6.2[d]	6 144
Beer	8 785	6.0[e]	52 711	6.0[e]	52 711
Wine and champagne	1 024	4.8[f]	4 913	9.0[f]	9 212
Total	20 057		87 923		90 625

[a] Anon. (1979).
[b] Doedens and Noda (1977).
[c] Schobinger (1978).
[d] Beuthe (1977).
[e] ATV (1977).
[f] Beuthe (1975).

drinks industry must not be forgotten. In 1976, for example, the production of drinks accounted for DM12.5 × 10^9: in other words, 32% of the gross product of the total food industry.

A very large proportion of the pollution caused by the beverage industry arises from production wastes during the processing and bottling of particular beverages. An overall view of these wastes for individual drinks is presented in Table 2.4. The wastewater from the processing of fruit and vegetables is also included, together with data for the production of fruit juice.

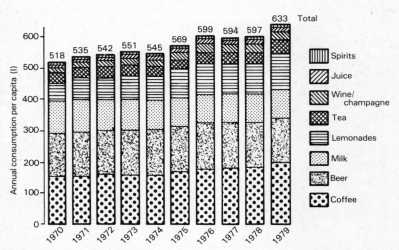

Fig. 2.1 Increase in per capita consumption of beverages in the Federal Republic of Germany from 1970 to 1979

Table 2.2 Consumption of beverages per capita in EEC countries in 1977 (Anon., 1979/80; Anon., 1979; UNO, 1977)

EEC country	Coffee (kg)	Tea (kg)	Milk (without cream) (l)	Soft drinks (l)	Mineral water (l)	Juices (l)	Beer (l)	Wine and champagne (l)	Spirits (l)	Total (without coffee, tea) (l)
Belgium	4.12	0.11	83(a)	5.775	4.464	0.350	134.6	17.5	2.24	—
Luxembourg	—	—	—	—	—	—	121.9	49.3	4.7	—
West Germany	4.18	0.20	81	40.273	20.722	8.264	147.7	23.4	8.35	329.7
Denmark	7.64	0.49	152	2.134	0.329	0.320	114.3	11.7	4.69	285.5
France	4.08	0.12	87	11.9	25.2	1.184	44.8	100.9	5.94	276.9
Great Britain	0.69	2.82	142	18.592	0.130	1.413	119.0	5.4	3.93	290.5
Republic of Ireland	—	3.65	211	24(b)	—	—	126.2	4.3	2.2	—
Italy	2.56	0.07	78	11.4	12.0	2.295	13.7	93.5	3.92	214.8
Netherlands	5.76	0.82	136	3.268	0.307	2.180	76.9	11.7	3.48	233.8

(a) Belgium and Luxembourg combined.
(b) 1976 value.

Table 2.3 Consumption of beverages (litres per capita) in West Germany (Breitenacher, 1980)

Type of beverage	1970	1971	1972	1973	1974	1975	1976	1977	1978	1979
ALCOHOLIC DRINKS (a)	165.4	172.5	172.6	175.5	174.8	178.8	183.6	179.6	178.3	177.9
Beer	141.1	144.4	145.3	146.7	146.9	147.8	150.9	148.7	145.6	145.1
Wine	15.3	17.8	18.2	18.9	17.6	20.5	20.5	20.3	20.7	20.3
Champagne	1.9	2.2	2.4	2.6	2.7	2.9	3.4	3.5	3.7	4.1
Spirits	7.1	8.1	6.7	7.3	7.6	7.6	8.8	7.1	8.3	8.4
NON-ALCOHOLIC BEVERAGES (b)	69.6	81.2	83.7	94.1	91.1	103.6	115.5	112.5	116.7	124.6
Refreshment drinks	59.7	69.6	71.4	80.1	79.5	90.1	101.6	99.2	102.4	108.6
Mineral water	13.6	16.2	16.9	20.3	21.2	26.1	32.6	34.0	36.6	39.6
Soft drinks	46.1	53.4	54.5	59.8	58.3	64.0	69.0	65.2	65.8	69.0
Juices	9.9	11.6	12.3	14.0	11.6	13.5	13.9	13.3	14.3	16.0
OTHER DRINKS (e.g. coffee, tea, milk) (c)	282.8	281.6	285.5	281.7	279.0	286.3	299.4	301.4	302.3	330.6
Total (a + b + c)	517.8	535.3	541.8	551.3	544.9	568.7	598.5	593.5	597.3	633.1

Table 2.4 Some analytical data for various beverages

Type of beverage	BOD$_5$ (g l^{-1})	COD (g l^{-1})	Perm. consumed[a] (g l^{-1})	Reference
Milk (cream 3.5%)	100–143	183	88–175	Doedens (1968), (1978), Viehl (1961), Rüffer and Rosenwinkel (unpublished results)
Skim milk	75– 90	147	70– 90	Doedens (1968), (1978), Rüffer and Rosenwinkel (unpublished results)
Buttermilk	61	134	155	Doedens (1978), Rüffer and Rosenwinkel (unpublished results)
Evaporated milk (cream 7.5%)	271	378	519	Doedens (1978)
Whey	33– 50	70	63–105	Damerow and Gerhold (1976), Viehl (1961)
Sour whey	35	60	75	Doedens (1978)
Apple juice	57– 95	85–155	280–514	Damm (1964), Dittrich (1965), (1975), Rüffer and Rosenwinkel (unpublished results)
Orange juice	60– 67	83– 95	237–297	Dittrich (1975)
Grape juice	78– 99	115–117	252–452	Damm (1964), Dittrich (1975)
Pear juice	94–109	175	315–427	Dittrich (1965), (1975)
Mixed juice	65	91	228	Dittrich (1975)
Grapefruit juice	70	117	354	Dittrich (1975)
Apricot juice	105	194	568	Dittrich (1975)
Peach juice	100	174	335	Dittrich (1975)
Banana juice	110	184	483	Dittrich (1975)
Cherry juice	110	162	461	Dittrich (1975)
Blackcurrant juice	60–114	111	324–491	Damm (1964), Dittrich (1965)
Red beet juice	50	93	—	Rüffer and Rosenwinkel (unpublished results)
Vegetable juice	20	50	—	Rüffer and Rosenwinkel (unpublished results)
Sour cherry juice	69	—	382	Damm (1964)
Plum juice	81	—	390	Damm (1964)
Beer	80– 90	136	57	Dittrich (1965), Schumann (1977), Putz (1977), Rüffer and Rosenwinkel (unpublished results)
Must	150–155	270	—	Weller (1979), Fischer and Coppik (1980)
Wine	120–150	216–225	85	Schmittel (1977), Weller (1979), Fischer and Coppik (1980), Rüffer and Rosenwinkel (unpublished results)
Soft drinks	47– 52	90–114	—	Schumann (1975b)

[a] Perm. consumed/4 ≈ oxygen consumed.

Table 2.5 Minimum requirements of §7a WRPA for water discharged into surface waters in West Germany

Source of waste	Specific waste-water	Settle-able solids grab sample	COD settled 2 h average	COD settled 24 h sample	BOD$_5$ settled 2 h average	BOD$_5$ settled 24 h sample
	(m^3 m^{-3})	(ml l^{-1})	(mg l^{-1})	(mg l^{-1})	(mg l^{-1})	(mg l^{-1})
Soft drink production and mineral water filling		0.3	160	110	35	25
Breweries	8.0	0.3	95	85	25	20
	6.0	0.3	120	100	30	25
	4.0	0.3	175	150	35	30
Dairies		0.5	170	160	35	30
Juice production, vegetable and fruit		0.3	250	200	60	45
Communities >600 kg BOD$_5$ d^{-1}		0.3	140	100	30	20

Table 2.4 gives a clear indication of the very high degree of pollution involved when the production wastes pass directly into the sewerage system. In the case of many drinks, as well as the high organic load, the very low pH value (pH 3–4) must also be taken into account. This is usually attributable to organic acids. The special problems connected with the biological degradation of these acids are discussed in Section 2.2.

In West Germany the quality of discharged wastewater is regulated by the Water Resources Policy Act. According to these regulations, wastewater may be discharged only if the degree of pollution is considered to be no higher than certain critical concentrations which are related to the available treatment methods. The Federal Government has issued general administrative directives for the implementation of these laws, in relation to the minimum requirements for the disposal of wastewater, which correspond to the generally recognized technical regulations. In Table 2.5 the minimum requirements for the disposal of wastewaters from the beverage industry are compared with the disposal of treated sewage from communities of more than 10 000 p.e.

As well as complying with the statutory proscriptions, the observance of the minimum requirements has an economic significance. The rate charged for the disposal of wastewater from all direct sources in West Germany was reduced by 50% from January 1981, by the observance of these minimum requirements. Thus, there is an economic incentive for the construction and use of wastewater plants.

Firms that discharge wastewater to common sewers have to observe concen-

Table 2.6 Standards recommended for wastewater discharge to common sewers in West Germany (ATV, 1980)

Parameter/pollutant	Permitted maximum
Temperature	35°C
pH	6.5–10.0
Settleable solids, where settling is prescribed	10 ml l^{-1} after 0.5 h
Saponifiable oil and grease	250 mg l^{-1}
Inorganics (dissolved and suspended)	
arsenic (As)	1 mg l^{-1}
lead (Pb)	2 mg l^{-1}
cadmium (Cd)	0.5 mg l^{-1}
chromium VI (Cr)	0.5 mg l^{-1}
chromium, total (Cr)	3 mg l^{-1}
copper (Cu)	2 mg l^{-1}
nickel (Ni)	4 mg l^{-1}
mercury (Hg)	0.05 mg l^{-1}
zinc (Zn)	5 mg l^{-1}
tin (Sn)	5 mg l^{-1}
aluminium (Al) } iron (Fe) }	no limit if no difficulties
ammonia (NH$_4^+$ and NH$_3$)	200 mg l^{-1}
cyanide, easily decomposed (CN$^-$)	1 mg l^{-1}
cyanide, total (CN$^-$)	20 mg l^{-1}
fluoride (F$^-$)	60 mg l^{-1}
nitrite (NO$_2^{2-}$)	20 mg l^{-1}
sulphate (SO$_4^{2-}$)	600 mg l^{-1}
sulphide (S^{2-})	2 mg l^{-1}
Organics	
steam-distillable phenols (C$_6$H$_5$OH)	100 mg l^{-1}

tration limits set by local authorities. The limits effective in West Germany are set out in Table 2.6.

2.2 Juices — production processes, the wastewater and its treatment

The products of the fruit juice industry include nectars and fruit juice drinks as well as pure fruit juices and musts. The composition and the minimum proportion of fruit in the drinks is regulated by law. Table 2.7 shows the minimum proportion of fruit juice according to the statutory requirements of several EEC countries together with the requirements of the EEC directive itself.

International standards for fruit juices are set by the FAO and the WHO. Musts are not included in Table 2.7 as they are listed partly under fruit juices and partly under nectars. As a general rule, ready-to-drink fruit juices are taken to comprise those obtained from berries, stone-fruits or wild fruits, with added sugar and water.

Table 2.7 Statutory minimum content of fruit juice for various beverages in several countries of the EEC (Schobinger, 1978)

Type of beverage	EEC directive (%)	National directives				
		FRG (%)	Switzerland (%)	Austria (%)	Italy (%)	France (%)
Fruit juice	100	100	100	100	100	100
Nectar	25–50	40–50	40–50	>40	40–50	—
Fruit juice drink	—	6–30	—	50–60	>12	8–12
Soft drinks	—	—	>4	—	—	—

2.2.1 Juice production

The technical procedures pertaining to the production of fruit juices can be divided into the processes of washing, crushing, juice extraction, preliminary clarification, heating, refining and bottling. In some cases, the liquid is concentrated before the bottle-filling operation and is subsequently diluted. In the concentration process the aroma of the juices is partly isolated and must be restored after redilution. The various stages in the production of fruit juices are presented schematically in Fig. 2.2. This also shows the locations of the fresh water inlet and the wastewater outlet as well as the possibilities for recycling the water within the processing plant itself.

In the main procedural step, namely the extraction of the juice from the fruit, three different methods are used: pressing, extracting and liquidizing. In pressing, the cell walls are mechanically broken down and the raw juice is pressed out through the filter-effect of the draff, the solid matter being retained. Extraction utilizes the destruction of the semi-permeability of the cell membranes. The juice is extracted with hot water and relies on quasiosmotic processes and hydrostatic pressures, which only slightly damage the cell walls. In liquidizing, the cell walls and membranes are dissolved by enzymatic breakdown, the solid matter which has not been broken down is sieved off and the raw juice can be collected.

2.2.2 Wastewater characteristics

In fruit juice production, rinse-water, condensates and cooling-water, together with water used in the filling and cleaning of bottles, pass into the effluent stream. In addition, at all stages of the processing, wastewater results from the rinsing and washing of surfaces, rooms, machines and containers. A considerable part of the pollution arising in the fruit juice industry comes from the production wastes of the processed juices and fruit. Data showing the degree of pollution have already been given in Table 2.4. Further analytical data are shown in Table 2.8.

Table 2.9 shows the composition of the wastewater at different process stages in two factories producing juice from berries and from stone-fruits. Characteristics of wastewater from the production of apple juice are given in Table 2.10.

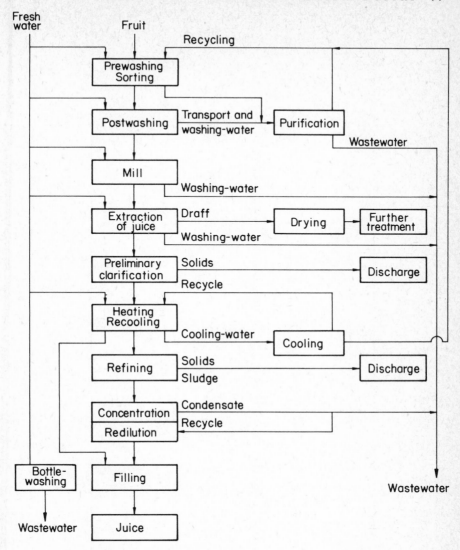

Fig. 2.2 Stages of production of fruit juices

These refer to factories which eliminate solids and use only new bottles. The production of other fruit juices yields similar wastewater flows and loads (Table 2.11). The juice yield is about 80%.

The wastewater volume and load from one of the factories quoted in Table 2.9 are broken down into the various process stages in Table 2.12. The specific figures for the wastewater generated by this factory, which has an annual output

Table 2.8 Analytical data for various fruit juices (Damm, 1964; Dittrich, 1965, 1975)

Type of juice	pH	Sugar content range (g l^{-1})		BOD$_5$ range (g l^{-1})		Perm. consumed[a] range (g l^{-1})		COD range (g l^{-1})		TOC range (g l^{-1})	
		min	max	min	max	min	max	min	max	min	max
Apple juice, pure	3.4	85.0			57.0		363.0				
Apple juice	3.2–3.4	92.0	106.0	61.5	94.5	280.7	514.0	85.0	155.0	32.1	60.0
Orange juice	3.6–3.8			60.9	85.0	237.0	303.4	83.4	134.6	38.2	54.5
Grapefruit juice	3.3				70.0		353.9		116.8		48.1
Pear juice	3.2–3.6			88.0	109.2	315.0	472.0	72.3	175.3	62.5	69.3
Apricot juice	3.5				105.0		568.8		194.0		77.5
Peach juice	3.6				100.0		335.0		174.2		70.5
Mixed juice	3.2				65.0		227.6		90.5		
Banana juice	4.1				110.0		483.0		184.1		53.6
Grape juice, white	3.2		170.0	80.0	89.0	252.8	445.0		116.8		46.8
Grape juice, red	2.9		170.0	78.0	99.0	368.0	452.0		115.6		51.5
Blackcurrant juice	3.0–3.2		92.0	60.0	113.5	324.0	491.0	110.8	158.8	44.5	62.0
Cherry juice	3.2				110.0		461.4		162.4		64.7
Sour cherry j. must	3.2		97.0		69.0		382.0				
Plum must	3.3		103.0		81.0		390.0				
Red beet juice	4.3				50.0				92.5		33.8
Vegetable juice	3.8				20.0				50.0		18.0

[a] Perm. consumed/4 ≈ oxygen consumed.

Table 2.9 Wastewater volumes and loads at various process stages in two fruit juice factories (Bauer and Brunner, 1976)

Process stage	Specific wastewater volume ($m^3 m^{-3}$)	pH	Settleable solids after 2 h ($ml l^{-1}$)	BOD_5 ($mg l^{-1}$)	COD ($mg l^{-1}$)
Pressing	0.82–1.42	5.8–6.0	26.4–26.6	2850–2870	4575–4685
Container cleaning	0.015–0.019	7.0–9.3	4.8–36.0	730–810	915–1270
Kieselguhr filtration	0.005–0.013	5.9–6.9	12.0–20.4	—	1490–7500
Refining solids[a]	0.06	—	—	67 500	147 000
Bottle cleaning	0.23–1.82[b]	8.4–9.4	—	52–290	82–410
Caustic solution	0.04–0.12[b]	—	16.6–23.5	740–2710	1700–4740
Bottle box washing	0.042[a]	—	0.1–15.0	210–245	300–550

[a] Only one factory.
[b] Per 1000 bottles.

Table 2.10 Specific wastewater volume and COD load from apple juice production (Béuthe, 1977)

Process stage	Wastewater volume ($m^3 t^{-1}$ raw juice)	COD (settled) Average ($kg t^{-1}$ raw juice)	Range ($kg t^{-1}$ raw juice) max	min
Storing, washing, pressing	2.0	3.0	5.0	1.5
Refining, separating, filtration, storing	1.0	1.5	2.0	1.0
Storing, out storing, mixing, filtration	1.0	0.7	1.5	0.5
Filling (new bottles)	2.0	1.0	1.5	0.5
Total	6.0	6.0	7.0	5.0

Table 2.11 Wastewater data from juice pressing operation (Dittrich, 1975)

Fruit processed	Wastewater volume ($m^3 t^{-1}$)	BOD_5 ($kg t^{-1}$)	COD ($kg t^{-1}$)
Apples	1.2	1.4	2.0
Pears	1.2	1.9	4.0
Cherries	5.0–8.5	2.8–4.5	3.9–4.5
Apricots	4.6	4.4	5.3
Plums	4.0	2.4–2.9	3.4

Table 2.12 Wastewater data for various stages of juice production from berries and stone-fruits (Bauer and Brunner, 1976)

Process stage	Wastewater volume ($m^3 m^{-3}$)	BOD_5 load ($kg m^{-3}$)	Ratio COD/BOD_5
Pressing	0.820	2.333	1.49
Tank cleaning	0.015	0.050	1.57
Filter cleaning	0.005	0.005	1.33
Solids, isolated by refining	0.060[a]	4.050[a]	2.17[a]
Draff	0.001	0.065	1.66
Bottle cleaning	1.820	0.528	1.39
Caustic solution	0.043	0.032	2.10
Bottle box washing	0.042	0.010	1.84
Total (without refining solids)	2.786	3.023	

[a] Not included in total.

of 4–5 million litres, are low in comparison with the findings of Beuthe (1975). This may be because this factory has developed all possible means to save water. In this case, the concentration of organics in the wastewater can be calculated as $1080 \, mg \, l^{-1}$ BOD and $1730 \, mg \, l^{-1}$ COD. In another investigation, Schobinger (1978) found BOD values of $2.6 \, g \, l^{-1}$ for the filling process

Table 2.13 Wastewater volumes and pollutional loads from juice production from various fruits

Product	Wastewater volume	Pollutional load[a]		Perm. consumed[b]
		BOD_5	COD	
	$(m^3 t^{-1})$	$(kg\,t^{-1})$	$(kg\,t^{-1})$	$(kg\,t^{-1})$
Apple juice	1.76	4.58	11.23	15.85
Berry juice (some frozen)	1.18	1.80	6.66	7.06

[a] Unsettled samples.
[b] Perm. consumed/4 ≈ oxygen consumed.

Table 2.14 Analytical data for the wastewater from apple processing (unsettled samples)

Parameter	Range		Average
	max	min	
pH	7.9	4.0	5.5
Temperature (°C)	32.0	21.0	27.0
BOD_5 $(mg\,l^{-1})$	6 785	985	2 880
COD $(mg\,l^{-1})$	17 200	3 190	6 374
Perm. consumed[a] $(mg\,l^{-1})$	19 118	3 900	10 556
Total P $(mg\,l^{-1})$	42.0	1.5	14.1
Kj-N $(mg\,l^{-1})$	148.8	19.3	47.7
Organic acids[b] $(mg\,l^{-1})$	2 101	780	1 354

[a] Perm. consumed/4 ≈ oxygen consumed.
[b] Volatile acids calculated as butyric acids.

of must and juices and values of $4.2\,g\,l^{-1}$ for pressing and filling when the centrifuged matter was discharged into the wastewater. A survey for the period 1979–1980 of a factory producing fruit and vegetable juice is shown in Table 2.13 (Rüffer and Rosenwinkel, unpublished results). Most of the output during the period of the survey was concentrated to semi-finished products. The concentration of several constituents of the total wastewater from this factory, specifically during the apple season, is shown in Table 2.14.

The major sources of problems in the treatment of raw effluents from the fruit juice industry are (a) low pH values, caused by the high content of organic acids, (b) imbalance of nutrients, caused by deficiencies in nitrogen and phosphorus, and (c) the very considerable fluctuations in the amount of effluent and waste matter produced. The last point is clearly demonstrated by the overall view of the seasonal amounts of wastewater given in Fig. 2.3.

2.2.3 Wastewater load reduction

In order to minimize the investment and the running costs involved in the installation of an effluent treatment system, it is essential to use all possible measures to reduce the quantities of wastewater resulting from the processing of fruit juices. Such measures include ensuring that musts, draff, turbid matter

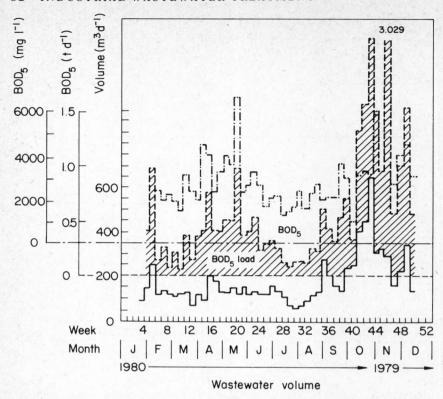

Fig. 2.3 Annual distribution of wastewater volume and load (mean weekly value 1979/80)

and filter residues, together with the sludge left over from clarification procedures, do not enter the effluent stream. The necessity for this is clear when it is realized that the solid matter from apple juice processing can represent, in suspension, a BOD concentration as high as 400 g l^{-1}. Thus, wherever possible the solid phase should be removed to reduce this load. Flocculation–sedimentation is one way of achieving this, the waste solids being dumped. Matter remaining in filters can be neutralized with lime and, after the fruitstones have been removed, can be put to agricultural use. Solid wastes can also be utilized in other ways — for example, liquid and muddy wastes can be digested or composted, apple cores can be despatched to factories producing pectin or used for feed for deer. A further reduction can be achieved by preliminary mechanical cleaning processes such as raking or sieving.

The installation of water-saving machines and processes such as, for example, drying, countercurrent washing and recycling (the reuse of condensate and unpolluted cooling-water) can bring about a significant reduction in wastewater loads. However, where the multiple use of water is practised, a mechanical treatment stage often has to be installed. Experience has also shown that training and instruction of factory personnel in the economic use of water, together with the installation and regular inspection of an adequate number of water meters in the factory for each individual production sector, will lead to definite additional economies in the use of water.

2.2.4 Wastewater treatment

Wastewater from the juice industry may be treated by physical, chemical and biological methods. However, it is always advisable to insert a flow equalization tank prior to the main treatment stage. This and an interval-aeration enables the bio-oxidation stage, particularly activated sludge, to be kept active during shutdown periods (holidays and breakdowns). The physical–mechanical methods that are used are screens, sieves, and, in the case of vegetable processing, sedimentation tanks. However, chemical treatment is employed only as a supplementary method to biological treatment. The techniques commonly employed are neutralization, using lime or sodium hydroxide, and flocculation, using alum, or ferrous or ferric salts. Experiments have shown, however, that the neutralization of organic acids is unfavourable for their biodegradation. Wastewaters from several fruit juice factories, together with juices and pure organic acids, have been examined in biodegradability tests and, when neutralized, were found to oxidize more slowly than the raw samples. Figure 2.4 shows an example of the oxygen consumption of the wastewater from apple processing, measured in a 'Saprocomp' unit.

The biological treatment methods usually adopted are irrigation, activated sludge and trickling filter plants, as well as oxidation ponds. Generally, the wastewater is readily biodegradable and the wastes can be treated alone or in mixture with sewage. Trickling filters can be 'overloaded' by 10–30% (in comparison with the normal permissible load) when fed with a mixture of fruit juice waste and sewage (Beuthe, 1977). The advantage of activated sludge plants is their buffer capacity for shock loads (e.g. pH, organics). However, bulking of activated sludge has occasionally occurred when fruit wastewater exceeded 50% of the applied load (Beuthe, 1977) and the daily volumetric load was about 1 kg BOD m^{-3}. When lower loads were used (<0.8 kg m^{-3} d^{-1}) the sludge volume index was maintained below 150 mg g^{-1}. The development of bulking sludge can also be avoided by the addition of ferrous salts (20 g Fe m^{-3}) to the aeration basins. On those occasions when the sludge volume index rose in spite of iron treatment, an increased dosage of 40 g Fe m^{-3} was effective in reducing the bulking.

When the aeration basins were filled initially, foam (presumably caused by proteins) developed, but this disappeared as the concentration of mixed liquor solids increased. The loading rates used (Dittrich, 1975) were:

volumetric loading   0.5–0.7 kg BOD m^{-3} d^{-1}
sludge loading (SLR) 0.1–0.2 kg BOD kg^{-1} MLSS d^{-1}

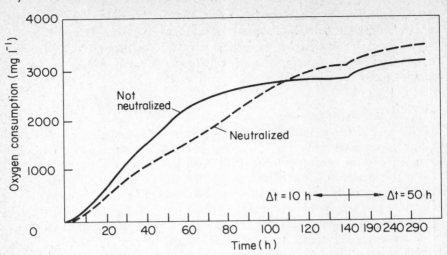

Fig. 2.4 Oxygen consumption in a Saprocomp facility with and without neutralization of wastewater from juice production

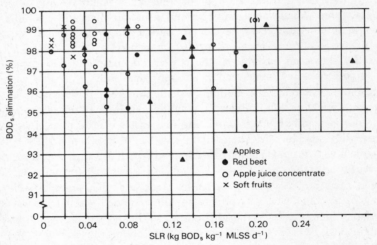

Fig. 2.5 Influence of SLR on BOD reduction for fruit juice wastewater in a single-stage activated sludge plant with the aeration tank divided into 3 cascades

Härig (1980), however, recommended the following rates for plants treating effluents from the fruit and vegetable industry:

volumetric loading 0.3 – 0.5 kg BOD m^{-3} d^{-1}
SLR 0.05 – 0.1 kg BOD kg^{-1} MLSS d^{-1}

For aerated lagoons, a volumetric loading of 0.025 – 0.03 kg BOD m^{-3} d^{-1} was recommended.

Figure 2.5 illustrates the effect of sludge loading on BOD removal and shows that there is little effect. The removal efficiency is normally better than 95% and, when the SLR is less than $0.05 \text{ kg kg}^{-1} \text{d}^{-1}$, is better than 97%. The BOD:COD ratio in the effluent is as high as 8:1. It is necessary to add some P- and N-salts to obtain a BOD:N:P ratio of 100:4:0.5, which has been found optimal for this type of wastewater.

A survey in the Federal Republic of Germany has shown that about 9.3% of fruit juice factories treat their wastewater in-plant, all using activated sludge systems, about 7.2% discharge their wastewater to irrigation, while the rest discharge it, after pretreatment, into common sewers.

2.2.5 Examples

The three examples given all use the activated sludge process.

(a) *Fruit juice factory in West Germany*
Production: fruit juice, vegetable juice, juice concentrates
Wastewater volume: $280-850 \text{ m}^3 \text{ d}^{-1}$
Pollutional load: $800-2500 \text{ kg BOD d}^{-1}$
Treatment plant: one hopper-bottom tank ($V = 12.5 \text{ m}^3$)
    one aeration tank, three cascades (variable $V = 200-333 \text{ m}^3$)
    aeration by air diffusion (foam plastic diffusers)
    three rotary piston blowers (each $1100 \text{ m}^3 \text{ h}^{-1}$)
    one final settling tank, two hoppers ($V = 118-230 \text{ m}^3$)
    sludge dewatering with sludge disposal to land at solids concentration 6%
Effluent: $\text{BOD} \leqslant 50 \text{ mg l}^{-1}$ (see Fig. 2.5)

(b) *Bottling station in West Germany (Jäppelt, 1977)*
Production: bottling of fruit juice
Wastewater volume: $279 \text{ m}^3 \text{ d}^{-1}$
Pollutional load: $\sim 255 \text{ kg BOD d}^{-1}$
Treatment plant: two aerated oxidation ponds (cascade) ($V = 5733 \text{ m}^3$, $A = 2271 \text{ m}^2$)
    surface aeration
Effluent: three unaerated lagoons ($V = 6.730 \text{ m}^3$, $A = 3.905 \text{ m}^2$)
    $\text{BOD} < 59 \text{ mg l}^{-1}$

(c) *Fruit juice factory, Yakima, Washington, USA (Esvelt and Hart, 1970)*
Production: pear, peach, apple and tomato juice
Wastewater volume: $9500 \text{ m}^3 \text{ d}^{-1}$
Pollutional load: $13\,775 \text{ kg BOD d}^{-1}$

Treatment plant: one aerated oxidation pond ($A = 7225 \text{ m}^2$; $V = 22\,700 \text{ m}^3$)
surface aeration
five centrifugal aerators, 13 W m^{-3}
addition of P- and N-salts
sludge loading $0.56 \text{ kg BOD m}^3 \text{ d}^{-1}$
one final settling tank
sludge reaeration by centrifugal aerators (22.5 kW)

Effluent: $\text{BOD} \leq 10 \text{ mg l}^{-1}$ (filtered)
$\text{COD} \leq 56 \text{ mg l}^{-1}$

2.3 Mineral waters and soft drinks—the wastewater and its treatment

Mineral waters are taken from wells, often from deep wells. They contain minerals and carbon dioxide and some contain hydrogen sulphide. Most mineral waters contain dissolved ferrous ions in high concentration and these must be removed, with a consequent loss of carbon dioxide. The carbon dioxide is normally recovered and redissolved before bottling. If necessary, sulphides may be removed by filtration through activated carbon (Knopf, 1975). The majority of table waters consist of tap water bottled after the addition of carbon dioxide.

Soft drinks consist of water, fruit juice, sugar, organic acids, natural flavour and colours and dissolved carbon dioxide. The same beverage without carbon dioxide is called a 'cold drink' (Schobinger, 1978).

2.3.1 Wastewater characteristics

Härig (1979) and Schumann (1975b) have reported the wastewater characteristics which result from these processes. The main source of organic load is the bottle-washer, causing up to 80% of the BOD_5. However, there are great differences in the waste volume and load between different factories. The proportion of juice and soft drink to mineral water and table water bottled has considerable influence on the wastewater characteristics. The relation between new and multiple-use bottles and the nature of the labels (durable or gummed) are also important.

Analysis by Härig (1979) of the wastewater of eight factories producing 50–60% mineral water and 40–50% soft drinks gave the data shown in Table 2.15. The bottling machines at these factories had a capacity of 10 000 to 15 000 multiple-use bottles per hour. The volume of wastewater was 14–35 $\text{m}^3 \text{h}^{-1}$ or 122–400 $\text{m}^3 \text{d}^{-1}$. Table 2.16 illustrates the wastewater quality corresponding to 1 m^3 of mineral water and soft drinks (Härig, 1979). Bottle-washer effluent and washdown water were analyzed separately.

Table 2.15 Range of analytical data for wastewater from mineral water bottling (Härig, 1979)

Parameter	Range min	max
Temperature (°C)	20	35
pH	6.5	8.5
Settleable solids (ml l^{-1})	0.1	1.8
Settleable solids (mg l^{-1})	8	54
Suspended solids (mg l^{-1})	45	114
Total suspended solids (mg l^{-1})	53	168
Perm. consumed[a] (mg l^{-1})	485	1135
COD settled (mg l^{-1})	350	835
BOD$_5$ settled (mg l^{-1})	180	370
Sulphate (mg l^{-1})	80	110
Chloride (mg l^{-1})	120	170
Nitrogen (mg l^{-1})	6	40
Phosphorus (mg l^{-1})	6	9

[a] Perm. consumed/4 ≈ oxygen consumed.

Table 2.16 Specific wastewater analysis of eight mineral water and soft drink factories — values correspond to 1 m^3 of drink (Härig, 1979)

Parameter	Range
Wastewater volume (m^3 m^{-3})	1.18–2.70
Settleable solids (l m^{-3})	0.09–0.33
Settleable solids (kg m^{-3})	0.03–0.06
Suspended solids (kg m^{-3})	0.14–0.15
Total suspended solids (kg m^{-3})	0.17–0.21
COD settled (kg m^{-3})	0.6 –1.5
BOD$_5$ settled (kg m^{-3})	0.4 –0.7
COD/BOD$_5$	1.7 –2.2

Table 2.17 Wastewater volume and pollutional load for the bottling of table water and sweet drinks (Schumann, 1975b)

Product	Wastewater volume (m^3 m^{-3})	BOD$_5$ load (kg BOD$_5$ m^{-3})
Table water and sweet drinks (one factory)	2.98	2.40
Sweet drinks (two factories)	2.68–2.72	3.0–3.1

Schumann (1975b) found a higher average load from similar factories. Table 2.17 shows data from a factory producing mineral water and soft drinks and from two firms producing only soft drinks. The organics originate mainly from remnants of sweet drinks in multiple-use bottles and from labels and their adhesives (Porges and Struzeski, 1961; Sidwick, 1973). The COD/BOD$_5$ ratio of 1.7–2.2 indicates a good biodegradability and toxins could not be found in the TTC test (triphenyltetrazolium chloride test). There were sufficient nitrogen and phosphorus compounds that, although admixture with sewage was not necessary, it was advantageous.

2.3.2 Wastewater load reduction
Measures taken within the factory can reduce the waste load significantly. In principle these measures are the same as those discussed above for juice factories (Voss and Rottke, 1979; Heller, 1979):

All processes
- Separation of cooling- and production-water and condensate.
- Use of small-diameter hoses.
- Use of 'water-saving' valves.
- Avoidance of product loss.
- Recycle of low-pollution-level water for service-water, floor rinsing, etc.
- Complete emptying of containers before cleaning.
- Cleaning of containers only once daily.
- Recycle of cooling-water, including recooling.
- Staff training in water saving.

Bottle-washing
- Use of pulsating jets.
- Provision of automatic water stop in case of production stoppage.
- Extended use of caustic solution by periodic recleaning and concentrating.
- Bleeding of used caustic solutions into the wastewater stream without neutralization.

Container-washing
- Recycle and storage of cleaning solutions, e.g. 'cleaning-in-place' (CIP) apparatus.

2.3.3 Wastewater treatment
In the majority of factories pH control is required. If caustic solutions are discharged intermittently, balancing tanks should be used to retain the discharge for reuse. This practice reduces acid consumption and avoids a needless increase in inorganics in the receiving waters. Neutralization can be carried out either as a batch or as a continuous process, using hydrochloric acid, sulphuric acid or carbon dioxide. Sulphuric acid is the cheapest but its use may be limited by restrictions on sulphate concentrations which are necessary to minimize corrosion. Carbon dioxide is used as a gas (Mette, 1977) or in the

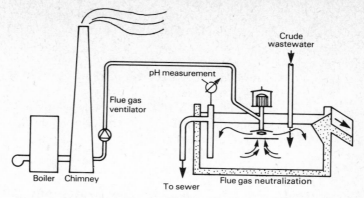

Fig. 2.6 Scheme for neutralization with flue gas (Pöpel and Mangold, 1979)

carbo-method (Krüger, 1977). To neutralize 1 kg of sodium hydroxide requires:

1.22 kg H_2SO_4 available as 1.29 kg conc. H_2SO_4 (95%); or
0.91 kg HCl available as 2.38 kg conc. HCl (38%); or
1.10 kg CO_2.

In the beverage industry carbon dioxide is the most convenient reagent for neutralization, as it is available on-site. The reaction produces sodium bicarbonate. Neutralization plants are available as proprietary units and incorporate loop reactors. Gaseous neutralization works more economically if the gas is diffused in fine bubbles (Pöpel and Mangold, 1979) (see Fig. 2.6). Surface diffusers can also be used but will increase the sulphite and sulphate concentrations in the water, as well as the temperature (Mette, 1977).

The carbo-method uses waste carbon dioxide from the impregnation plant and the bottling machine. The gas is taken to the bottle-washer, where it reacts with the sodium hydroxide rinsing-water in a mixing device (Krüger, 1977). Using waste carbon dioxide is very economical: the waste gas can be supplemented automatically by carbon dioxide from the reservoir. The pH of the rinsing-water is a function of the sodium hydroxide concentration and the hardness of the water.

The wastewater can be treated by conventional biological processes, provided that the BOD:N:P ratio is acceptable.

2.3.4 Example

FRG, Harzer–Grauhof–Brunnen, Grauhof/Goslar (Sierp, 1967)
Production: 67% soft drinks, 33% table water
              225 000 bottles d^{-1}, about 85% multiple-use bottles.

Wastewater volume: 425 m³ d⁻¹ (included 15 m³ domestic sewage d⁻¹)
Pollutional load: 276 kg BOD_5 d⁻¹ about 6 ml l⁻¹ settleable solids
pH ≃ 8.5 (average)
Treatment plant: screening, grit chambers
activated sludge basin ($V = 250 \text{ m}^3$, SL = 1.1 kg BOD_5 m⁻³ d⁻¹, MLSS ≃ 6 g l⁻¹, SLR = 0.18 kg BOD_5 kg⁻¹ MLSS d⁻¹, surface aerator)
final settling tank ($V = 100 \text{ m}^3$, Dortmund type)
sludge thickener ($V = 80 \text{ m}^3$)
Effluent: ~ 98.8% BOD_5 removal

2.4 Beer — production processes, the wastewater and its treatment

Beer is an alcoholic and carbonated beverage which in West Germany, because of purity requirements, is prepared exclusively from malt (e.g. dried and germinated barley), hops, yeast and water. In other countries, other sugars are used to replace those in malt and other flavouring agents are permitted.

The different kinds of beer are distinguished chiefly by varying concentrations of malt and hops, and by the types of malt and yeast from which they are made. The distinctions between the various types of malt (e.g. chit malt, caramel malt, burnt and acid malt) depend on the temperature of the malting process, while yeast distinction depends upon the desired fermentation process (e.g. bottom or top fermentation).

The beer consumption per capita in the countries of the EEC has already been given (Table 2.2). Chronological development of the beer output in these countries and of the numbers of their commercial breweries are given in Tables 2.18 and 2.19 respectively. Although beer output increased by 87% in the years from 1957 to 1978, the number of commercial breweries decreased by 50%. Between 1977 and 1979 several countries experienced in beer production a decrease which also affected some other areas of beverage production. Nevertheless, there is a general trend towards an expansion of beer production.

In 1957 the annual production of the average EEC brewery was 3.3 million litres, but that figure increased to 12.5 million litres by 1978. In a few EEC countries, such as the UK and the Republic of Ireland, the breweries have a much higher mean production than the EEC average. A brewery is considered large if it has an annual output of 100 million litres or more. Large breweries generate a high pollutional load which requires on-site treatment, possibly coupled with treatment in municipal systems.

2.4.1 Beer production
The beer manufacturing processes, illustrated in Fig. 2.7, include the wastewater and trub substances flow. The supplied malt is bruised and then mashed in the

Table 2.18 Beer barrelage of EEC countries (million m³) for the years 1950–1978 (Anon., 1980a)

Country	1950	1957[a]	1970	1976	1978
West Germany	1.82	4.55	8.71	9.57	9.17
France	0.91	1.47	2.03	2.39	2.28
Belgium–Luxembourg	1.05	1.06	1.37	1.53	1.41
Netherlands	0.14	0.27	0.87	1.39	1.47
Italy	0.15	0.17	0.60	0.73	0.80
Great Britain	3.80	4.21	5.51	6.56	6.64
Republic of Ireland	0.30	0.31	0.51	0.57	0.58
Denmark	0.30	0.35	0.71	0.89	0.81
Total EEC	8.47	12.39	20.31	23.63	23.16

[a] Foundation of the EEC.

Table 2.19 Number of industrial brewery plants in EEC countries for the years 1938–1978 (Anon., 1980a)

Country	1938	1957[a]	1960	1970	1977	1978
West Germany	3408	2369	2218	1815	1490	1447
France	824	281	225	114	70	67
Belgium–Luxembourg	1167	510	425	241	145	141
Netherlands	78	43	38	23	20	20
Italy	37	28	29	38	33	32
Great Britain	948	390	358	177	139	135
Republic of Ireland	16	10	8	7	7	7
Denmark	204	128	112	28	25	25
Total EEC	6680	3779	3413	2443	1929	1874

[a] Foundation of the EEC.

mashing tun with warm water. During the mashing process, the insoluble components (starches) and the high-molecular protein compounds are converted or decomposed by enzymes into soluble compounds (sugar, dextrine). In the 'Dekoktions' method, the thicker flowing portion is broken down separately at higher temperatures in the mashing pan; using the 'infusion' method, the whole mashing process takes place in the mashing tun.

Separation of wort from insoluble elements (grains) takes place in a lauter tun, a mashing filter or a strainmaster (Knopf, 1975). The concentration of dissolved materials decreases with the removal of the wort, and the liquid is referred to as flushing water when this concentration is less than 0.5%. The flushing water can be used again for mashing, and the grains can eventually be used as animal feed even if some filter aid is used to assist grain-separation, provided that neither carbon nor asbestos is used.

After the addition of hops, the wort is cooked in a boiler where it is concentrated (i.e., adjusted to its original mass), enzymic activity is stopped, the bitter particles of the hops are extracted, and the proteins are separated. The

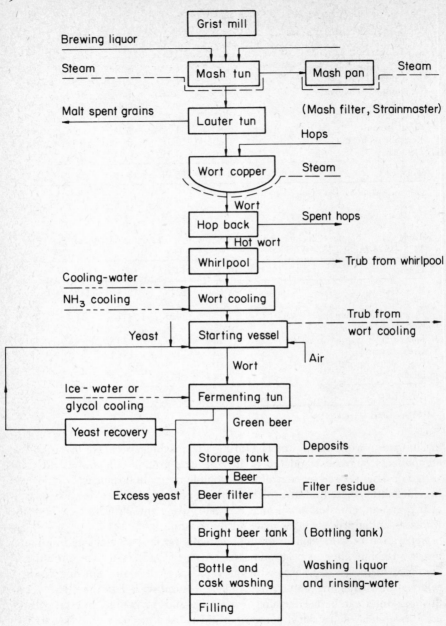

Fig. 2.7 Scheme for beer processing (ATV, 1977)

fumes generated pass to the atmosphere by way of an emission control system, or pass through a vapour condenser.

When using whole hops, the hot wort must be separated from the spent hops before hot break separation begins. The trub (hot or coarse) is then separated, often by means of vortex separators. After separation of hot trub, the hot wort is cooled to the temperature of fermentation, as the cool or fine trub separates. The cooling is carried out in plate coolers or cooling pipes and the cool trub can be removed by settlement, flotation, filtration or centrifuging.

After the cooled wort and yeast have been combined, the fermentation process begins in the fermenting tanks. For lagers, primary fermentation is carried out at $5°-10°C$. Covered cylindrical tanks are used for the primary fermentation, and the carbon dioxide which is evolved is recovered. The duration of this fermentation is between seven and ten days. The secondary fermentation takes place in substantially smaller tanks at temperatures around $0°C$, and it is at this stage that carbon dioxide is dissolved in the beer, and the beer becomes clear and matures. Some breweries store the beer to allow for discharge of more cooling trub particles before bottling at temperatures around $-2°C$. During fermentation an average of 20–30 litres of liquid yeast with ~12% dry solids per m^3 of beer output is produced. Approximately 75% of this comes from the primary fermentation (Anon., 1980a). After the fermentation process the yeast settles to the bottom of the tanks.

The healthy seed yeast is salvaged and some is used again in a new brew. The protein-rich yeast that is not reused can be utilized as animal or human food supplement. Waste beer recovery from the yeast sediment, using centrifugation or filter pressing, results in higher beer production and a reduction in the pollutional load. The remainder of beer from the deposits or sediments of the secondary fermentation can also be reclaimed.

Before bottling, the beer must be filtered to improve its storage properties and to give it a bright, clear appearance. Pulp filters, kieselguhr filters, layer filters and centrifugal filters are commonly used. Kieselguhr removed from the filters is not allowed to be discharged into sewers, because it would add considerably to the pollutional load, and would settle out in the sewerage system. The conditions for the addition to the grain (no activated carbon and no asbestos) also apply at this stage. The spent kieselguhr can be disposed of as a soil conditioner/fertilizer. After the filtering process the beer is placed in a carbon dioxide pressure tank before packaging in bottles, casks or cans. Possible sources of pollution at this stage include beer losses (spilt beer, broken bottles, etc.), contamination of supplied bottles and casks, and soluble labels and dissolved label glue from the bottles.

2.4.2 Wastewater characteristics

The water consumption in a brewery was measured by Pöhlmann (1980); data from this study, showing the amounts of water needed in the various sections of

Table 2.20 Fresh water consumption in a brewery (Pöhlmann, 1980)

Location of water consumption	Water consumption ($m^3\,m^{-3}$ marketable beer)
Brew house with vortex separator	1.8 –2.0
Wort cooling	–2.2
Fermenting room	0.4 –0.7
Storage cellar	0.3 –0.6
Filter and bright beer tank room	0.1 –0.4
Bottling (70% bottled beer)	0.9 –1.9
Racking hall (30% draught beer)	0.1 –0.2
Boiler	0.1 –0.3
Air compressors	0.15–0.5
Other cleaning (office, truck fleet, canteen, social rooms)	1.0 –2.5
Total	4.85–11.3

the brewery, are shown in Table 2.20. The average water consumption in West Germany in 1980 was approximately $8.0\,m^3\,m^{-3}$ marketable beer. However, not all of this becomes wastewater. Losses and utilization amount to about $2.1-2.5\,m^3\,m^{-3}$ (in special instances, $2.7\,m^3\,m^{-3}$) marketable beer, so that the least amount going to wastewater treatment is about $2.35\,m^3\,m^{-3}$. Even with extreme in-plant measures for waste reduction, smaller volumes of wastewater can rarely be achieved. Moreover, excessive recycling of the water might cause the quality of the beer to suffer.

Pollution figures for brewery wastewaters are given in Table 2.21. These include the specific wastewater volumes and, for comparison, pollution load figures. Wastewater analyses from other breweries give the following ranges:

$2.4-9.0\,m^3$ wastewater m^{-3} beer
$2.3-16.5\,kg\,BOD_5\,m^{-3}$ beer

The $BOD_5:COD$ ratio lies between 1:1.3 and 1:1.8 (ATV, 1977). With careful management and removal of all solids (grain, hops, yeast), as well as cooling-water circulation, the following values should be attainable (ATV, 1977):

$4.0-6.0\,m^3$ wastewater m^{-3} beer
$5.0-6.0\,kg\,BOD_5\,m^{-3}$ beer
$BOD_5/COD \approx 0.67$

The analytical results from 22 breweries (Winzig, 1975) are listed, for breweries with an annual production of up to $15\,000\,m^3$ beer, in Table 2.22.

The wastewater characteristics of several different process stages, for a brewery with an annual production of $110\,000\,m^3$ beer, are set out in Table 2.23. The pollution and specific volume of wastewater in each part of a brewery are heavily dependent upon the preventive measures taken within that section, and therefore may vary considerably.

The observed volume of wastewater amounted to $8.2\,m^3\,m^{-3}$ (see Table 2.21).

Table 2.21 Brewery wastewater characteristics (daily averages)

Parameter	According to Seyfried (1969)	Brewery in West Germany according to Seyfried and Rosenwinkel (1981)	According to Gehm and Bregman (1976)
Specific wastewater volume ($m^3\,m^{-3}$)	2.66	8.2	8.3
Specific pollution load (kg $BOD_5\,m^{-3}$)	6.0	5.3	11.8
pH	3.9–12.8	3.7–12.1	7.6
Settleable solids after 2 hours ($ml\,l^{-1}$)	4.7–13.5	4.32	—
Insoluble solids ($mg\,l^{-1}$)	561–914	117	722
BOD_5 ($mg\,l^{-1}$)	1080–4270	775	1622
COD ($mg\,l^{-1}$)	—	1220	2944
Total nitrogen ($mg\,l^{-1}$)	69.2	19.2	—
Total phosphate ($mg\,l^{-1}$)	80.0	7.6	—

Table 2.22 Volumes and pollutant concentrations (BOD$_5$ and settleable solids) produced by breweries (Winzig, 1975)

Plant no.	Annual production Beer (m^3)	Annual production Soft drinks (m^3)	Specific wastewater volume (m^3 m^{-3})	BOD$_5$ Original (mg l^{-1})	BOD$_5$ Settled (mg l^{-1})	Settleable solids (mg l^{-1})
1	10 000	—	6.0	1550	1390	480
2	10 000	1000	6.2	980	940	450
3	10 000	2500	4.9	1260	1150	227
4	4 000	—	7.3	730	670	73
5	2 000	500	7.3	775	710	208
6	1 000	50	7.6	830	780	132
7	3 000	—	5.2	1450	1360	313
8	5 500	—	4.3	1610	1440	95
9	2 000	—	3.9	755	640	368
10	13 000	—	5.5	1400	1120	734
11	11 000	2500	6.2	1420	1330	277
12	3 000	200	5.1	1540	1370	76
13	6 000	—	4.1	1050	970	266
14	15 000	3000	4.8	1890	1820	284
15	5 000	—	6.2	1580	1520	164
16	12 000	2000	3.4	1840	1760	306
17	2 000	400	6.6	650	610	172
18	700	200	2.6	1030	930	128
19	10 000	1400	2.0	1290	1180	108
20	8 000	2000	5.0	1510	1440	180
21	5 000	—	4.9	710	660	59
22	5 000	—	3.1	940	920	65
Mean			5.1	1220	1120	—

Table 2.23 Analytical data for wastewaters from various sites within a brewery (room cleaning not included) (Seyfried and Rosenwinkel, 1981)

Location of production	Settleable solids (ml l^{-1})	BOD$_5$ (settled) (mg l^{-1})	COD (settled) (mg l^{-1})	Perm. consumed[a] (mg l^{-1})
Engine and boiler room	—	3	28	16
Brewhouse	7.2	470	1791	1728
Pitching cellar	15.4	1045	3179	2190
Fermenting room	—	476	948	974
Storage cellar	—	280	539	274
Filter room	—	185	464	212
Bottle and cask filling	3.7	773	1262	754

[a] Perm. consumed/4 ≈ oxygen consumed.

The pollutant distribution, and the possible in-plant measures for waste control taken to reduce the pollution load, are shown in Fig. 2.8. With the implementation of these measures (which have been placed in order of priority by Schumann (1980)), the related range of values remains unchanged.

Fig. 2.8 Example of waste load distribution in a brewery with no in-plant measures for waste reduction, but without discharging draff and press-water (Schumann, 1980)

2.4.3 Wastewater load reduction

Before discussing the treatment of brewery wastewater the options for in-plant waste reduction in the various areas of the brewery should be mentioned (Pöhlmann, 1980; Putz, 1977; Schumann, 1980; Narziss and Meyer, 1978):

The whole operation
- Installation and control of water meters for all points in the operation.
- Staff training.
- Acquisition of dependable hose fittings (ball cocks), use of small-diameter hoses and high-pressure cleaners.
- Use of recycled water for prewash purposes.
- Reduction in quantity of wash-water and rinse-water.
- Use of automatic CIP installations for pipes and tanks.
- Reuse of cleaning solvents.
- Separate drainage or recirculation of cooling-water.

Brewhouse
- Reduction of the last runnings losses by exact dosages of second worts to the lauter tun, from $0.3-0.4 \, m^3 \, t^{-1}$ to $0.1 \, m^3 \, t^{-1}$ malt (10 litres last runnings = 1 p.e. with 1% wort capacity).
- Use of compressed air to clean the mashing filter.
- Use of last runnings and wash-water from the lauter tun underside cleaning ($BOD_5 = 12\,000-15\,000 \, mg \, l^{-1}$) for mashing (after pretreatment by sedimentation, centrifugation or activated carbon).
- Use of spent hops and malt, and compressed spent grain ($BOD_5 = 12\,000-15\,000 \, mg \, l^{-1}$) as feed. The spent grain yields = $125 - 130$ wet kg (≈ 30 dry kg) for every 100 kg malt, and its composition is 28% protein, 8% fat, 41% nitrogen-free extract substances.

Wort pipes
- Removal of the hot trub from the spent grain. The hot trub is usually separated by means of vortex separators, but some centrifugation and wort kieselguhr filtration is also used. About 600–800 g m^{-3} of extract-free dry substance becomes available, of which 50–70% is protein, 15–30% is bitter particles, and 20–24% is tannin, mineral and other organic material.
- Reuse of cool water from the plate cooler.
- Cleaning of deflection vessels with cleansing and disinfection agents only once a week, using hot water at other times.
- Removal of cooling trub with the left-over yeast or with spent grain (sediment must be removed and collected in yeast collection vessels when flotation is used); separation of cooling trub is the result of settling, floating, filtering and cold-centrifuging processes — approximately 150–300 g cooling trub per m^3 of hot wort becomes available, of which about 50% is protein and 20–30% is high molecular weight carbohydrate.

Fermentation and storage cellar
- Collection of left-over and deposited yeast for sale as cattle or pig feed or for sale to other breweries. The yield of yeast comprises about 15 litres from left-overs and 3–4 litres from deposits per m^3 beer (BOD$_5$ = 170 000–200 000 mg l^{-1} in fresh condition, ≤500 000 mg l^{-1} in stale condition). On a dry solids basis, the yeast contains:

    50–60% proteins
    15–35% carbohydrates (glycogen)
    2–12% fat (lipids, lipoides)
    6–12% ash (minerals), in addition vitamin B

- Waste beer recovery from deposits and draw-off beer. This is achieved by separation through a settler, yeast press, separator or vacuum-drum filter (through utilization of yeast, the relative amount of extract loss can be reduced from 100% to 15%).
- As the yeast is cleaned over the vibrating filters, the catching and removal of filtrate above the yeast collection tanks.
- Settling of the wash-water from fermentation and supply tanks; eventual use of the sediment in mashing.
- Installation of larger fermentation and supply tanks.

Filtration
- Emptying of the filter with compressed CO_2 to reduce the volumes of forerun and postrun wastewater.
- Removal of fines from the filter with beer in circulation (after filling with water and emptying with CO_2). This beer can be used twice, but must go through the fermentation process again (eventually problems can be caused by foam and the distribution of kieselguhr).

- At the onset of turbidity, after filtration and emptying with carbon dioxide, collection of the beer in the dregs tank.
- The weighing-out of kieselguhr as dry powder rather than as slurry, using compressed air (kieselguhr addition $1.00-1.50\,\text{kg m}^{-3}$ beer $\approx$ 5 litres settled matter of kieselguhr per m^3).
- Intake and transportation of the kieselguhr by tip-carts, transport spirals, eccentric screw pumps and containers; use with the spent grain, yeast, or for bottom loosening or deposition.
- Hot water sterilization in circulation, or steam sterilization.

Cask washing and filling
- Use of second washing-water for prewashing.
- Steaming of the casks.
- Precise rinsing dosages; eventually intermittent rinsing.
- Excess beer collection.
- Precise adjustment of the cask fillers to avoid beer loss.

Bottle-washing and filling
- Testing of the nozzles in the bottle-cleaning machines with a view to use of smaller-diameter nozzles (fresh water pressure $\leq$ 1 bar).
- Installation and control of a water meter at every machine.
- Installation of solenoid valves to control water flow, adjusted to the operating times of the washing machine, filler and cleaning machine.
- Installation of a metal sheet under the fillers to collect spilt beer (2.5 ml beer per bottle = $400\,\text{g BOD}_5\,\text{m}^{-3} \approx 10\,\text{p.e. m}^{-3}$); this collected beer can be saved, or added to the yeast in the collection tank and sold.
- Optimum utilization of the liquor in bottle-washing and eventual separation through settling, flocculation, filtering.
- Rapid sifting out of labels, broken glass and foils.
- Use of the water from bottle-washing to wash the casks, or in malting.
- Sorting out and emptying of unfilled bottles.
- Use of liquor-proof labels and reduction of the amount of the glue (strip or point gumming uses less glue than overall gumming).

Neutralization of the predominantly alkaline wastewater is seldom needed if intermittent discharge of the cleansing agent is avoided. For the same reasons as given when discussing juice, a neutralization stage can be avoided. It is much better to lead the residue liquor through a bypass in very small dosages over a long period of time. Slightly alkaline sewage is not harmful to the wastewater treatment and is neutralized by acidity produced during biological oxidation. Where more acidic wastewater (pH $\leq$ 6.5) is present it will require neutralization by the controlled addition of alkaline liquors.

2.4.4 Wastewater treatment

After the application of in-plant measures for waste reduction, there is not enough settleable material in the wastewater to warrant preliminary treatment for BOD_5 removal.

Equalization basins with aerated mixing are useful for the wastewater pretreatment. The aeration is necessary because of the dangers of acidification, sludge deposits and odour emissions. With partial biological treatment, a north German brewery achieved a 50% reduction in BOD_5 in its wastewater. The effectiveness of this type of treatment can be improved substantially by the addition of excess sewage sludge (Seyfried, 1980) and, provided that mixing and equalization tanks are available, this is best carried out at municipal treatment works.

A totally biological treatment of brewery wastewater is possible. However, Seyfried (1969) noted a nitrogen deficiency during the initial start-up of a brewery treatment plant. This was remedied by the addition of calcium nitrate. The phosphorus requirement for biological treatment can be met by using the cleaning water and bottle-washing liquors.

The use of conventional effluent treatment plants to treat domestic sewage with a proportion of brewery wastewater is well established. Where the total BOD_5 load comes predominantly from breweries, the following methods have proved successful (ATV, 1977):

(a) Activated sludge process with a daily sludge loading of $SLR \leq 0.1$ kg BOD_5 kg^{-1} MLSS. At higher loads bulking sludge is reported with SVI values of up to 270 ml g^{-1}. The bulking sludge can sometimes be controlled by using higher nitrogen/BOD_5 ratios.
(b) Aerobic oxidation ponds (max. daily space loading = 20 g BOD_5 m^{-3}) with a final clarification pond ($t_R \geq 1$ d).

Activated sludge plants can be constructed under the majority of soil conditions, and need only a moderate amount of space. Oxidation ponds, on the other hand, require substantially more space (over 1 m^2 per p.e.) and demand a less porous subsoil. As long as the land is available and suitably priced, the choice is self-evident; oxidation ponds call for little investment, they have low operational costs, and the waste sludge amounts are relatively small. Sludge removal from the ponds is needed only every two to three years (ATV, 1977).

Flocculation of brewery wastewater containing high concentrations of suspended solids is considered to be economical (Heyden and Kanow, 1975), although the reduction of suspended solids by means of in-plant measures for waste reduction may be more significant.

A summary of the results of an anaerobic treatment of brewery wastewater has been given by Sixt (1979). The purification capacity, in terms of COD, ranges from 72% to 90% and the digestion process takes between four and ten days with the daily volumetric loading reaching 5.55 kg $BOD_5\,m^{-3}$. At the Anheuser Busch Inc. brewery in St Louis, Missouri, the left-over yeast, deposits, spent hops and kieselguhr as well as the wastewater have been digested. However, because of large amounts of solids in the effluent together with residual BOD_5 and hydrogen sulphide, a subsequent aerobic treatment was necessary. Plastic media filtration with daily volumetric loading >2 kg $BOD_3\,m^{-3}$ is suitable.

Sludge treatment can be by either aerobic or anaerobic digestion. The stabilized sludge from either of these processes usually needs to be concentrated using drying beds, flotation or mechanical operations (filters, presses or centrifuges). Kieselguhr or waste sludges from water treatment can be used as conditioner. The amount of excess sludge is given by Steiger (1973) as 0.3–0.5 kg dry substance for every kg of decomposed BOD_5.

Erling *et al.* (1977) report the use of single-stage and double-stage pilot plants with pure oxygen to treat brewery wastewater. Effluent values have been achieved for the BOD_5 of less than 20 mg l^{-1} at SLR ≤ 0.45 kg BOD_5 kg^{-1} MLSS d^{-1} in single-stage plants and SLR ≤ 0.62 kg BOD_5 kg^{-1} MLSS d^{-1} in double-stage plants. In the case of single stage plant, the development of filamentous bacteria at the higher sludge loading was observed. The production of excess sludge was 0.1–0.3 kg MLSS kg^{-1} BOD_5 decomposition, and the specific oxygen requirement was 0.5–0.6 kg O_2 kg^{-1} BOD_5 removed. The removal of ammoniacal nitrogen was 50–65% (Erling *et al.*, 1977).

2.4.5 Examples

(a) *France: activated sludge treatment plant at the brewery Kronenbourg, Obernai (Heyden and Kanow, 1975)*

Production: 240 000 m³ beer yr^{-1} (1974–75)
Wastewater volume: 8500 m³ d^{-1}
Pollutional load: 9350 kg BOD_5 d^{-1}
Treatment plant: one coarse screen to eliminate labels, closures, etc.
    one buffer tank ($V = 810$ m³)
    one holding tank for the alkaline wastewater before the bottle inspection; pH regulation and aeration
    one static prethickener ($V = 400$ m³)
    two aeration tanks ($V = 9600$ m³ in total), each divided into three separate sections (SL = 0.97 kg BOD_5 m^{-3} d^{-1}) with surface aerators
    two secondary clarifiers (circular) with special suction scraper
    one twin thickener for sludge preparation
    two vacuum filters ($A = 28$ m² per filter)
Effluent: $BOD_5 = 30$–40 mg l^{-1}

(b) *West Germany: activated sludge treatment plant for a brewery (Seyfried, 1980)*

Production: 60 000 m^3 beer yr^{-1}
Wastewater volume: 2.5–3.0 m^3 m^{-3} beer
Pollutional load: BOD$_5 \approx$ 500–2000 mg l^{-1}
COD $\approx$ 1000–4000 mg l^{-1}
(peaks are higher for a short time)
Treatment plant: two aeration tanks ($V = 2 \times 850$ m^3, $V_{min} = 2 \times 500$ m^3), single-stage aeration with simultaneous sludge stabilization, operation with storage capacity, operation as cascade of both aeration tanks
Effluent: BOD$_5$ = 4–8 mg l^{-1}
COD = 65–105 mg l^{-1}

(c) *West Germany: oxidation pond treatment plant for the Bernreuther brewery, Pyras (near Roth) (Meixner et al., 1975)*

Production: 3000 m^3 beer yr^{-1} (1975)
Wastewater volume: 50–114 m^3 d^{-1}
Pollutional load: 15–95 kg BOD$_5$ d^{-1}
Treatment plant: two aeration ponds, operated simultaneously
($A_1 = 1950$ m^2, $V_1 = 4880$ m^3, line aerator, energy density 0.8 W m^{-3}; $A_2 = 1012$ m^2, $V_2 = 2840$ m^3, line aerator)
secondary clarification pond ($A = 500$ m^2, $V = 400$ m^3)
Effluent: BOD$_5 < 20$ mg l^{-1} (filtrated)

(d) *West Germany: oxidation pond treatment plant in D-8891 Kühlbach (near Aichach-Friedberg) (Anon., undated)*

The treatment plant is used both by the brewery (2400 p.e.) and by the local community (1600 p.e.).
Wastewater volume: 550 m^3 d^{-1}
Pollutional load: 240 kg BOD$_5$ d^{-1}
Treatment plant: one oxidation pond for brewery wastewater
($A = 2400$ m^2, $V = 4800$ m^3, SL = 0.03 kg BOD$_5$ m^{-3} d^{-1})
one oxidation pond for community wastewater
($A = 1800$ m^2, $V = 3450$ m^3, SL = 0.028 kg BOD$_5$ m^{-3} d^{-1})
one oxidation pond for brewery and community wastewater ($A = 1250$ m^2, $V = 2600$ m^3, SL = 0.02 kg BOD$_5$ m^{-3} d^{-1})
one secondary clarification pond ($A = 2800$ m^2, not aerated)
Effluent: BOD$_5 \approx 19$ mg l^{-1}

2.5 Wine — production processes, the wastewater and its treatment

The leading group of countries with a high annual per capita consumption of wine and champagne includes France, with 100.9 litres in 1977, and Italy, with 93.5 litres (see Table 2.2). In 1975 the area planted with vines throughout the world was 100 000 km², 73 300 km² being in Europe. The global volume of wine produced in the same year was 31 660 800 m³, with Europe producing 24 870 000 m³ (Vogt et al., 1979). In terms of volume of wine production, Italy and France are followed by the USSR, Spain, Argentina and the USA. In West Germany the area given over to wine production was 882 km² in 1979, 68% of this in Rheinland-Pfalz. The average must-yield was 1184 m³ km^{-2} vine cultivation area.

2.5.1 Wine production

Wine production in West Germany is subject to the EEC directives and the wine law of 1971. The main reaction in wine production is the fermentation process, converting glucose and fructose of the must into ethanol, carbon dioxide and energy (heat):

$$C_6H_{12}O_6 \rightarrow 2C_2H_5OH + 2\ CO_2 + \text{energy}$$

The harvested grapes are transported to the winery, the stalks are removed, and the grapes are mashed using rubber rollers. The mash is then usually stored after the addition of 50 mg SO_2 l^{-1} (in the form of sodium bisulphite) as a preservative. Separate storage is provided for each grape type and often for grapes from different locations within the vineyard. After extraction of the juice, the mash is pressed and the draff (kernels, pulp, skins) separated. The must is clarified by sedimentation or separation of the solids (dust, pulp, etc.). For white wines, the must is fermented and after six to eight weeks the young wine is decanted and pumped into another tank. The settled material (yeast, etc.) is then concentrated. The wine is polished by the addition of flocculating agents which include bentonites, kieselguhr, tannin, gelatin, proteins, wine-yeast and potassium-ferrocyanide (ATV, 1981). The main difference in the preparation of red wines is that the mash is heated without the prior separation of the grapeskins so that the red colour of the skins (anthocyane) is dissolved. Hence the preseparation of the draff from the must is omitted.

After the second decantation, the wine is filtered, sweetened if necessary (with non-fermented filtered must) and bottled. Schematic flow-sheets for white and red wine production are shown in Figs 2.9 and 2.10 (Weller, 1979).

2.5.2 Wastewater characteristics

The wastewater from wineries is a seasonal product. As a rule in the Northern hemisphere wine producers work from September till March. The pollutional

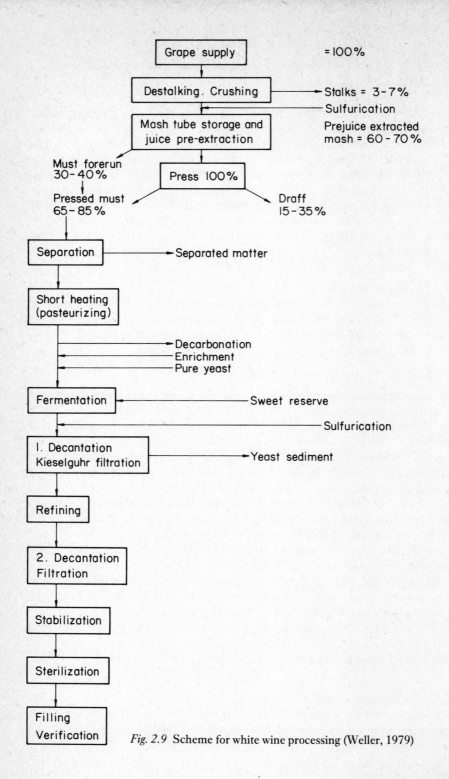

Fig. 2.9 Scheme for white wine processing (Weller, 1979)

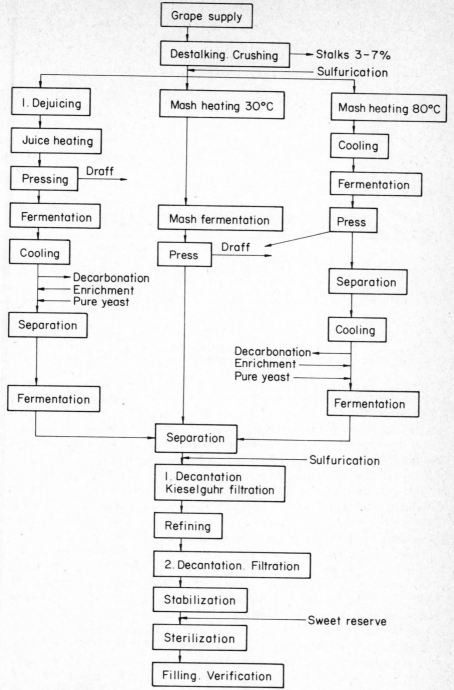

Fig. 2.10 Scheme for red wine processing (Weller, 1979)

Table 2.24 Wastewater data for rinsing-water and washing-water from vinification (based on 1 km² of vineyard) (Klotter and Hantge, 1965)

Location of wastewater flow	Wastewater volume (m³ d⁻¹)	BOD$_5$ concentration (mg l⁻¹)	BOD$_5$ load (kg d⁻¹)
Chopper cleaning	1.0	150	0.15
Destalking	0.1		0.50
Mill cleaning	0.1		0.50
Tank cleaning	8.6	270	2.32
Press cleaning	5.0	300	1.50
Room cleaning	5.0	150	0.75
Total	~20.0		5.72

Table 2.25 Influence of barrel volume on the pollutional load and volume of rinsing-water (Klotter and Hantge, 1965)

Barrel volume	Rinsing-water volume	Pollutional load (BOD$_5$)	
		Slime removing	First racking
(m³)	(l m⁻³ bar. vol)	(kg BOD$_5$ m⁻³)	(kg BOD$_5$ m⁻³)
1	87	0.322	0.521
30	28	0.105	0.170
100	19	0.071	0.115
300	14	0.049	0.080

load consists of product losses, yeast, draff and pulp, together with washing- and cooling-water.

An analysis of the various wastewater fractions — rinsing- and washing-water from the bottle-washer, cleansing-water from machines, pipes, tanks and buildings — is given in Table 2.24. These data have been analyzed for 1 km² of vine cultivation area by Klotter and Hantge (1965) (the unit km² has been retained for convenience). As already mentioned in Section 2.4, waste volume is related to container size. Table 2.25 gives information on the washing of different barrels after separation of the settled draff and yeast: the bigger the barrel, the smaller is the relative volume of waste.

In white wine production the juice is pre-separated from the draff, which consists of dust and ground particles, pulp, etc. The volume of this is approximately 30 to 120 litres m⁻³ must and its BOD$_5$ can reach 115 000 mg l⁻¹.

After the first racking of white wine there will remain about 2–3.5% by volume of settled yeast, and about 1% by volume of separator-yeast; for red wine, the figures are 3–5% and 1% respectively (Weller, 1979). The BOD$_5$ concentration of this yeast-squash is very high: 170 000–500 000 mg l⁻¹. The substances settled during the polishing process can amount to 2.5% by volume of the wine produced and the BOD$_5$ of this concentrate is less than 100 000 mg l⁻¹.

One of the main in-plant measures to reduce the waste load is thus the retention of draff, yeast and settled substances. Backwashing water from kieselguhr filters and separators and sulphur dioxide wash-water can be minimized. Cooling-water, however, should be recycled and should not increase the wastewater volume (ATV, 1981). If no in-plant measures are adopted, the pollutional load will reach the values shown in Table 2.26 (Beuthe, 1975).

The concentration for wastewater and draff from wineries is given for the various stages of production in Table 2.27 (Fischer and Coppik, 1980).

Applied to a vine cultivation area of $10\,000\,m^2$, the retention of draff, etc., reduces the pollutional load by 90 p.e. to a value of 9–16 p.e. (Hantge, 1977/78). The methods and objectives of in-plant measures, summarized by Bauer (1978), are given in Table 2.28.

Alkaline water from the bottle-washer should be neutralized only if the sewer conditions make it necessary. It is preferably discharged continuously as a small flow. Thus, in some cases, the use of a flow-equalizing tank may be advantageous.

For the bottling of $1\,m^3$ of wine the following chemicals are needed (assuming 65% multiple-use bottles) (Beuthe, 1975):

caustic soda (NaOH)	2.0 kg
phosphate cleanser	0.5 kg
sulphur-dioxide gas (SO_2)	1.0–2.0 kg
sulphur-dioxide solution	0.2–0.4 kg

When sterilized with SO_2 gas, the combined wastewater can contain up to $400\,mg\,l^{-1}$ SO_2. This can be reduced to $100-150\,mg\,l^{-1}$ SO_2 if SO_2 solution is used.

Weller (1979) found $350\,mg\,l^{-1}\,N$ and $68\,mg\,l^{-1}\,P$ in wine and $870\,mg\,l^{-1}\,N$ and $166\,mg\,l^{-1}\,P$ in must. An analysis of the wastewater load according to source within the winery (Bauer, 1978) is shown in Fig. 2.11.

The average wastewater volume from wine villages was found by Weller (1979) to be $16-30\,m^3\,km^{-2}\,d^{-1}$ and that from large wineries to be $4-20\,m^{-3}\,km^{-2}\,d^{-1}$. Fischer and Coppik (1980), however, report a value of $15-25\,m^3\,km^{-2}\,d^{-1}$. Hantge (1977/78) analyzed the effluent of wine villages (total retention of the draft) as follows:

villages with small and medium-sized wineries	$54-97\,kg\,BOD_5\,km^{-2}\,d^{-1}$
large wineries	$32\,kg\,BOD_5\,km^{-2}\,d^{-1}$

If draff is not retained, the daily pollutional load is increased by $540\,kg$ $BOD_5\,km^{-2}$.

Fischer and Coppik (1980) published a load hydrograph (Fig. 2.12), expressed as average weekly p.e., for the sewage from the village of Herxheim (population = 600) with a $2.7\,km^2$ vine cultivation area. The dependence of the wastewater load on the season is very clear. The daily hydrograph shows that

Table 2.26 Average wastewater volume and load from wineries without in-plant measures (after Beuthe, 1975)

Process stage	Operation period	Duration (d)	Production losses (l m^{-3})		Pollutional load (kg O$_2$ m^{-3})				Wastewater volume (m^3 m^{-3})
			White wine	Red wine	White wine BOD$_5$	COD	Red wine BOD$_5$	COD	
Grape receiving, mashing, pressing	End Sept.–early Oct.	15–40	4–10	6–12	0.78	1.20	1.04	1.60	0.2–0.5
Desliming and start of fermentation	End Sept.–early Nov.	15–40	13	—	1.70	2.60	—	—	0.2–0.3
Racking with separation and filtration	End Oct.–end Dec.	30–50	27 (24–35)	38 (32–45)	3.50	5.40	4.90	7.60	0.5–1.0
Clearing, separation and filtration	Mid. Dec.–end Mar.	60–80	14 (12–18)	14 (12–18)	1.80	2.80	1.80	2.80	0.2–0.5
Storing, filling, cooling, filtration	Year-round	100–250	10 (8–15)	10 (8–15)	1.30	2.00	1.30	2.00	1.8–2.5
Total production			61–91	58–90	9.08	14.00	9.04	14.00	2.9–4.8

Table 2.27 Organic load in wastewater and draff from wineries (Fischer and Coppik, 1980)

Source of pollutant	BOD_5 (mg l^{-1})	COD (mg l^{-1})	COD/BOD_5
Apparatus-cleaning	100–700	220–1540	2.2
Barrel-washing, first racking	6 000	12 000	2.0
Barrel-washing, removal of slime	3 500	7 000	2.0
Slime concentrate	110 000	198 000	1.8
Fresh yeast	170 000	306 000	1.8
Stale yeast	500 000	900 000	1.8
Refinery draff	100 000	180 000	1.8
Separator draff, first racking	82 000	147 600	1.8
Separator draff, slime removal	54 000	97 200	1.8
Wine	120 000	216 000	1.8
Must	150 000	270 000	1.8

Table 2.28 In-plant measures for waste reduction/treatment in wineries (after Bauer, 1978)

Objective	Method
Separation of coarse settleable solids in the press house (e.g. skins and stalks)	Sieve Rotary sieve
Dewatering of concentrates	Low solids concentration: centrifuge High solids concentration: filter
Reduction of settleable substances and COD	Partial treatment by flocculation Full treatment by mechanical–biological plant
Neutralization of caustic solution	HCl, CO_2, waste flue gases or continual discharge

the main pollutional load occurs between 6 p.m. and 12 p.m. and represents up to 8200 p.e. If the BOD_5 load from the population is subtracted from the daily average of 3400 p.e., the results for specific waste load from vineyard operation are:

daily average load of 1000 p.e. km^{-2} = 54 kg BOD_5 km^{-2}
daily maximum load of 3200 p.e. km^{-2} = 172 kg BOD_5 km^{-2}

The sizing of wastewater treatment plants for wine villages is proposed in the following way (Menkens, quoted in Weller, 1979):

500 p.e. km^{-2} for a village with separate wineries
250 p.e. km^{-2} for a village with a cooperative winery
250 p.e. km^{-2} for bottling works only, calculated on the basis of 750 m^3 wine per km^2 vine area.

Weller recommended, however, that actual wastewaters should be analyzed. Fischer and Coppik (1980) suggest 1000 p.e. km^{-2} vine cultivation area.

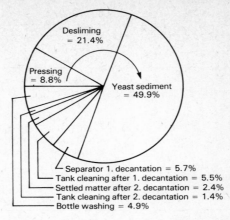

Fig. 2.11 Distribution of waste load for wine processing (Bauer, 1978)

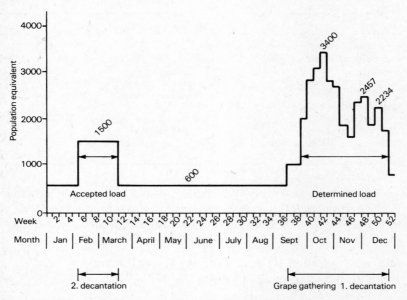

Fig. 2.12 Average weekly waste load at Herxheim (Fischer and Coppik, 1980)

2.5.3 Wastewater load reduction

All these observations point to the necessity of adopting in-plant measures to reduce the polluting loads. The target of these should be to avoid product losses as far as possible and to retain byproducts such as draff, yeast, pulp and other settled material.

It is possible to decrease product losses to 1.5–2% of the wine production.

Filtration and centrifugation coupled with chemical flocculation are
for solids retention; the solids can be composted or dumped. In all c
essential to cover the material if odour emission is to be avoided. Ta
gives an overview of in-plant measures suitable for wineries.

2.5.4 Wastewater treatment

There is little published information about the biological treatment of winery wastewaters as most of the discharges are into municipal sewers. Normal biological treatment results in a 93% BOD_5 removal, when a SLR of 0.05 kg BOD_5 kg^{-1} MLSS d^{-1} is adopted. However, only double-stage biological plants can produce an effluent of less than 20 mg l^{-1} BOD_5 (Weller, 1979).

Fischer and Coppik (1980) found a satisfactory treatment efficiency with mechanical–biological plants when the vineyard effluent was combined with municipal sewage, as long as there was an effective retention of solids in the winery. The combined treatment is enhanced by the addition of nitrogen and phosphorus.

The capacity of the treatment plants should be designed for a volume load < 0.5 kg BOD_5 m^{-3} d^{-1} during the processing season and the oxygen provision should be calculated on the basis of oxygen capacity/load = 4. During the campaign it is possible to increase the effectiveness of the treatment by additional chemical flocculation and filtration of the effluent (Fischer and Coppik, 1980). In the case of trickling filters, Weller (1979) proposes a daily loading rate of 0.3–0.4 kg BOD_5 m^{-3}.

2.5.5 Example — combined treatment of winery wastewater and municipal sewage

West Germany: activated sludge treatment plant in Pfalz (Anon., 1980b)

Wastewater volume: combined treatment plant for sewage and winery wastewater during campaign (dimensioning for 7000 p.e.)
Pollutional load: $BOD_5 = 150–1200$ mg l^{-1}
Treatment plant: one-stage activated sludge plant, Carrousel system (SL = 0.2 kg BOD_5 m^{-3} d^{-1}, SLR = 0.05 kg BOD_5 kg^{-1} MLSS d^{-1})
Effluent: $BOD_5 = 3–15$ mg l^{-1}
COD $= 50–60$ mg l^{-1}

2.6 Spirits — distillation, the wastewater and its treatment

Distilleries produce liquors with a high alcohol content. These may or may not be for human consumption, and the distinction is usually made by their

Table 2.29 Fruits and other substances that may be used as feedstock for distillation processes under West German law

potatoes	plums	rowanberries
rye	prunes	strawberries
barley	mirabelles	elderberries
wheat	cherries	mountain cranberries
rice	apricots	roots of:
maize	apples	gentian
turnips	greengages	ginger
millet	quinces	calumus
kaffir corn	raspberries	wine
buckwheat	blackberries	yeast
oats	blueberries	draff
dried fruits	haws	must
fruit marrow	juniper	molasses
fruit pulp	sloes	stone-fruit draff
Fruit flour	currants	must yeast
pears	gooseberries	

particular production processes: fermentation and synthesis. A difference is also made between the fermented alcohols from agricultural and from non-agricultural sources (Weller, 1979). In the Federal Republic of Germany distilleries are regulated by means of laws such as those on 'Spirits Monopoly' and on 'Food and Requisites'. The type of alcoholic content and the annual alcohol quotas are set for each distillery class and structure by the law governing spirits monopoly. Only distilleries producing fermented alcohol from agricultural products are discussed here.

The highest annual per capita consumption of spirits in the EEC is in West Germany with 8.35 litres. This is followed by France with 5.94 litres. In the operating year 1975/76, 296 422 m^3 of pure alcohol was produced in distilleries in West Germany (Weller, 1979).

Distilleries can be classified according to the types of raw material used. These include potato, molasses, corn, wine, wine yeast, fruit and must. The fruits and substances which may legally be used for distillation in West Germany are listed in Table 2.29.

2.6.1 Distillation

The production processes in a distillery consist of mashing, fermentation and distillation, but if the substances for distillation already contain alcohol (e.g. wine), only the latter process is necessary. A schematic illustration of potato, grain, and molasses distillation is shown in Figs 2.13–15. The substances for distillation are mashed and fermented after the initial cleaning and grinding processes.

Potatoes are transported mechanically or by means of cooling-water over the fluming canals into the washer, and then are steamed at a pressure of 8 atm. After stripping by air in the mashing tank, the mash is cooled to the sugaring

THE BEVERAGE INDUSTRY 113

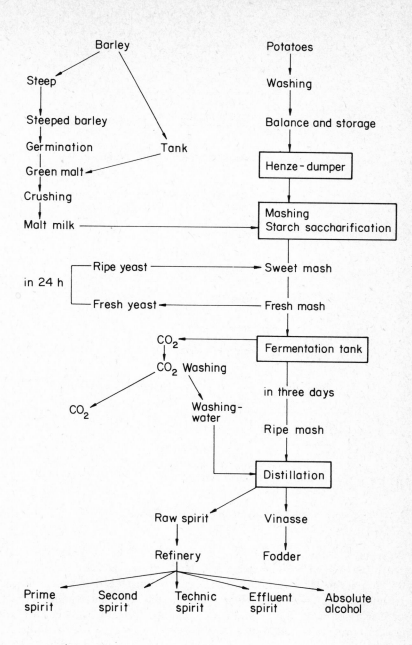

Fig. 2.13 Scheme for potato distillation (Kreipe, 1972)

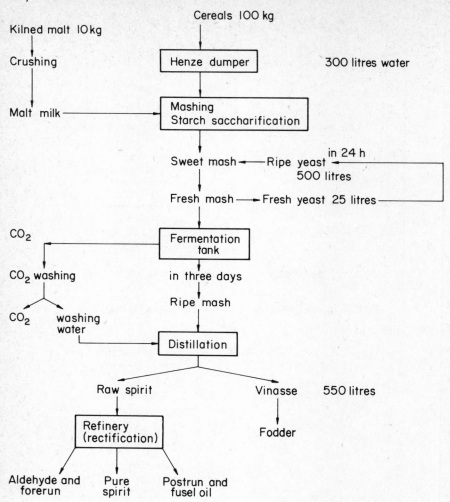

Fig. 2.14 Scheme for cereals distillation (Kreipe, 1972)

temperature (around 60°–70°C). The sugaring itself is effected by the addition of malt and enzymes. The mash is cooled further, added to the yeast, and allowed to ferment for 72 hours. The fermented mash is then passed through a continuously operated distillation column. The residual distiller's mash is stored without prior cooling and is used as food for cattle. In a grain distillery the same process is followed after the steaming, except that washing is not needed. Sometimes the mash is acidified with sulphuric acid and heated before the fermentation. In a molasses distillery a preliminary dilution with water usually takes place. Sheehan and Greenfield (1980) state that stillage can be

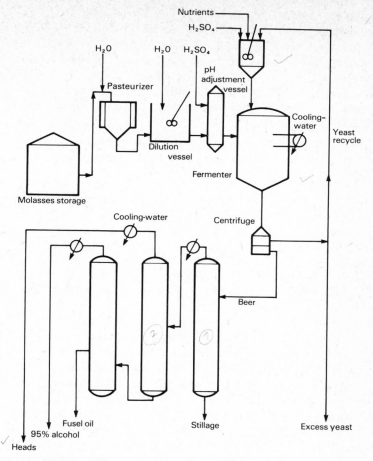

Fig. 2.15 Scheme for molasses fermentation and distillation (Sheehan and Greenfield, 1980)

used for up to 20% of the molasses dilution. The water and energy consumption (especially for steam) can be substantially limited by these in-plant measures and a reduction of the wastewater load can be achieved.

After the fermentation process, the alcohol-rich mash is passed through the distillation column. The first distillate contains 20–30% alcohol, while in the second distillate 50–80% alcohol is extracted (Weller, 1979). The distillation of the alcoholic mash proceeds until all of the alcohol is in the distillate, and the alcohol-free mash is left as stillage. The watery or pasty stillage is known as spent mash. The so called 'low wines' become available at the rectification process of the first to the second distillation. When the distillation is continuous, so that the second distillate is produced in one stage, a spent mash–low wines mixture results (Weller, 1979).

2.6.2 Wastewater characteristics

The wastewater from distilleries consists of wash-water, steaming-water, stillage, low wines, cleaning-water and cooling-water. The wash-water used for cleaning the potatoes contains between 5 and 20% sand and dirt and amounts to as much as $10\,m^3\,t^{-1}$ potatoes. The BOD_5 can vary between 400 and $1600\,mg\,l^{-1}$ or 100–400 p.e. Potato distilleries with washing arrangements and water circulation that are functioning well, and with normal raw material quality, have the following p.e. values (Ruf, 1971):

23–72 p.e.   potato wash-water without fluming system
27–75 p.e.   potato wash-water with fluming system

When farm fruits with high starch contents are processed in 'Henze-dampers', steam and blanching-water is produced from the dampers. The amount of wastewater from this process is $0.3\,m^3\,t^{-1}$ potatoes (Ruf, 1971), and the average organic content is $15\,410\,mg\ BOD_5\,l^{-1}$, the range being $5800–24\,600\,mg\ BOD_5\,l^{-1}$. With larger amounts of wastewater (blanching-water), the value is $3100\,mg\ BOD_5\,l^{-1}$ (Weller, 1979). In addition to sugar, dextrine and starch, rubber and the toxic solanine from the potato skins are present in the steam and blanching-water (Sierp, 1967). The steam-water can be led back into the mashing tanks and reused (Seyfried, 1980).

The spent mash contains large amounts of organic acids, and it is subject to a rapid anaerobic fermentation because of the high concentrations of BOD_5 and COD. In addition, spent mash is nutritionally unbalanced and this leads to the formation of hydrogen sulphide during the decomposition of the sulphur constituents. Spent mash from fruit also has a high solids content (seeds, stems, grit, and fruit and grape skins).

The amount of spent mash resulting from the potato process is about $1.4\,m^3\,t^{-1}$ potatoes (Seyfried, 1980), and from the cereals process is about $5\,m^3\,t^{-1}$ cereals (Weller, 1979). The use of double-column distillation will produce a stillage volume equal to 91% of that resulting from the use of a single-column system. However, the latter product will be predominantly low wine, so that its alcohol content will be lower (Kreipe, 1980). The consumption of water and energy is, however, substantially higher with the double-column system. The amount of spent mash from molasses distilleries has been reported as being $3.6\,m^3\,t^{-1}$ molasses (Bronn, 1975), while for grape juice operations it is 300–500 litres m^{-3}, and for red wine 500–600 litres m^{-3} (Montens, 1965). The composition of various spent mashes is shown in Table 2.30.

Low wines are heavily polluted with organic material. The quantities of low wines produced and their organic content are given in Table 2.31. According to Weller (1979), the concentration is $20\,850–22\,550\,mg\ BOD_5\,l^{-1}$ for grape draff and $21\,440\,mg\ BOD_5\,l^{-1}$ for apple draff.

Wastewater from the cleaning of equipment is of secondary importance where operation is continuous. The amount produced at potato distilleries is only $0.6\,m^3\,t^{-1}$ with the organic load varying from 100 to $3000\,mg\ BOD_5\,l^{-1}$,

Table 2.30 Pollutional data for various stillages (Weller, 1979)

Origin of stillage	BOD$_5$ (mg l^{-1})	Perm. consumed[a] (mg l^{-1})	pH
Potato	15 000–25 000	7 000– 10 000	4–5
Rye	15 000–25 000		
Cereals	12 000–34 000		
Malt	25 000	8 000	3.5–4
Corn-rye	40 000		
Molasses	18 000–22 000	26 000– 81 000	
Pip-fruit	37 326	80 000	3.5
Stone-fruit	51 078	158 000	3.5
Plum	24 733–58 830	143 780–175 380	3.8
Pear	57 000	200 625	3.4
Wine	5 600–32 000		

[a] Perm. consumed/4 ≈ oxygen consumed.

Table 2.31 Pollutional load and volume of low wines from various feedstocks

Feedstock	Volume of low wines	BOD$_5$ (mg l^{-1})	pH	Reference
Potatoes	0.3 (m^3 t^{-1})	250–710		Ruf (1971)
Barley malt	0.15 (m^3 t^{-1})	800–1200		Hemming (1970)
Wine (Brennwein)	0.35 (m^3 m^{-3})	600		Supperl (1976)
Red wine		26 700–29 600	3.4	Weller (1979)
Molasses		1280	3.5	Seyfried (1980)

Table 2.32 Analytical data of total wastewater (without stillage) from a potato distillery (Ruf, 1971)

Parameter	Wastewater without recycling	Wastewater with 60% recycling	Wastewater with 80% recycling
Settleable wet volume (ml l^{-1})	6.5	18	22
Dissolved organic carbon (mg l^{-1})	164	245	304
Dissolved organic nitrogen (mg l^{-1})	7.3	20	24
Perm. consumed[a] (mg l^{-1})	830	930	970
BOD$_5$ (mg l^{-1})	470	770	1030
pH	6.9	6.9	6.0
Specific volume (m^3 t^{-1})	6.2	3.2	2.2

[a] Perm. consumed/4 ≈ oxygen consumed.

depending on the production site. Cooling-water has a temperature of between 30° and 60°C and should be kept in circulation. The analytical results of the total wastewater from a potato distillery are shown in Table 2.32. They are dependent upon the circulation of the flushing-water; the stillage is not included (Ruf, 1971). The specific amounts of wastewater and pollutional load from potato distilleries are given in Table 2.33.

Table 2.33 Specific wastewater volumes and pollutional load for potato distillation

Feedstock	Specific wastewater volume ($m^3 t^{-1}$)			Specific pollution load ($kg\, BOD_5\, t^{-1}$)		
	Range min	max	Average	Range min	max	Average
Potatoes (Rübelt, 1971)	6.00	7.50	6.50	0.465	0.669	0.598
Potatoes exclusive of cooling-water (Seyfried, 1980)	0.40	5.70	4.00	0.140	11.400	1.500

For a cereals distillery with the stillage used as cattle feed, Seyfried (1980) quotes values of

0.29 m^3 wastewater t^{-1} cereals
3.4 kg BOD_5 t^{-1} cereals

Molasses distilleries have a pollutional load of about 324 kg $BOD_5 m^{-3}$ molasses processed, with a wastewater concentration (stillage) of 18 000–22 000 mg $BOD_5 l^{-1}$ (Seyfried, 1980). The amounts of wastewater (stillage) from fruit distilleries with an annual production of 50–300 litres of pure alcohol lie between 0.5 and 0.8 $m^3 d^{-1}$ with a pollution load of 6.0–35.0 kg $BOD_5 d^{-1}$ (Viehl and Mudrack, 1970). Analytical results of wastewater from cereal, molasses and wine distilleries are shown in Table 2.34 (Sheehan and Greenfield, 1980). These values, by comparison with the analytical values for the stillage and total pollutional load, appear to include stillage.

2.6.3 Wastewater load reduction
In-plant measures that can reduce the wastewater load from distilleries include:

- Installation of a single-column distillation apparatus for potato processings.
- Installation of a wash-water circulation system for potato washing.
- Removal of steam-water from the mashing tanks.
- Recycle of distillation residue for the molasses dilution.
- Circulation of cooling-water.
- Utilization of the spent mash.

2.6.4 Wastewater treatment
The problem of distillery wastewater treatment can be reduced essentially to the treatment and removal of the spent mash and the low wines, although at potato distilleries the wash-water must be treated.

At small distilleries (those with an annual production of ⩽ 300 litres pure alcohol) spent mash can be disposed of:

- by transportation to agricultural areas
- by use as compost

Table 2.34 Analytical data for wastewater from various distilleries (Sheehan and Greenfield, 1980)

Parameter	Grain			Molasses			Wine		
	Range		Average	Range		Average	Range		Average
	min	max		min	max		min	max	
pH	3.8	7.5	5.4	3.5	5.7	4.2	3.9	4.5	4.1
Temperature (°C)	42	95	73	80	105	94	—	—	—
Total solids (g l^{-1})	20.5	47.3	33.8	21	140	78.5	24	125	62
Volatile solids (g l^{-1})	24	36	29.5	40	100	58.9	—	—	29.5
Suspended solids (g l^{-1})	—	—	11.4	1	13	5.1	0.2	0.9	0.55
Dissolved solids (g l^{-1})	—	—	—	25	110	56.9	—	—	22
BOD$_5$ (g l^{-1})	15	340	22.2	7	95	35.7	—	—	12.3
COD (g l^{-1})	—	—	—	15	176	77.7	—	—	—
Total nitrogen (g l^{-1})	0.2	1.9	0.98	0.6	8.9	1.78	0.4	1.0	0.69
Phosphorus (P^{5+}) (g l^{-1})	0.039	0.087	0.063	0.026	0.326	0.168	—	—	1.17
Chloride (Cl$^-$) (g l^{-1})	—	—	—	0.68	7.39	3.79	—	—	1.34
Sulphates (SO$_4^{2-}$) (g l^{-1})	—	—	—	1.56	6.60	4.36	—	—	3.64

- by delivery to a central treatment plant
- by direct dumping.

In West Germany, agricultural grain and potato distilleries are required to utilize their stillage as feed for cattle. This cannot be done with the stillage from fruit, wine and molasses distilleries. Agricultural use of the spent mash from commercial distilleries is usually impossible or difficult.

Spraying and flush irrigation with wastewater diluted with cooling-water produces excellent yields of grass, turnips, corn, hemp and other crops. Ruf (1971) reports that the annual addition of 100 mm of wastewater to root crops also gives good results. However, odours and the danger of nematode transmission are potential disadvantages of irrigation. The latter can be prevented by the addition of chlorine ($0.1-0.3$ g m^{-3}) prior to spraying. In addition, neutralization of the wastewater is recommended by Bronn (1975).

The treatment of distillery wastewater in fish ponds requires a large land area and the load should be no more than 5 g BOD_5 m^{-2} pond area d^{-1} (Ruf, 1971). Meixner et al. (1975) have reported on artificially aerated ponds and suggest that they offer a useful solution in rural areas.

A degree of treatment can be achieved by purely mechanical means (i.e. sedimentation); three-stage clarifiers that can hold all of one day's wastewater are used. The efficiency of sedimentation can be improved by the addition of flocculants (iron chloride, alum sulphate, lime) giving a BOD reduction of 45%.

Conventional aerobic biological processes can be used for stillage, provided that it is diluted. Neutralization is usually necessary as the wastewater is usually strongly acidic (pH $3.0-4.5$). The separate treatment of distillery effluents is more difficult than combined treatment with domestic sewage at dilutions of 1:50 or more. The addition of nutrient salts is necessary if the nutritional balance, BOD_5:N:P, is less than 100:5:1. The potential problems of operating an activated sludge plant, mainly the build-up of bulking sludges, are described by Sierp (1967). Furthermore, the waste sludge is highly active and must be handled so as to avoid odour nuisances. Ruf (1971) reported on the purification of wastewater from a potato distillery in a laboratory-scale aerated biofilter filled with pumice stones, coke and pebbles. His results are shown in Table 2.35. The possible load was given as 1 m^3 distillery wastes for every $0.5-0.8$ m^3 of filter volume, although a substantially lower degradation of 'septic' wastewaters was observed. Sheehan and Greenfield (1980) have described the performance of biofilters for the elimination of organic materials in distillery wastes. The results were based on bench-scale work and show a COD reduction of up to 91% with a retention time of 43 hours.

Vent condensate can be biologically purified in activated sludge plants after an anaerobic pretreatment. Sixt (1979) reports details of the anaerobic purification of distillery wastes. The first large European treatment plant with modern engineering techniques for an anaerobic and alkaline digestion system

Table 2.35 Performance data of a pilot filter treating distillery effluent (Ruf, 1971)

Wastewater	BOD$_5$ Influent (mg l^{-1})	Effluent (mg l^{-1})	Percentage elimination
Barley steep-water	1350	107	92
Rinsing-water	970	40	94
Low wines	250	23	90
Potato wash-water	360	32	91

was built in 1957 by Ruhrverband for a distillery and a yeast factory in West Germany. Seyfried (1980) reports that the treatment plant has two digestion tanks with connected settling tanks and sludge recirculation. The effluent is then treated aerobically. The theoretical digestion period lasts 3.7 days. Arrangement in parallel gave an efficiency of only 40%, whereas an efficiency of over 80% with respect to BOD$_5$ could be achieved by arrangement in cascades (spillways). This difference stems from the effect of high hydrogen sulphide production in the anaerobic stage, which hinders both reactors during the parallel operation. In the cascade operation, the hydrogen sulphide production is restricted to the first digestion stage, with only partial decomposition taking place. The results of a long research programme gave the following average values (Seyfried, 1980):

	Influent	*Effluent*	*Efficiency*
Permanganate consumption	9 600 mg l^{-1}	2700 mg l^{-1}	72%
BOD$_5$	10 800 mg l^{-1}	2200 mg l^{-1}	80%

Retention time 90 h
Space loading 3.2 kg BOD$_5$ m^{-3} d^{-1}

Data for the digestion of stillage are also available from a treatment plant for the purification of wastewater of 'Bühl and environs' Bühl/Baden (Weller, 1979). This plant, as reported in the first stage of construction, is limited to 45 000 residents and 40 000 p.e. It will treat domestic and agricultural wastewater, as well as the industrial wastewater from distilleries. Since November 1976, up to 23 m^3 spent mash has been fed daily to the digestion tank, which has a working volume of 3200 m^3. The spent mash has a pH of 3.3 and a BOD$_5$ of 20 000 mg l^{-1}. Under standard operating conditions, 8 m^3 of methane gas per 1 m^3 mash is produced during this process (Weller, 1979). Since 1977, 7 kg lime for every m^3 of spent mash has been used to maintain the pH value between 8.0 and 10.0.

Sixt (1979) reports the following data for the anaerobic purification of distillery wastes: a volume loading of 1.12–4.0 kg BOD$_5$ m^{-3} d^{-1} (with one exception of SL = 9.9 kg BOD$_5$ m^{-3} d^{-1}) and a BOD$_5$ reduction of between 60 and 99.6%. The digestion period is between 5.6 and 10 days.

2.6.5 Examples

(a) *Canada: activated sludge treatment plant for a distillery (EPS, 1979)*
Production: Spirits, industrial alcohol (feedstock: corn, rye, barley)
Wastewater volume: $1000 \, m^3 \, d^{-1}$
Pollutional load: $800 \, kg \, BOD_5 \, d^{-1}$ ($1000-1500 \, kg \, COD \, d^{-1}$)
Treatment plant: pumping pit (dosage of urea and sodium dihydrogen phosphate)
one aeration tank ($V = 19\,000 \, m^3$), aeration by two floating surface aerators (each 19 kW, energy density = $4 \, W \, m^{-3}$)
$SL = 0.8 \, kg \, BOD_5 \, m^{-3} \, d^{-1}$ (retention time = 24 h)
one secondary settling tank
one lagoon (retention time = 5 d)
sludge draining beds
Effluent: $COD \leq 50 \, mg \, l^{-1}$.

(b) *West Germany: aerobic ponds for treatment of wastewater from Schönsee distillery association, Landkreis Schwandorf (Meixner et al., 1975)*
Production: potato distillery ($150 \, m^3 \, yr^{-1}$, 1750 t potatoes per season)
Wastewater volume: $2600 \, m^3$ per season, $22 \, m^3 \, d^{-1}$ (cooling-water is drained off separately)
Pollutional load: $40 \, kg \, BOD_5 \, d^{-1}$ ($2.7 \, kg \, BOD_5 \, t^{-1}$ potatoes)
Treatment plant: one settling tank ($V = 240 \, m^3$, three chambers)
one aerated wastewater pond ($A = 300 \, m^2$, $V = 750 \, m^3$), aeration by subsurface aerator, oxygen entry = $12.5 \, kg \, O_2 \, h^{-1}$)
two unaerated wastewater ponds ($A = 4000 \, m^2$)
one unaerated settling pond
Effluent: good purification

2.7 Conclusions

The range of beverages manufactured or bottled includes mineral waters, which involve mainly a bottling operation, to spirits, which can involve fermentation, distillation and bottling. The wastewater pollutional load tends to reflect the complexity of these operations. Soft drinks are manufactured by admixture of sugar syrup, flavouring, acid, carbon dioxide and water (EPA, 1977). There is little handling of the product, hence although the drink itself will have a BOD in excess of $50\,000 \, mg \, l^{-1}$ and more (Sidwick, 1973), the majority of the wastewater is due to bottle- and plant-cleaning and bottling; effluent BOD_5 values usually are less than $600 \, mg \, l^{-1}$ (Härig, 1979; EPA, 1977;

Porges and Struzeski, 1961). The production of spirits from potatoes involves the transport and preparation of the potatoes, fermentation, distillation and bottling (Ruf, 1971; Weller, 1979). The BOD of these wastewater streams is rarely less than $1000\,\mathrm{mg\,l^{-1}}$ and can exceed $20\,000\,\mathrm{mg\,l^{-1}}$.

As with many industries, an improvement in plant housekeeping operations can reduce both the volume and the pollutional load produced in the beverage industry. For breweries which involve several production processes a range of in-plant operations can be adopted (Pöhlmann, 1980; Putz, 1977; Schumann, 1980; Narziss and Meyer, 1978). Even for soft drink bottling operations, improvements in plant operations can reduce wastewater loads by up to 50% (EPA, 1977).

As beverages tend to be produced from natural products such as sugar, malt and fruits, the wastewaters usually can be treated by conventional biological processes. The effluents contain a relatively high proportion of carbohydrates and are prone to exhibit a nutrient imbalance which requires the addition of nitrogen and/or phosphorus for effective biological treatment. The effluents can contain solid materials such as waste fermentation products or washings, but in many cases the major loads are soluble or colloidal and physical/mechanical separators are only of limited use. In addition to the flow and load variability which reflects seasonal and operational processes, effluents from the beverage industry tend to show a very variable pH value. Many beverages are acidic, for example soft drinks usually have a pH value of $\geqslant 3$, while cleaning solutions are strongly alkaline (Porges and Struzeski, 1961). The use of equalization tanks, flow attenuation and neutralization (Krüger, 1977; Mette, 1977) may be necessary to protect both mechanical equipment from corrosion and biological treatment processes from the inhibitory effects of variable pH.

Therefore, while there are some wastewaters from the beverage industry that are difficult to manage, these tend to be confined to the more complex manufacturing processes such as the production of spirits. Other wastewaters can be managed relatively easily provided that due attention is given to the problems of flow, load and pH variability and the wastewater is adjusted to correct any nutrient imbalances. In-plant measures to reduce loss of product and to conserve water can make significant improvements to the wastewater quality. This latter point, which stresses the close integration of manufacturing and wastewater treatment operations, is best illustrated by the soft drink industry. Spillage or loss of the concentrated syrups used to formulate the beverages should be minimized; the syrups can have a BOD in excess of $\frac{1}{2}$ million $\mathrm{mg\,l^{-1}}$ and so should be retained in the product, not released to the wastewater.

References

Anon. (1979) *Statistisches Jahrbuch für Ernährung, Landwirtschaft und Forsten*, Landwirtschaftsverlag, Hiltrup, Münster.

Anon. (1979/80) *European Marketing Data and Statistics*, 16th edn, Euromonitor Publications, London.
Anon. (1980a) *Brauwelt-Brevier*, Brauwelt-Verlag, Nürnberg.
Anon. (1980b) Technical information, Firma Esmil.
Anon. (undated) Publicity leaflet, Firma Universal, Mölln.
ATV (1977) Abwassertechnische Vereinigung Fachausschuss 7.2, Arbeitspapier 1.18 Brauereien (1977), *Korresp. Abwasser*, **24**, 182–186.
ATV (1981) Abwassertechnische Vereinigung Fachausschuss 7.2, Arbeitspapier 1.13 Weinbereitung (1981), *Korresp. Abwasser*, **28**, 656–660.
ATV (1982) *Hinweise für das Einleiten von Abwasser in eine öffentliche Abwasseranlage*, Abwassertechnische Vereinigung Arbeitsblatt A 115 (Entwurf).
Bauer, H. (1978) Abwasserabgaben in Weinkellereien, *Deutsche Weinbau*, **33**, 595–603.
Bauer, H. and Brunner, L. (1976) Abwasseruntersuchungen in Fruchtsaftbetrieben nach dem Abwasserabgabegesetzentwurf und zu erwartende finanzielle Belastungen, *Flüssiges Obst*, **43**, 79–83.
Beuthe, C.G. (1975) Die Abwasserverhältnisse in Wein- und Fruchtsaftkeltereien, *Weiterbildung Gewässerschutz, 10. Lehrgang*, Ministerium für Ernährung, Landwirtschaft und Umwelt, Baden-Württemberg, 243–286.
Beuthe, C.G. (1977) Abwasserprobleme in Süßmostereien und bei der Gemüse- und Obstkonservenindustrie, *Weiterbildung Gewässerschutz, 12. Lehrgang*, Ministerium für Ernährung, Landwirtschaft und Umwelt, Baden-Württemberg, 355–423.
Breitenacher, U. (1980) Ungebrochene Verbrauchsentwicklung bei alkoholfreien Getränken in der Bundesrepublik, *Flüssiges Obst*, **47**, 219–220.
Bronn, W.K. (1975) Die Abwasserbelastung bei Hefefabriken und Melassebrennereien, *Branntweinwirtsch.*, **115**, 388–396.
Damerow, G. and Gerhold, E. (1976) Umweltfreundliche Beseitigung von Molke und Gewinnung von Molkenprotein und Verwertung der enteiweissten Molke durch Fermentation, *Deutsche Milchwirtsch.*, **27**, 796–803.
Damm, H. (1964) Abwasseraufgaben und -probleme in der Süssmosterei, *Flüssiges Obst.*, **31**, 498, 548, 609.
Dittrich, V. (1965) Abwasserentstehung und Abwasserreinigung in der Mineralbrunnen-, Süssmost- und Obstgetränkeindustrie, *IWL-Forum*, 263–284.
Dittrich, V. (1975) Abwasserbeseitigung in der Fruchtsaftindustrie — Technische Möglichkeiten und Kosten, *Flüssiges Obst*, **42**, 42–52.
Doedens, H. (1969) *Neue Verfahren zur biologischen Reinigung von Molkereiabwasser*, Institut für Siedlungswasserwirtschaft der Technischen Universität Hannover, 32, Eigenverlag.
Doedens, H. (1978) Kenndaten für Molkereiabwasser, *Deutsche Milchwirtsch.*, **29**, 236–239.
Doedens, H. and Noda, N. (1977) *Über den Anfall von Molkereiabwasser in Abhängigkeit von der Produktion sowie über die bei der Reinigung von Molkereiabwasser entstehenden Schlammengen*, Gutachten des Institutes für Siedlungswasserwirtschaft der Technischen Universität Hannover, 2nd edn, Eigenverlag.

EPA (1977) *State of the Art Study of Water Pollution Control from the Beverage Industry, Bottled and Canned Soft Drinks*, Environmental Research and Applications, Wilton, Connecticut.

EPS (1979) *Biological Treatment of Food Processing Wastewater — Design and Operations Manual*, Report EPS 3-WP-79-7, Water Pollution Control Directorate, Environmental Protection Service, Environment Canada.

Erling, F., Gregor, C.H. and Neubert, D. (1977) Vollbiologische Kläranlage auf kleinstem Raum, *Brauwelt*, **117**, 1080–1083.

Esvelt, L.A. and Hart, H.H. (1970) Treatment of fruit processing waste by aeration, *J. Water Pollut. Control Fed.*, **42**, 1305–1308.

Fischer, H. and Coppik, L. (1980) Abwasserbeseitigung in Weinbaugemeinden, *Weinwirtsch.*, **22**, 614–617.

Gehm, H.W. and Bregman, J.I. (1976) *Handbook of Water Resources and Pollution Control*, Van Nostrand Reinhold, New York.

Härig, H-J. (1979) Abwasseruntersuchungen in Mineralbrunnenbetrieben, *Mineralbrunnen*, **29**, 73–77.

Härig, H-J. (1980) *Nahrungsmittelindustrie, Branchenspezifische Darstellung der Industrieabwasserreinigung*, Vortrag anlässlich des Fortbildungskurses der Abwassertechnischen Vereinigung (ATV) für Wassergütewirtschaft, Abwasser- und Abfalltechnik, Laasphe.

Hantge, E. (1977/78) Beseitigung und Verwertung von Trubstoffen einschliesslich Hefe aus Winzerbetrieben und Brauereien, *Ber. Landwirtsch.*, **55**, 683–690.

Heller, R. (1979) Frischwasser und Abwasser im Betrieb für alkoholfreie Getränke, *Erfrischungsgetränk*, **32**, 602–606.

Hemming, U.C. (1970) Experience in the treatment of effluents from the manufacture of malt whisky using high rate biofiltration and plastic media, *Int. Congress on Industrial Wastewater*, Stockholm.

Heyden, W. and Kanow, P. (1975) Zur Aufbereitung von Brauereiabwässern, *Brauwelt*, **115**, 1398–1402.

Jäppelt, W. (1977) Abwasser und Abwasserreinigung bei der Obst- und Gemüseverarbeitungsindustrie, *Hannoversche Industrieabwasser-Tagung am 21.3.1977*, Institut für Siedlungswasserwirtschaft, Universität Hannover.

Klotter, H.E. and Hantge, E. (1965) Bemessungsgrundlagen für Abwasser aus Winzerbetrieben, *Vom Wasser*, **32**, 310–346.

Knopf, K. (1975) *Lebensmitteltechnologie*, Ferdinand Schöningh, Paderborn.

Kreipe, H. (1972) *Technologie der Getreide- und Kartoffelbrennereien*, Hans Carl, Nürnberg.

Kreipe, H. (1980) Einsäuliger oder zweisäuliger Maischedestillierapparat in Kartoffelbrennereien, *Branntweinwirtsch.*, **120**, 178–180.

Krüger, R. (1977) Abwasserneutralisation nach dem Carbo-Verfahren, *Mineralbrunnen*, **27**, 73–76.

Lohmann, U. (1977) Abwasserneutralisation im Mineralbrunnen, *Mineralbrunnen*, **27**, 68–69.

Meixner, G., Lindenmaier, M. and Sixt, K. (1975) Abwasser aus Gewerbebetrieben im ländlichen Raum, *Wasser Boden*, **27**, 279–281.

Mette, M. (1977) Abwasserneutralisation mit CO_2-Gas, *Mineralbrunnen*, **27**, 70–72.
Montens, A. (1965) Organisch hochbelastete Abwässer, ein technisches Problem, *IWL-Forum*.
Narziss, L. and Meyer, L. (1978) Die Auswirkung der Heisstrubabscheidung bei Whirlpools und Zentrifugen, *Brauwelt*, **118**, 1191–1200.
Pöhlmann, R. (1980) *Innerbetriebliche Massnahmen zur Abwasser- und Abfallverminderung in Brauereien, Massnahmen im Sudhaus, auf dem Würzeweg, im Gär- und Lagerkeller*, Fachveranstaltung Nr. F-7-912-09-0, Haus der Technik, Essen.
Pöpel, F. and Mangold, S. (1979) Neutralisation und Flotation alkalischer Abwässer mit submers eingetragenem Rauchgas, *Lebensmit.-Tech.*, **12**, 16–19.
Porges, R. and Struzeski, E.J. (1961) Wastes from the soft drink industry, *J. Water Pollut. Control Fed.*, **33**, 167–175.
Putz, M. (1977) Recycling des extraktschwächeren Filtrationsvor- und -nachlaufs, *Brauerei-J.*, **74**, 484–490.
Rosenwinkel, K.-H. and Rüffer, H. (1981) Reinigung von Abwässern der Fruchtsaftindustrie, dargestellt an Beispielen, *Vom Wasser*, **57**, 243–262.
Rübelt, Ch. (1975) Die Bewertung von Brennereiabwässern in Schadeinheiten und Einwohnergleichwerten, *Branntweinwirtsch.*, **115**, 209–213.
Ruf, G-D. (1971) Zusammensetzung, Bewertung und Abbaufähigkeit von Abwässern der Kartoffelbrennerei, *Branntweinwirtsch.*, **111**, 49–56.
Rüffer, H. (1980) *Low Waste Technology in Food Industries, Vol. 52*, Pergamon Press, 2005–2015.
Schmittel, H.W. (1977) Abwasserprobleme in Sektkellereien, *Weinwirtsch.*, **19**, 688–690.
Schobinger, U. (1978) *Handbuch der Getränketechnologie, Frucht- und Gemüsesäfte*, Eugen Ulmer, Stuttgart.
Schumann, G. (1975a) Senkung der Abwasserfracht durch innerbetriebliche Massnahmen in Brauereien und Mälzereien, *Tagesz. Brauereien*, **137/138**, 750–754.
Schumann, G. (1975b) Hinweise für die alkoholfreie Erfrischungsgetränke herstellende Industrie zum bevorstehenden Abwasserabgabengesetz, *Erfrischungsgetränk*, **28**, 248–255.
Schumann, G. (1977) Abwasserteilfrachten aus Reinigungsverfahren und ihre Umweltprobleme, *Monatsschr. Brauereien*, **9**, 319–323.
Schumann, G. (1980) *Innerbetriebliche Massnahmen zur Abwasser- und Abfallverminderung in Brauereien, Auswirkungen der Frachtverminderung auf die Abwasser-Belastungsgrössen und -Kostenlage*, Fachveranstaltung Nr. F-7-912-09-0, Haus der Technik, Essen.
Seyfried, C.F. (1969) Erfahrungen mit Brauereikläranlagen, *Münchner Beiträge zur Abwasser-, Fischerei- und Flussbiologie*, **16**, 258–267, Oldenbourg, München/Wien.
Seyfried, C.F. (1980) *Getränkeindustrie*, Vortrag anlässlich des Fortbildungskurses der Abwassertechnischen Vereinigung (ATV) für Wassergütewirtschaft, Abwasser- und Abfalltechnik, Laasphe.
Seyfried, C.F. and Rosenwinkel, K.-H. (1981) Abwässer aus Brauereien, Winzereien und aus der Fruchtsaftheustellung, *Wissenschaft und Umwelt*, **3**, 89–107.

Sheehan, G.J. and Greenfield, P.F. (1980) Utilisation, treatment and disposal of distillery wastewater, *Water Res.*, **14**, 257–277.

Sidwick, J.M. (1973) Review of the problems of treating and disposing of liquid waste from breweries and soft drink plants, *Int. Bottler Packer*, **47**, 64–72.

Sierp, F. (1967) *Die gewerblichen und industriellen Abwässer*, 3rd edn, Springer, Berlin–Heidelberg–New York.

Sixt, H. (1979) *Reinigung organisch hochverschmutzter Abwässer mit dem anaeroben Belebungsverfahren am Beispiel von Abwässern der Nahrungsmittelherstellung*, Institut für Siedlungswasserwirtschaft der Universität Hannover, 50, Eigenverlag.

Steiger, E. (1973) Möglichkeiten der Abwasserreinigung für Brauereiabwässer, *Doemensianer*, **13**, 30–42.

Supperl, W. (1976) Brennereiabwasser, *Berichte der Abwassertechnischen Vereinigung e.V. (ATV), Nr. 28, 1975*, Gesellschaft zur Förderung der Abwassertechnik (GFA), Bonn.

UNO (1977) *Statistical Yearbook*, United Nations Organization, New York.

Viehl, K. (1961) Die Abwasserfrage bei Molkereien, *Molk. Käsereiz.*, **12**, 369–375.

Viehl, K. and Mudrack, K. (1970) Stand und Entwicklung der Abwasserbehandlung bei der Nahrungs- und Genussmittelindustrie in der Bundesrepublik Deutschland, *Industrieabwasser*, Juni, 37–45.

Vogt, E., Jakob, L., Lamperle, E. and Weiss, E. (1979) *Der Wein*, 8th edn, Eugen Ulmer, Stuttgart.

Voss, W. and Rottke, A. (1979) Das Abwasserabgabengesetz und die Berechnung der Abwasserabgabe, *Mineralbrunnen*, **29**, 160–170.

Weller, G. (1979) Abwasserbeseitigung in Brennereien und Winzerbetrieben, *Münchner Beiträge zur Abwasser-, Fischerei- und Flussbiologie*, **31**, 107–137, R. Oldenbourg, München/Wien.

Winzig, K. (1975) Grundlagen und neuere Aspekte auf dem Sektor Umweltschutz/Abwasser, *Brauer Mälzer*, **60**, 34–41.

3 The management of wastewater from the meat and poultry products industry

S E Hrudey, *Department of Civil Engineering, University of Alberta*

3.1 Introduction

The meat and poultry industry is operative throughout the world because of its central role in food production. Considerable differences occur from one nation to another in traditional diets and these differences are reflected in the organization and performance of local meat and/or poultry processing. In North America, Australia, New Zealand and parts of Europe a tendency towards centralization of meat and poultry processing has resulted in the development of relatively large facilities. Such facilities provide the best documentation of wastewater management. Moreover, the scale of their operations requires the application of sound environmental engineering practice because of the potentially large pollutant load which they are capable of generating. Hence, this chapter will focus on the wastewater problems and solutions encountered by medium-to-large centralized facilities. The problems faced by small, local processors are magnified by economic constraints but the relatively small total pollutant load which they generate often allows simple agricultural solutions (i.e. landspreading of concentrated wastes) which may not be practical for the large generator.

3.1.1 Relevant water pollutant parameters

The meat and poultry industry is primarily involved in the conversion of live animals to human food products and animal feed. The slaughtering and processing operations carried out in this conversion release materials which can cause adverse conditions if discharged to natural watercourses. These materials are mainly organic in nature owing to their animal origin.

The pollutants which are characteristic of this industry are generally classified according to the nature of their effects on receiving waters rather than their specific identity as chemical compounds.

Biochemical oxygen demand — BOD_5

Organic matter which can be degraded by micro-organisms in an aerobic water medium will exert an oxygen demand upon that medium in order to support the respiratory requirements of the degrading micro-organisms. Readily biodegradable organic components are characteristic of meat, poultry and rendering plant wastewaters.

BOD_5 is used widely in the design of biological treatment systems for meat and poultry industry wastewaters. Such usage occurs because BOD_5 is representative of the food, or substrate, which is to be provided to the active micro-organisms which perform biological treatment.

Unlike many industrial wastewaters, those from the meat and poultry industry can be analyzed for BOD_5 without the use of a microbial seed. The latter is unnecessary because these wastewaters are already rich in heterotrophic microbial flora derived from the animals being processed. The possibility of inhibition of the BOD_5 test is generally remote, but cannot be totally ignored. Use of sanitizing chemicals in plant cleanup could conceivably lead to occasional anomalies in BOD_5 measurements on meat and poultry industry wastewaters.

Suspended solids

Matter which is essentially insoluble in water, but which is carried by the water medium, constitutes suspended solids. Such matter is contributed from many sources and includes both organic and inorganic constituents. Where wastewaters have received biological treatment, suspended solids in the effluent are predominantly micro-organisms which have failed to coalesce into a floc and hence do not settle in a clarifier following the biological reactor.

Fats, oils and greases — FOG

Animal oils and fats are esters of long chain fatty acids and glycerol and are major components of the biochemical group known as lipids. Naturally, meat, poultry and rendering plant wastewaters can be expected to contain such materials which collectively may be termed fats, oils and greases or FOG. The insoluble nature of FOG causes such materials to form a scum on water unless emulsified by some physical means. This coating action can create physical problems in sewerage systems and treatment plants as well as in the receiving environment. Clogging of pipes, pumping stations and screens will inevitably occur if FOG levels are not adequately controlled. FOG derived from animal sources is generally biodegradable and exhibits extremely high specific BOD, more than 2 g BOD_5 per g of lipid (Hrudey, 1981c).

FOG may be determined empirically by several standard procedures (APHA, 1980), but the basis for all the procedures is the extraction of the lipid compounds into a nonpolar organic solvent followed by determination of the dry weight of extract. Unfortunately, most such procedures are tedious at best and inaccurate at worst (Hrudey, 1981b).

Ammonia

Nitrogen, in the form of ammonia, occurs in most meat, poultry and rendering wastewaters because of the breakdown of proteinaceous wastes into amino acids and, ultimately, ammonia. Ammonia is directly toxic to fish and other aquatic life, primarily as the unionized form (NH_3 being 300–400 times more toxic than NH_4^+, Thurston et al., 1981). The nature of the ammonia species present is a function of pH; at higher pH values unionized ammonia is the predominant form. Ammonia is measured by several standard methods (APHA, 1980), most of which are based on the reaction of specialized reagents with ammonia to produce coloured complexes.

Hydrogen ion concentration — pH

Substances which readily accept or donate hydrogen ions will tend to alter the pH of water. Meat, poultry and rendering wastewaters are generally well buffered (resistant to rapid pH change), but pH extremes may occur as a result of careless handling of chemicals or because of wastewater treatment processes which require or produce pH extremes. For example, protein precipitation requires very acid conditions and ammonia stripping requires alkaline conditions, while anaerobic biological processes may produce acidic effluents. The latter would result from formation of organic acids in the anaerobic decomposition process, if subsequent methanogenesis of the organic acids was incomplete.

Pathogenic micro-organisms

Micro-organisms which can cause disease in humans and animals are often found in meat, poultry and rendering wastewaters. In particular, bacteria of the genus *Salmonella*, which are common causative organisms in food poisoning (Hobbs, 1974) are found in meat and poultry packing wastes (Vanderpost and Bell, 1976). The overall bacterial count of raw wastewaters is normally very high, but pathogenic bacteria constitute only a very small percentage of these organisms.

The isolation and identification of pathogenic micro-organisms is normally a very difficult and time-consuming task. In the case of domestic sewage bacterial monitoring, the so-called 'faecal coliforms', which are representative of intestinal bacteria, are used as indicators of faecal contamination. The confirmed presence of faecal coliforms indicates only the possible presence of intestinal pathogens, while the absence of faecal coliforms indicates only the probable absence of intestinal pathogens.

3.1.2 Regulatory controls for the industry

The specific philosophy for regulatory control differs widely from one jurisdiction to another. In North America, there has been a trend towards specific effluent standards for given industry categories. In Europe, the regulatory pattern has been oriented to site-specific requirements based upon receiving water criteria. Although the industry-specific regulation development exercises in Canada and the USA have been responsible for the generation of a large quantity of data which is useful in understanding process waste generation and control, the actual regulations and guidelines developed for these industries have become quite complex. Presentation of regulated values without adequate explanation of the rationale for their development and use could be misleading. On the other hand, presentation of the detail necessary to understand the specific regulations and guidelines for various jurisdictions would pre-empt the primary purpose of this chapter, that is, consideration of water pollution control technology. Accordingly, discussion of regulatory considerations will be limited to broader concepts.

Pollutants selected for regulatory control have been drawn primarily from those considered in the foregoing introduction. Canada's regulations under the Federal Fisheries Act specify limits for BOD_5, suspended solids, grease (FOG) and pH. Furthermore, regulated plants are required to monitor ammonia levels and to conduct periodic acute lethality bioassays with rainbow trout. The US Federal regulations under the Clean Water Act cover all these parameters except fish toxicity but they include an additional regulation for faecal coliforms.

With the exception of pH and faecal coliform standards, the North American regulations are specified in terms of mass of pollutant per unit of production. In the case of the US regulations, limits are specified in kg of pollutant per tonne of live weight killed (LWK). Canadian regulations have been specified in terms of kg of pollutant per tonne of finished product. In either case, by limiting the mass of pollutant rather than concentration, any unintentional incentive for dilution of wastewater prior to monitoring and discharge is eliminated. Additionally, the relation of limits to units of production recognizes the larger waste generation potential of larger plants.

In many instances meat and/or poultry plants are located near enough to urban centres that they are able to connect to municipal sewerage systems. Pollutants produced by the industry are basically compatible with municipal sewerage and treatment systems, with the exception of floatable FOG. As noted earlier, such materials can cause physical problems in municipal sewerage and treatment systems. The remaining parameters represent an additional organic loading upon the municipal treatment plant rather than a source of treatment upset. Consequently, most municipalities levy sewer surcharges on meat and poultry plants based upon the additional loading which they contribute to the treatment plant.

3.2 Industry characterization

The meat and poultry industry, like most industries, cannot be uniquely categorized in a totally unambiguous fashion. However, in keeping with the stated focus on medium-to-large centralized facilities, the industry generally can be subdivided into:

(1) red meat products;
(2) poultry products; and
(3) inedible rendering.

Each of these major categories will be described, first in terms of the characteristic industrial processes, then, in the next major section, in terms of in-plant water pollution control measures which can be undertaken.

3.2.1 Red meat products

Within this category three common subcategories may be encountered, according to type of operation:

(i) *Slaughterhouses* or *abattoirs* perform only the killing and dressing of animals with or (more commonly) without an on-site rendering operation.
(ii) *Integrated plants* perform killing and dressing, as well as processing (cutting, cooking, canning, curing, pickling, sausage-making, smoking and/or retail packaging). They frequently incorporate on-site rendering.
(iii) *Processing plants* process meats (see above) but perform no slaughtering.

An understanding of the individual processing steps which occur in this industry is best gained from consideration of an integrated plant which, as noted above, generally conducts the full range of processes. Figure 3.1 (EPA, 1974a) provides an overview of the various processing steps which will each be considered in turn.

In the subsequent discussion, waste generation will be expressed as kg BOD_5 per tonne LWK. However, it must be stressed at this point that raw waste load data are very approximate since they are often generated by extrapolation and deduction from limited data at specific plants. Hence, the reader must regard such data cautiously and use them only to gain an appreciation of the relative magnitude of major waste-generating activities in meat industry plants.

Pens holding livestock

Animals are held in storage areas for a few hours prior to slaughter. They are watered, but will not generally be fed unless holding exceeds one day. Pollutants are derived from manure and urine, feed, dirt borne by livestock, and sanitizers and cleaning agents which may be used in pen washdown. These pollutants will reach the sewer by means of water emanating from overflow of water troughs, rain and snowmelt runoff, and pen washdown water. Provided that solid contaminants are initially removed in a dry cleanup, the sewered raw waste has been estimated at 0.25 kg BOD_5 per tonne LWK (Wells, 1970).

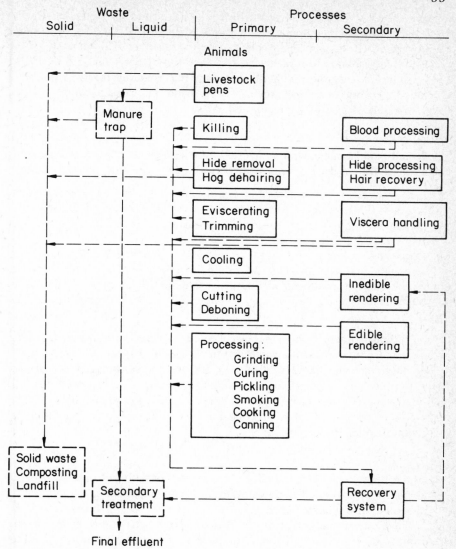

Fig. 3.1 Process waste sources for an integrated red meat products plant (after EPA, 1974a)

Killing floor
Animals are driven from the holding pens to the slaughtering area where they are first stunned, then suspended from an overhead rail by their hind legs, and finally are stuck and bled over a collecting trough. The major pollutant from the slaughtering operations is blood which has a BOD_5 of between 150 000 and

200 000 mg l^{-1} (Witherow *et al.*, 1973). Blood comprises 5% of the liveweight of a typical steer and amounts in total to about 23 kg. Of this total, only approximately 16 kg is normally collected during bleeding. The remaining 7 kg of blood corresponds to a wasteload of 3.0 kg BOD$_5$ per tonne LWK (EPA, 1974a) which could appear in the slaughter and hide removal area wash-water. If no blood recovery is practised, the raw waste load is increased proportionally to approximately 10 kg BOD$_5$ per tonne LWK.

Blood processing
Collected blood is most commonly dried for use as an animal feed supplement or fertilizer. Blood drying is normally performed by direct or indirect heating.

1. *Indirect heating.* As the name implies, the blood is dried by external steam heating. This process generates only a small volume of bloodwater, corresponding to a raw waste load of approximately 0.3 kg BOD$_5$ per tonne LWK (EPA, 1974a).
2. *Direct heating.* Direct steam sparging is used to coagulate blood followed by screening and centrifuging. As would be expected, larger quantities of bloodwater are produced in this process option. The corresponding raw waste load is estimated at approximately 1.3 kg BOD$_5$ per tonne LWK (EPA, 1974a).

In European practice, blood may be collected, stabilized with an anticoagulant, refrigerated and shipped for direct livestock feeding (Akers, 1973). Alternatively, the blood may be allowed to coagulate and the separated blood plasma used in processed meat production or sold to bakeries as a substitute for egg albumin. One German firm freezes plasma at $-30°C$ to produce flakes which are subsequently used as a processed meat additive. Blood processing is discussed further in Section 3.2.3.

Hide removal
After bleeding, cattle are skinned while hogs are normally dehaired. However, some plants have adopted skinning of hogs to eliminate scalding and dehairing. In the medium-to-large plants relevant to this discussion, hide removal is performed mechanically. The largest source of BOD$_5$ from this operation will be from blood and tissue which falls to the floor. External contamination of the hide with dirt and manure provides a secondary source of pollutants. Sewered waste load is generated from the cleanup operations in this area. It is highly variable, depending on cleanup practices; no specific estimate of raw waste load has been located.

Hide processing
Prior to processing by a tannery, hides must be defleshed, washed and covered in brine. The meat products plant may either perform these operations on-site, thereby recovering the fleshings for inedible rendering, or they may pack the

unwashed hides in salt for shipment to a tannery for processing. Although the former will increase byproduct recovery for the meat products plant, it will create additional raw waste load in terms of cleanup wastewaters and brine-curing solution overflow. Raw waste load from cleanup of this operation will depend upon the extent of dry cleanup performed. Brine solution overflow under controlled conditions has been estimated to provide approximately 1.5 kg BOD_5 per tonne LWK (EPA, 1974a). Alternatively, dry packing of unwashed hides will transfer any potential raw waste loading to the tannery operation, except perhaps for salt spills or similar unintended occurrences.

Another operation which may be practised at meat products plants is cattle hide dehairing. This process, which is commonly performed at tanneries, produces a high-strength organic liquor which may be complicated by the presence of surfactants, lime and sulphides. An estimated raw waste load from sewering hide-dehairing liquor is approximately 3.8 kg BOD_5 per tonne LWK, according to data provided by Ramirez et al. (1977).

Hog scalding, dehairing and hair recovery

As mentioned above, hogs are usually scalded to loosen their hair before the hair is removed. The scald tank operates at temperatures of 45° to 65°C and will collect blood, dirt, manure and hair. Intentional overflow from the scald tank produces an estimated raw waste load of approximately 0.15 kg BOD_5 per tonne LWK (EPA, 1974a). After scalding, the hogs are mechanically dehaired by abrasion. Hair is carried away by fluming and is recovered by screening the spent flume-water. This wastewater carries some residual hair, blood and dirt which provides an estimated raw waste load of approximately 0.4 kg BOD_5 per tonne LWK (EPA, 1974a). Hog hair recovered at this stage may then be simply disposed of as solid waste, washed and baled for direct marketing, or hydrolyzed by pressure cooking for marketing as a feed supplement. In the case of washing and baling, a raw waste load of 0.7 kg BOD_5 per tonne LWK is estimated, while hydrolyzing is assigned a raw waste load of 1 kg BOD_5 per tonne LWK (EPA, 1974a). Finally, dehaired hogs are usually singed in a gas flame to complete the hair removal process. They are then cooled by a water spray which makes a negligible contribution to the plant raw waste load.

Evisceration

Following hide or hair removal, carcasses are beheaded and cut open and the viscera are pulled out for inspection. Approved carcasses are eviscerated and trimmed prior to subsequent washing and cooling stages. Carcass trimmings and heads are collected for inedible rendering, but blood and smaller scraps of fat and other tissue will reach the floor during this operation. Specific estimates of raw waste load from this operation have not been located.

Viscera handling

Edible products such as heart, liver and kidneys are generally recovered at this stage. Lungs may be separated for inedible rendering or processing as pet food. Next, the paunch (the first stomach of ruminants) must be handled. This contains an estimated 27 to 40 kg of 'paunch manure' which is partially digested feed (FWPCA, 1967). Paunches may be handled in four different ways corresponding to varying degrees of concern for raw waste load generation.

1. *Total dumping:* the entire paunch contents are flushed directly to sewer, contributing an estimated 2.5 kg BOD_5 per tonne LWK of raw waste load (EPA, 1974a).
2. *Wet dumping:* the paunch contents are washed out and the wet slurry is screened for gross solids removal. The screened effluent carries dissolved organics, suspended solids and FOG, contributing an estimated raw waste load of 1.5 to 2 kg BOD_5 per tonne LWK (EPA, 1974a).
3. *Dry dumping:* the paunch contents are dumped for subsequent rendering or disposal as solid waste without needless water flushing. Consequently, the estimated raw waste load contribution is only 0.2 kg BOD_5 per tonne LWK.
4. *Whole paunch handling:* the entire paunch may be removed, intact, for rendering or disposal as solid waste. Although this option eliminates any contribution to the plant raw waste load, it also eliminates the paunch as a recovered byproduct. Paunches are marketable as pet food (after washing) or as tripe (after washing and bleaching). Washing alone will create a raw waste load of approximately 0.6 kg BOD_5 per tonne LWK while bleaching by cooking in a hot alkaline solution will contribute an additional 0.4 kg BOD_5 per tonne LWK (EPA, 1974a).

Intestine handling

Intestines may be rendered directly, hashed and washed prior to rendering, or processed (in the case of hog intestines) for further use. Intestines for processing must be deslimed prior to thorough washing. Large hog intestines may be sold as chitterlings while diminishing markets exist for small intestines as sausage casings or surgical sutures. Desliming and casing washing contributes FOG and an overall estimated raw waste loading of 0.6 kg BOD_5 per tonne LWK (EPA, 1974a).

Washing and cooling

To deter spoilage, carcasses are washed and chilled to temperatures between 0.5° and 1.5°C for at least 24 hours. The sanitary value of whole carcass washing, which markedly increases water consumption, is currently being debated. At this point, the carcass will normally receive final inspection and grading.

Cutting and deboning

Although many large-scale plants ship only whole graded carcasses to retail market, most will perform some on-site processing to produce retail cuts. Cutting and deboning operations will generate trimmings, blood, bones and bone dust, but need not contribute significantly to plant raw waste load, depending on the quality of cleanup operations.

Meat processing

Meat processing covers a variety of operations including curing, pickling, smoking, cooking and canning. These generate raw waste load from blood, tissue and fats which reach the sewer during cleanup. Curing solutions containing sugar and salt are additional potential waste load generators if significant quantities of curing solution are spilled to sewer. In such cases both BOD_5 and chloride concentrations will be raised. Total raw waste loading for meat processing plants (including cutting and deboning) have been estimated at between 5.7 and 6.7 kg BOD_5 per tonne of product (Pilney *et al.*, 1968; Dept H.E.W., 1965).

Edible and inedible rendering

Rendering is a heating process for meat industry wastes whereby fats are separated from water and protein residues. The products may be either edible lards or inedible (for human consumption) tallows and dried protein residues. In order to produce edible lards, the same standards required of all other meat products for human consumption are enforced. Consequently, the edible rendering industry is almost exclusively associated physically with a meat or poultry packing operation because of the need for the feedstock to be transferred freshly from the edible (inspected) process areas. Inedible rendering is practised both on-site at meat and poultry packing operations and at independent off-site rendering operations. Because few independent edible rendering operations exist, the relevant processes will be discussed here in conjunction with red meat products operations. Processes for inedible rendering are discussed briefly here, but will be dealt with in more detail in Section 3.2.3.

Two basic processes are used for edible rendering in the red meat and poultry industry:

1. *Wet rendering.* As the name implies, this process employs direct steam heating of the feedstock to promote separation of water, fat and protein. Differences in the physical properties of these materials are then used to separate the floatable fat, then the insoluble proteinaceous residue from the tankwater. The latter, if sewered, would generate a raw waste load of approximately 2 kg BOD_5 per tonne LWK (EPA, 1974a). Evaporation of tankwater to recover more residual organic matter can further reduce the raw waste load. However, this process will still generate condensates which carry a BOD loading. Furthermore, inadequate control of evaporation processes can result in overflow or foaming of tankwater into the condensers. Consequently the raw waste load where tankwater evaporation is practised may range from 0.5 to 1.0 kg BOD_5 per tonne LWK (EPA, 1974a).

2. *Low temperature continuous rendering.* This process represents an advance from conventional wet rendering technology since it produces a better quality lard. The major differences relative to older technology are the use of finely ground feedstock, lower temperatures (slightly above the fat melting point) and continuous feed. The process, unfortunately, still generates tankwater which, if sewered, would produce a waste load similar to that from batch wet rendering. If, alternatively, tankwater were evaporated, the raw waste load would be similar to that reported above for that option.

Inedible rendering is now carried out almost exclusively by means of dry rendering processes. These may be either batch or continuous processes, and are described in more detail in Section 3.2.3. In both cases, indirect heating is used. As a result, the process waste load is limited to organic materials contained in condensates of the vapours drawn off from the cooking vessels. Some Canadian meat processors have analyzed this condensate and found it to contain in the order of 200 mg l^{-1} of BOD_5 (Hrudey, 1980). The total raw waste load from dry rendering operations originates from spillage, cleanup and vapour condensates. This has been estimated at 0.5 kg BOD_5 per tonne LWK (EPA, 1974a).

General cleanup

Because of the inspection requirements associated with the production of food, the meat products industry has tended to emphasize cleanup without concomitant concern for water use and waste load generation. Thus, for example, hot water cleanup is regarded by some operators as essential. Yet, prolonged hot water washing can produce fogging which may effectively blind the operator, cause very inefficient water use and even a poor overall cleanup. Washup personnel have been known to remove floor drain covers and use a high-pressure hose to wash scraps into the sewer. Such practices will occur if the objectives of cleanup are one-sided, with no consideration of water use or waste loading.

Given the very wide range of cleanup practices employed, a meaningful estimate of raw waste load from this operation is difficult to establish. However, it should be clear that this step can significantly add to the total raw waste load if processing operations are sloppy and compensatory cleanup is achieved by flushing everything to sewer.

Total process waste load

Having considered the relative waste load contributions of the various processes, it is instructive to consider the total process waste load for overall plant operations. The United States Environmental Protection Agency (US EPA) gathered data (EPA, 1974a) for waste loading following byproduct recovery (gross solids and floatable FOG removal). They subdivided the industry into simple and complex slaughterhouses and low and high processing

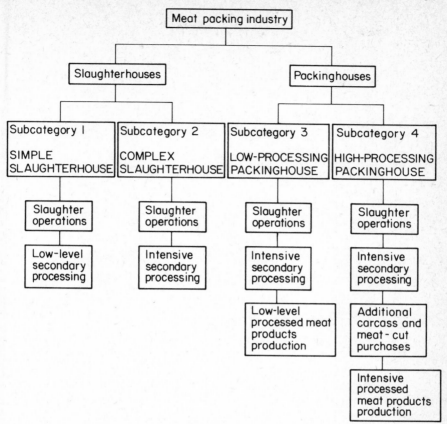

Fig. 3.2 Red meat products industry categorization (after EPA, 1974a)

packinghouses. This classification is illustrated schematically in Fig. 3.2. Waste load data for these various subcategories is summarized in Table 3.1.

3.2.2 Poultry products

The poultry products industry is engaged in the slaughter and evisceration of domestic birds, primarily chickens, turkeys and ducks. Further operations may include cutting and packaging as well as canning and related processing operations. This discussion will address integrated poultry products plants which both slaughter and process. Although independent processing operations exist, their raw waste loading is generally considerably lower than that of plants which perform slaughtering. Figure 3.3 summarizes the major processing steps in an integrated poultry products plant.

Although it would be desirable, for comparison with the red meat products industry, to utilize kg BOD_5 per tonne LWK, literature data in this format are

Table 3.1 Raw waste load from red meat products plants (after EPA, 1974a)

Base	Flow (m³ t⁻¹ LWK)	Kill (t d⁻¹)	BOD$_5$ (kg t⁻¹ LWK)	Suspended solids (kg t⁻¹ LWK)	FOG (kg t⁻¹ LWK)	Kjeldahl nitrogen as N (kg t⁻¹ LWK)	Chlorides as Cl (kg t⁻¹ LWK)	Total phosphorus as P (kg t⁻¹ LWK)
SIMPLE SLAUGHTERHOUSES								
(Number of plants)	(24)	(24)	(24)	(22)	(12)	(5)	(3)	(4)
Average	5.328	220	6.0	5.6	2.1	0.68	2.6	0.05
Standard deviation	3.644	135	3.0	3.1	2.2	0.46	2.7	0.03
Range, low–high	1.334–14.641	18.5–552	1.5–14.3	0.6–12.9	0.24–7.0	0.23–1.36	0.01–5.4	0.014–0.086
COMPLEX SLAUGHTERHOUSES								
(Number of plants)	(19)	(19)	(19)	(16)	(11)	(12)	(6)	(5)
Average	7.379	595	10.9	9.6	5.9	0.84	2.8	0.33
Standard deviation	2.718	356	4.5	4.1	5.7	0.66	2.7	0.49
Range, low–high	3.627–12.507	154–1498	5.4–18.8	2.8–20.5	0.7–16.8	0.13–2.1	0.81–7.9	0.05–1.2
LOW-PROCESSING PACKINGHOUSES								
(Number of plants)	(23)	(23)	(20)	(22)	(15)	(6)	(5)	(4)
Average	7.842	435	8.1	5.9	3.0	0.53	3.6	0.13
Standard deviation	4.019	309	4.6	4.0	2.1	0.44	2.7	0.16
Range, low–high	2.018–17.000	89–1394	2.3–18.4	0.6–13.9	0.8–7.7	0.04–1.3	0.5–4.9	0.03–0.43
HIGH-PROCESSING PACKINGHOUSES								
(Number of plants)	(19)	(19)	(19)	(14)	(10)	(3)	(7)	(3)
Average	12.514	350	16.1	10.5	9.0	1.3	15.6	0.38
Standard deviation	4.894	356	6.1	6.3	8.3	0.92	11.3	0.22
Range, low–high	5.444–20.261	8.8–1233	6.2–30.5	1.7–22.5	2.8–27.0	0.65–2.7	0.8–36.7	0.2–0.63

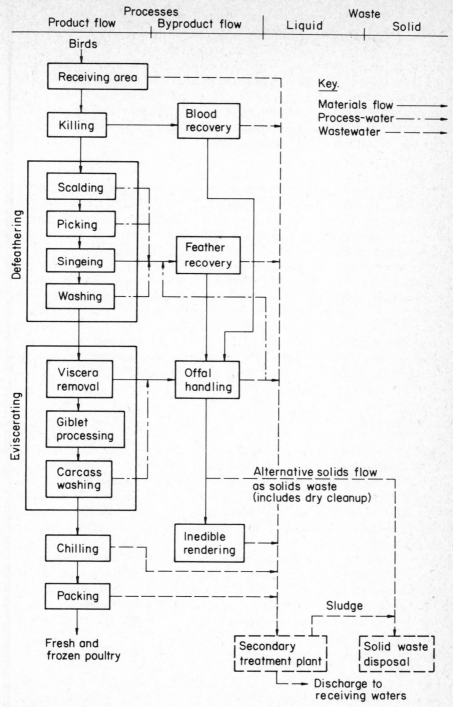

Fig. 3.3 Process waste sources for an integrated poultry products plant (after EPA, 1975a)

limited. Much of the available data is presented in the form of kg BOD_5 per 1000 birds processed. Unfortunately, simple conversions to mass of LWK based on average bird mass are not reliable since relative waste loadings differ for different types of birds processed. Hence, data will be presented in whatever form they were originally reported (except for metric conversion). This is not a serious drawback for our purposes of gaining an understanding of relative magnitude for waste load generating processes, since we have earlier noted that raw waste load data are very approximate.

Receiving areas
The inlet to the plant is equipped to accommodate and equalize fluctuations in bird deliveries, since the processing line has a fixed maximum capacity. Birds are intermittently unloaded into the holding areas and, in a relatively constant flow, are removed and attached by their feet to a continuous conveyor travelling to the slaughter area.

Waste loading to the receiving area can be extremely variable since it is derived from dirt, manure and feather deposits which vary with the length of holding time. Sewered waste load will, of course, depend upon the nature of cleanup, which ranges from total wet flushing into the sewer to total dry cleanup. Dry cleanup is complicated by loose feathers and dust and some form of vacuum system is advantageous. The most common compromise involves extensive dry cleanup followed by wet washdown. This approach is claimed to be capable of reducing raw waste loads from 0.45 to 0.09 kg BOD_5 per 1000 chickens processed (Porges and Struzeski, 1962).

Slaughter area
Birds suspended from the conveyor are killed and bled by cutting their jugular veins. Without prior stunning, the birds flap and writhe after cutting so that blood is splattered over a wide area. To contain this blood, the conveyor usually travels through a blood collection tunnel with a preselected bleeding travel time. However, excessive blood splattering will coat the birds' feathers, leaving this blood to be lost in subsequent scalding and defeathering operations. Hence, utilization of stunning prior to bleeding can reduce overall blood losses for the plant.

Blood collected in the bleeding tunnel can be gathered by continuous drains or intermittent scrape-down. Given the extremely high BOD_5 of blood and the fact that blood represents 6 to 8% of poultry body weight, efficient blood collection is essential to minimize raw waste load. The total potential raw waste load from uncontrolled blood sewering has been estimated at 7.9 kg BOD_5 per 1000 chickens processed (Porges and Struzeski, 1962).

Scalding
Prior to defeathering, the birds must be scalded to loosen the feathers. This is accomplished at water temperatures ranging from 50° to 65°C. In order to

maintain scald water temperature and quality, the tank receives a makeup flow of approximately 8 litres per bird. Scald tank overflow has been reported to generate raw waste loads ranging from 1.0 to 5.4 kg BOD_5 per 1000 birds processed (Stanley Assoc., 1981). Scald tank dumping at intermittent intervals will cause high BOD shock-loading because of the settleable residues that collect in the tank.

Defeathering

Feathers are mechanically abraded from the scalded birds, usually by rotating rubber fingers. Removed feathers drop to underlying troughs which flume the feathers to screening devices for recovery. The organic loading contributed by defeathering is difficult to estimate for most plants because the feather flume often receives makeup from other sources. However, the BOD_5 of feather flume-water is normally several hundred $mg\,l^{-1}$ and the feather flume is thought to contribute a greater waste load than the evisceration trough (EPA, 1973a; Stanley Assoc., 1981). Recovered feathers may be disposed of as solid waste or may be pressure cooked to hydrolyze the otherwise low nutritional value protein, keratin. Hydrolyzed feather meal can provide one-third of the dietary protein requirements of young chicks, if supplemented with the amino acids methionine and lysine (Blair, 1974).

Pinfeathers are more troublesome for automated processing; although singeing is effective for pinfeather removal from chickens and turkeys, geese and ducks usually require hand or wax stripping.

Whole bird washing

Prior to evisceration, the defeathered carcasses receive a spray wash. One study reported data corresponding to a raw waste load of approximately 0.1 kg BOD_5 per 1000 birds processed (EPA, 1973a).

Evisceration

Evisceration is a closely controlled step in poultry processing since it is essential to avoid any possibility of contaminating the meat. The carcasses are cut open manually and the viscera are pulled out for inspection. Any abnormality observed at this stage results in condemnation of the carcass and its removal for inedible rendering. Approved carcasses then have their head, feet and viscera removed. The head and feet become part of the waste or offal which is collected for disposal or inedible rendering. Viscera are sorted to recover the heart, liver and gizzard, the remaining inedible viscera being added to the offal.

Medium-to-high capacity poultry plants have generally adopted continuous flowaway fluming for offal removal rather than dry batch offal collection. Despite subsequent screening to recover offal solids, offal flume-water is contaminated with blood, fat, tissue and intestinal micro-organisms. It has been estimated that offal flume-water contributes a raw waste load of approximately one-third of total plant load (Stanley Assoc., 1981). Loadings for this source were reported to range from 3.0 to 22.9 kg BOD_5 per 1000 birds processed, with values near 6 kg most common. Although flowaway systems have provided quicker and more efficient processing for modern plants, a balance between the appeal of flowaway systems and wastewater treatment costs may encourage the development and utilization of automated dry viscera handling methods.

Final bird wash
Eviscerated carcasses still contain blood and loose tissue which must be removed by a final wash. Data from one survey corresponded to a raw waste load of approximately 1.3 kg BOD_5 per 1000 birds processed (EPA, 1973a).

Chilling
Because of the concern regarding bacterial propagation, particularly for the pathogenic *Salmonella* spp., the birds are chilled in cold water to reduce carcass temperature and, thereby, reduce bacterial reproduction rates. Chiller-water has final contact with the carcass and its bacteriological integrity is critical to avoid carcass contamination. Thus, although the concentration of organic contaminants will be considerably less than that of, for example, scald-water overflow, the chiller overflow rate is necessarily higher. Estimated raw waste loads ranging from 0.7 to 4.3 kg BOD_5 per 1000 birds are reported for this process (Stanley Assoc., 1981).

Grading, weighing and packing
The chilled birds are drained prior to weighing, grading and final packing. In medium-to-large plants the majority of the products are then cryogenically frozen for shipment to retail market. These operations do not involve any water use other than general area cleanup.

Plant cleanup
Waste load generation from plant cleanup can vary widely, depending on the extent of blood recovery and dry cleanup practised. Concern for microbial contamination of poultry products has tended to focus efforts toward final results of cleanup rather than the methods used. However, as with the tradeoffs concerning flowaway versus dry offal removal, rising wastewater treatment costs will require more consideration of raw waste load generation by plant management.

Total process waste load
As mentioned earlier, the process waste load in poultry plants is dependent upon the type of bird being processed. A summary of raw waste loads for chicken and turkey processing waste operations is provided in Table 3.2 (EPA, 1975a).

Table 3.2 Raw waste load from poultry products plants (after EPA, 1975a)

Base	Flow (m³/1000 birds)	Average live weight (kg/bird)	Production (birds/day)	BOD₅ (kg t⁻¹ LWK)	Suspended solids (kg t⁻¹ LWK)	FOG (kg t⁻¹ LWK)	Kjeldahl nitrogen as N (kg t⁻¹ LWK)	Ammonia as N (kg t⁻¹ LWK)	Chlorides as Cl⁻ (kg t⁻¹ LWK)	Total phosphorus as P (kg t⁻¹ LWK)
CHICKEN PROCESSORS										
(Number of observations)	(88)	(90)	(90)	(60)	(53)	(39)	(15)	(19)	(12)	(22)
Average	34.4	1.74	73 000	9.89	6.91	4.21	1.84	0.23	1.97	0.39
Range	15.9–87.0	1.45–1.97	15 000–220 000	3.26–19.86	0.13–22.09	0.12–14.03	0.15–12.16	0.0005–0.73	0.006–9.16	0.054–2.46
TURKEY PROCESSORS										
(Number of observations)	(34)	(34)	(34)	(15)	(13)	(10)	(5)	(5)	(4)	(4)
Average	118.2	8.3	12 100	4.94	3.17	0.89	0.94	0.15	2.49	0.098
Range	36.3–270.2	4.1–11.4	2000–20 000	0.96–9.1	0.57–10.89	0.34–1.81	0.38–1.89	0.064–0.37	0.38–5.41	0.034–0.18

Note: units are kg t⁻¹ LWK = $\mathrm{kg\,t^{-1}\,LWK}$

3.2.3 Independent rendering

Rendering operations can be primarily classified into edible and inedible processes. As previously mentioned, since edible rendering operations are almost exclusively associated with meat packing operations, these processes have been discussed in Section 3.2.1. Inedible rendering, however, may be conducted at a meat or poultry plant or at a separate independent plant. It is the latter which is considered in this section.

As with meat and poultry operations, BOD_5, suspended solids and FOG are important waste parameters for inedible rendering operations. The tendency for rendering operations to produce emulsified FOG, which is less amenable to gravity separation from water, makes FOG a particular concern in rendering wastewaters.

Pollutant loading may be expressed on the basis of kg per tonne of raw material (RM) or kg per tonne of finished product (FP). Both terms offer advantages and disadvantages for correlating waste loads from different plants. However, the moisture contents of different raw materials vary widely. Since raw material moisture is a prime source of wastewater in the inedible rendering process, expression of waste load as kg per tonne raw material provides a more relevant comparison among operations using different raw materials.

Rendering is a separation process whereby fat-containing materials are converted to tallow or lard and to dry proteinaceous materials (cracklings). Heat is used to melt the fats out of animal tissue, to coagulate cell proteins and to evaporate raw material moisture to varying degrees. Common raw materials include blood, feathers, offal, fat and meat scraps, bones, hog hair and condemned carcasses. The fundamental processes of an independent rendering facility are presented in the flow sheet (Fig. 3.4) and are described briefly below.

Raw material delivery

The raw materials used by an independent renderer are mainly collected from meat and poultry packing operations, restaurants and, in the case of dead stock, from livestock producers. Because of the putrescible nature of the raw material, rapid and efficient collection and processing are necessary to avoid nuisance conditions at the plant.

Plant reception facilities vary considerably, but normally provide separate facilities for:

1. offal, scraps, etc.;
2. feathers or hog hair; and
3. blood.

Studies have shown that a large waste load may be generated by the liquid drainage from the raw material, particularly for poultry offal and feathers (EPA, 1974b). In one case, the drainage amounted to 20% of the original raw material weight and contributed a waste load of 2.5 kg BOD_5 per tonne RM, some 43%

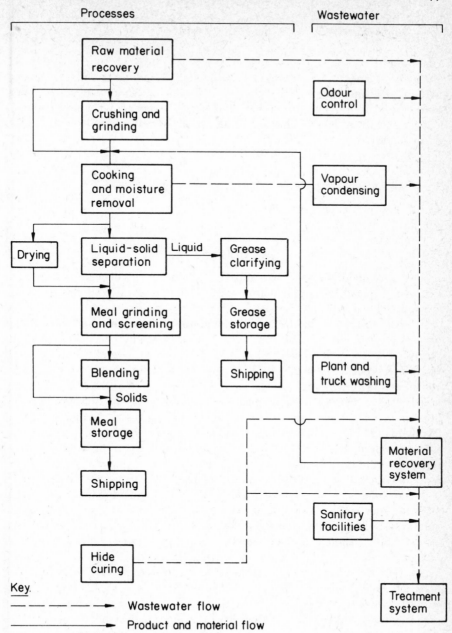

Fig. 3.4 Process waste sources for an independent rendering plant (after EPA, 1974b)

of the total plant waste load. In another case, the feather drainage alone produced 1.4 kg BOD_5 per tonne RM.

Grinding

Bulky raw materials are normally subjected to grinding to reduce the net particle size and increase the efficiency of cooking. Poultry offal does not normally require grinding; feathers are handled separately before cooking. Wastewater from the grinding area is the result of material spills and regular cleanup operations.

Cooking

The cooking process involves heating the raw materials:

(i) to melt the grease and allow phase separation;
(ii) to condition the proteinaceous solids to provide efficient fat release from the tissue;
(iii) to evaporate the raw material moisture.

The specific processes used and the product yield depend on the raw material. Typical product yield and moisture content for various raw materials are summarized in Table 3.3 (EPA, 1974b; Drynan, 1974).

Cooking processes for the inedible rendering industry are now exclusively dry processes (no direct contact between heating steam and raw material). Until recently, dry rendering has been a batch process, but several continuous dry rendering processes are now available. The two major process types will be considered in turn.

In batch cooking, the raw material is charged by screw conveyors or pressurized blow tanks into a steam-jacketed cooking vessel. A complete batch of raw material is fed before applying 270–610 kPa steam to the steam jacket. Cooling time is determined from experience, monitoring of product temperature and sampling of product. During this process, the raw material moisture is evaporated and is carried over to a condenser.

In most plants, poultry feathers and hog hair are handled in batch processes since these materials must be cooked under pressure (370 kPa) to hydrolyze the keratin to usable protein. This is followed by regular batch cooking. Blood is also normally handled in batch cookers. These materials are often dried in a ring or rotary drier following batch cooking rather than in the batch cooker itself. This is done because of the poor heat transfer characteristics of these materials as they reach low moisture levels and become viscous. The ring drier operates by reducing the particle size of the concentrate and blowing it into a vertical ring chamber where the particles are supported by hot gases. As they dry, their density drops and the particles are carried out with the hot gases and separated by a cyclone. The rotary drier employs a drum with steam-filled tubes running longitudinally. The material is cascaded over the heated tubes by rolling the drum.

Table 3.3 Product yield for inedible rendering by raw material (after Drynan, 1974; EPA, 1974b)

Raw material	Tallow and grease (%)	Cracklings with 10–15% fat (%)	Moisture (%)
Shop fat and bones	37	25	38
Dead cattle	12	25	63
Dead cows	8–10	23	67–69
Dead hogs	30	25–30	40–45
Dead sheep	22	25	53
Poultry offal	14	4	82
Poultry feathers	—	12	>75
Blood	—	12–14	86–88

The nutritive value of feed produced by drying blood is reduced with prolonged high temperature exposure. From this viewpoint, the ring driers apparently cause less nutritive loss, but they cause more of an odour problem than rotary driers. The latter may be attributed to higher temperature operation (shorter residence time) and higher volumes of air used.

Blood has also been processed by steam sparging to coagulate the albumen. The albumen and fibrin are separated from the bloodwater by screening and are further processed in a drier while the bloodwater is typically sewered. This blood process is undesirable from a waste load viewpoint because of the high BOD_5 (up to $16\,000\,\mathrm{mg\,l^{-1}}$) carried in the bloodwater. One plant sewering bloodwater added 16.3 kg BOD_5 per tonne blood processed to its raw waste load (EPA, 1974b). However, it was believed that the blood screening at this plant was inefficient and could be improved to reduce the BOD_5 loss with the bloodwater.

Continuous cooking processes have the advantage of increasing the plant throughput while decreasing the raw material and product residence time in the cookers, thereby improving product quality. Several continuous systems have been developed. In one process, a cylindrical cooking vessel is equipped with a rotating shaft and paddles which lift the material and move it through the cooker. Moisture vaporized in the cooker is drawn off and condensed.

A second process feeds the raw material to a fluidizing tank where recycled fat is used to suspend the incoming particles. From here the slurry passes to a disintegrator which reduces the particle size before the slurry passes to a shell and tube evaporator where steam flow through tubes, together with an applied vacuum, removes moisture vapour. The withdrawn vapour is again condensed.

Another process uses a series of batch-type cookers stacked one upon another with the continuous feed to the first cooker resulting in a continuous feed through the system. A common manifold collects the vapours from each stage and carries them to a condenser.

Condensing

In all the systems mentioned above, water is evaporated from the raw material during cooking. This vapour is then condensed, producing the only actual process-generated wastewater in dry rendering. As mentioned previously, one meat processor found condensates to contain about 200 mg l^{-1} BOD$_5$. A large rendering operation demonstrated condensate values from 1200 to 2700 mg l^{-1} BOD$_5$ and found that the process condensates made up roughly 25% of the plant BOD$_5$ waste load (Drynan, 1974). In another study (EPA, 1974b), undiluted condensed cooking vapours for 11 plants were found to average 1165 mg l^{-1} BOD$_5$ and to contribute, on average, 0.73 kg BOD$_5$ per tonne RM, some 30% of the total plant raw waste load.

Older installations mainly used barometric leg condensers which draw the vapours into the high velocity water flow of a water ejector, thereby condensing them. These devices are undesirable from a water use viewpoint as they dilute the condensed organics in a large ejector water flow and create large volumes of dilute wastewater. Air condensers or shell and tube heat exchangers are preferable in that only concentrated condensate is produced. However, the type of condenser used does not significantly affect waste loadings in terms of kg per tonne RM.

Liquid–solid separation
The desired product mixture from the cooker, whether batch or continuous, is usually discharged to drain pans where the grease is allowed to drain off the solids. In some systems, gravity draining is replaced by centrifuging. The solids from this process are then pressed mechanically to reduce further the grease content of the solids. Other than product spills, there is no wastewater generated in this process as dry cleaning of the area can normally be performed.

Grease processing
The grease recovered from the solids is normally clarified by screening, centrifuging and/or filtration. The solids recovered in the grease clarification are usually recycled to a mechanical grease separation stage. There is no normal wastewater generated in this process, but product spills may occur.

Solids processing (meat grinding, blending, etc.)
The solids cake produced from mechanical grease separation is screened and ground with a hammermill to produce protein meal. Protein meal from a single source (e.g. blood meal) will often be blended with meal from other sources (e.g. feather meal) to make up deficiencies in particular amino acids (Blair, 1974).

Spills and delivery truck cleanup
As mentioned previously, one survey (Drynan, 1974) reported that 75% of the plant waste load was produced by cleanup wastewaters. In another survey (EPA, 1974b), one plant was found to generate a waste load of 16.2 kg BOD$_5$ per 100 kg RM strictly from cleanup. Further investigation indicated that the

production area was washed with a continuous flow of hot water (52°C) which collected continuous spillage from leaking process equipment, thus producing the remarkably high waste load. In another plant studied in the latter survey, 43% of the plant BOD_5 load was attributed to plant cleanup.

Total process waste load
As with the red meat and poultry products industries, the US EPA conducted surveys of process waste loading from independent rendering operations. These data were obtained following byproduct (solids and floatable FOG) recovery and are summarized in Table 3.4.

3.3 In-process waste control

3.3.1 General concepts
In-process waste control should aim to reduce both the volume and the strength of the wastewater stream. A meat products plant producing a high quality product may quite conceivably be producing a poor quality effluent. Some industry and inspection personnel still hold the view: 'the more potable water used in the plant, the safer the product'. While standards of product quality may be maintained, indiscriminate water use increases the loss of organic materials to sewer. In the broader view, organic materials consisting of proteins and oils that are lost to the sewer are effectively lost product. If subsequently recovered from the wastewater, they are likely to have become lower grade products. Thus, as in any other highly competitive production business, optimization of maximum product quality should include minimized purchased resource (potable water) usage and minimized product loss. Only by considering the overall raw material–product balance can the optimal production strategy be achieved. These aims are entirely consistent with the process control goals of reducing wastewater flow and total pollutant loading upon final treatment.

For the first goal to be achieved, water consumption must be minimized. This will effectively reduce the size of treatment facilities required to achieve a given effluent standard. Such size reductions will be reflected in both lower capital and lower operating costs for on-site treatment. Alternatively, reduced flows will reduce sewer surcharge costs for off-site treatment. Moreover, whether water is purchased from a municipality or treated and produced on-site it represents an operating cost which can be reduced by intelligent water conservation.

Regarding the second goal, waste load reduction offers the following advantages. Firstly, reduction of raw waste load will also be reflected in reduced treatment capacity requirements for on-site treatment or reduced sewer surcharges for off-site treatment. Secondly, reduction of raw waste load implies greater recovery of organic byproducts which can be processed to provide an

Table 3.4 Raw waste load from inedible rendering plants (after EPA, 1974b)

Base	Flow (m³ t⁻¹ RM)	Raw material (t RM d⁻¹)	BOD₅ (kg t⁻¹ RM)	Suspended solids (kg t⁻¹ RM)	FOG (kg t⁻¹ RM)	Kjeldahl nitrogen as N (kg t⁻¹ RM)	Ammonia as N (kg t⁻¹ RM)	Chlorides as Cl⁻ (kg t⁻¹ RM)	Total phosphorus as P (kg t⁻¹ RM)
(Number of observations)	(47)	(48)	(29)	(26)	(18)	(17)	(16)	(14)	(17)
Average	3.26	94	2.15	1.13	0.72	0.476	0.299	0.793	0.044
Range	0.47–20.0	3.6–390	0.10–5.83	0.03–5.18	0.00–4.18	0.12–1.2	0.08–0.74	0.08–2.56	0.003–0.280

economic return. Successful implementation of in-plant controls is founded on a few simple premises.

- Downstream removal of waste materials is not as effective as their initial exclusion from the sewer because large solids break up into smaller solids which are difficult to remove, and FOG becomes emulsified.
- Drains throughout the plant should be equipped with threaded covers to discourage their removal. Once drain covers are removed, the washing of gross solids into the sewer is possible. Where solids loss cannot be prevented, recovery should be undertaken as near to the source as possible so that less BOD_5 will go into solution and less FOG will be emulsified.
- Use of potable water should be justified for each purpose for which it is used and quantities should be optimized to ensure that excess water is not being used.
- Waste streams should be segregated with regard to possible reuse of some streams and with regard to compatibility with individual treatment processes.
- For those areas where sanitary concerns are limited, non-potable water reuse should be considered.
- A conscientious dry cleanup procedure followed by a controlled and efficient wet cleanup should be used for all cleanup operations.

Efficient waste reduction and water conservation should become a measure of achievement for staff. Incentive and education schemes to encourage these goals are generally more successful than mere modification of physical facilities. The establishment of a waste reduction and water conservation programme should become the responsibility of one efficient staff member assigned reasonable powers for programme implementation. A programme should be initiated by determining where and how much water is used in the plant. Flow meters and pressure gauges should be installed in major water use areas. Pressure regulators should be installed to prevent major water pressure variations which can lead to inefficient water use.

If a plant is equipped with biological treatment, careful use of a biodegradable detergent may enhance cleanup efficiency and minimize water use. The magnitude and effect of a possible increase in emulsified FOG load as a result of detergent use must be evaluated before commitment to full-scale detergent application.

Specific measures to reduce waste loadings and minimize water use are discussed for the various processes in subsequent sections.

3.3.2 Red meat products

Process upgrading
Holding pens. Uncontrolled wastewater generation can occur in holding pens if rainwater (or snow) is allowed to reach the pen floor to create storm drainage.

This can be controlled most easily by roofing the pens to divert clean rainwater away from contact with the pen floor pollutant load. As well, the potential for waste discharge to sewer can be markedly reduced by dry cleanup prior to final pen washdown.

Slaughter area. Inefficient blood handling and recovery is undoubtedly the single largest factor in determining the raw waste load for a red meat products plant. Objectively viewed, the open bleeding trough, which is the most common system in use (in North America), is a relatively crude method for blood recovery.

Alternative blood collection technology (Akers, 1973) includes the following options:

1. A flat sticking knife equipped with a collection vessel at the base may be used. Blood is drained into the vessel and removed by vacuum.
2. A perforated tube attached to the sticking knife assembly may be forced into the incision with suction being applied to remove blood.

The efficiency of the above methods is enhanced by the application of anticoagulant to the blood being collected.

A segregated blood drain and collection tank should be provided for the concentrated blood. High concentrations of blood recoverable from the process line immediately following the slaughter area can also be directed to the blood collection drain prior to final wet washdown. Although the final washdown wastewater will still have a relatively high BOD, its high water content would devalue the collected concentrated blood. Therefore, such final washdown water is generally not a welcome addition to the blood collection tank.

Blood processing. Any direct steam method of blood processing will produce unacceptable quantities of bloodwater. Consequently, reduction of waste load requires either the use of indirect steam heating or bloodwater evaporation, either directly or as part of an inedible rendering process.

Hide processing. Provided that the waste load from this operation is not simply eliminated by dry salting and shipping of raw hides to off-site processing, waste load reduction must focus on wash-water, defleshing and curing. Since hide washing does not require potable water, a closed loop of hide wash-water recycle with solids removal may be employed. Defleshing waste load can be essentially eliminated by effective dry collection of fleshings prior to final wet cleanup. Finally, the hide-curing brine can be recycled, provided that solids are removed. Some system blowdown will be necessary. Also, intermittent tank draining for cleaning must be timed to avoid shock loading to downstream treatment.

Scalding tub. The tank for hog processing develops a sludge buildup throughout the day. This tub should be provided with a perforated riser pipe projecting

15 cm or more above the tub floor. The riser prevents draining of the sludge layer during tub draining. The sludge must then be removed through a side drain for rendering or dry disposal.

Paunch handling. Reduction of raw waste loading from this operation may be most effectively achieved by inedible rendering of whole paunches (thereby eliminating paunch manure dumping) or by dry dumping. The high BOD_5 of paunch manure (estimated at $50\,000\,\text{mg}\,l^{-1}$) makes it a major candidate for raw waste load reduction.

Inedible viscera handling. Washing of viscera prior to inedible rendering is a nonproductive step which only reduces the amount of organic feedstock to the inedible rendering process while increasing the overall plant raw waste load. Consequently, elimination of this washing step is desirable for waste load reduction.

Intestine utilization. Desliming of intestines produces high strength, high lipid content waste which is directly valuable to inedible rendering. Minimization of wash-water generation and maximization of FOG recovery for this step will effectively reduce raw waste loadings.

Meat processing. The major source of raw waste load from these operations is associated with pickling and curing operations. The salt and sugar content of these solutions, combined with the organic matter they collect from the meat being processed, provides the pollutant loading. Consequently, measures to reduce raw waste loading from these operations aim to avoid unnecessary discharge of these solutions. Possibilities for the above include solution reuse or alternative curing technologies.

On-site rendering. Waste load reduction from these operations will be discussed here in the context of edible rendering (wet rendering); dry rendering processes are discussed in Section 3.3.4 below. The major waste source for edible rendering is tankwater which can be dealt with by evaporation or by feeding to a dry inedible rendering operation. In such cases, waste load becomes associated with vapour condensates, which are also considered in Section 3.3.4.

Cleanup. As mentioned in Section 3.3.1, cleanup operations can play a pivotal role in determining plant raw waste load. Common sense measures such as extensive dry cleanup can markedly reduce raw waste load from this source. Avoidance of hot water fogging can prevent unnecessary waste discharges to sewer.

Water use reduction

Several rational plant and procedural modifications can be made to reduce water consumption. These options will be considered in turn.

1. Automatic level controls or intermittent timed valves can be used to control water supply for livestock in the holding pens, thereby reducing wastage associated with overflow.
2. Cleanup hoses and taps should be equipped with automatic shutoff valves to reduce careless water loss.
3. Drinking and handwashing water should be provided with foot-operated valves to avoid continuous operation. Significant advances have been made in recent years in the design of water-conserving fixtures which can improve washing efficiency.
4. Redesign of carcass-washing facilities can usually reduce the quantity of water required by optimizing the spray pattern and by providing shutoff valves tied to the progress of the carcass conveyor.
5. Side pan washers in an eviscerating line can be made intermittent by the use of a timer. Efficient cleaning can be achieved with much less water use.
6. Water consumption can be reduced by judicious water reuse. Although health regulations require the use of potable water in many processing steps, several viable reuse options remain. The major stipulation is that water must flow from cleaner areas to progressively dirtier areas. For example, carcass wash-water can be collected and reused in the hog dehairing operation. Also, refrigeration condenser water may be used for virtually any purpose for which potable water is not specifically stipulated.
7. Dry cleanup operations have already been proposed for raw waste load reduction. Clearly, such procedures can also reduce water consumption. Use of high-pressure nozzles can provide more water-efficient cleanup.

Waste segregation

Water reuse options mentioned above depend upon segregation of reusable, cleaner streams from more polluted wastes. Moreover, individual treatment processes are most effective for specific classes of pollutant. Consequently, it is generally desirable to segregate major waste streams carrying different classes of pollutant in order that individual smaller waste flows can receive individual treatment. Some examples of waste streams suitable for segregation are listed below.

1. *Manure sewer.* Holding-pen wastewater is generally contaminated with manure solids and dirt which are amenable to removal by screening and sedimentation. These materials are incompatible with grease which could otherwise be recovered for high quality tallow production.
2. *Blood drain and tank.* Low-water-content blood is a potentially valuable byproduct. However, separate collection of blood is necessary to obtain an

economic return on this material. Blood is also incompatible with recoverable grease because it will discolour the product tallow.
3. *Grease sewer.* Wastewaters from areas relatively low in blood but high in FOG, such as cutting, rendering, lard storage and meat processing, should be segregated for quality grease recovery by gravity and/or dissolved air flotation.
4. *Low grease sewer.* Wastewaters which are relatively low in FOG, such as those from the slaughter area, hog scalding, dehairing, evisceration and hide washing, do not necessarily require extensive FOG removal. Thus, the solids removal processes for this stream will not be unduly hampered by grease fouling.
5. *Sanitary wastes.* Health regulations normally preclude the mixing of human sanitary wastes with process wastewaters prior to byproduct recovery. Consequently, a segregated sewer system is necessary.
6. *Clean water sewer.* Water streams such as cooling-waters and steam condensates have normally been exposed to negligible contamination and are, therefore, amenable to reuse.

3.3.3 Poultry products

Process upgrading
Receiving area. Since waste load generation from the receiving area is a direct function of the length of time for which birds are held prior to slaughter, the load can be reduced by organization of deliveries to coordinate with processing capacity. Because of the dust and feathers in this area, some form of vacuum cleaning system is recommended for dry cleanup. Final wet washdown should use high-pressure hoses which consume less water for a given degree of cleanup.

Slaughter area. Spasmodic flapping by birds following the throat-cutting operation results in coating of their feathers with blood. Unlike blood spilled on the floors and walls of the slaughter tunnel, blood on the feathers cannot be recovered efficiently and will generate unnecessary waste load in the scald tank. Therefore, apart from improving blood collection within the slaughter tunnel, the most significant reduction in waste generation may be achieved by stunning the birds to prevent unnecessary splattering of blood in the slaughter area.

Scalding. Raw waste load from this operation can be reduced by providing a tank drain equipped with a screened or perforated riser as described for the scald tank in hog processing. The collected sludge at the bottom of the scald tank must be collected for disposal as solid waste, otherwise it will make a shock load contribution to the total raw waste load.

Defeathering. Some reduction in raw waste load may be achieved by screening and recycling the feather flume-water. Frequent checking of the screens is required to maintain them in effective working order, since they can easily clog. Alternatively, use of a vacuum feather-removal system effectively eliminates the raw waste load contribution from this source.

Evisceration. Medium-to-large poultry plants use the flowaway fluming system almost exclusively. This exerts a large water demand, a typical eviscerating flume using about 12 litres per bird, while gizzard washing, where practised, demands a further 12 litres per bird. This system tends to be favoured by health inspection officials, since the heavy water use rapidly removes the viscera from the carcass, thereby reducing the probability of contact and contamination. However, this practice is responsible for a major contribution to raw waste load and water use at poultry products plants. Solutions to this dilemma will require innovative schemes to utilize less water more efficiently or utilize mechanical removal to isolate the carcasses from the viscera.

Chilling. About 2.7 litres of water per bird are required for chilling. Raw waste load generated in this process is limited and the relatively clean chiller-water overflow is likely to be suitable for direct reuse in areas not required to use potable water. For example, recycle to the feather flume and the scald tanks might be considered.

Cleanup. Measures proposed for raw waste load reduction during cleanup of red meat processing facilities also apply here.

Water use reduction

Poultry operations offer several specific areas where significant reductions can be made in water consumption or where used water can be recycled. The comments made for red meat packing regarding the use of automatic shutoff and employee actuated control valves are equally applicable to poultry plants. In addition, the following water conservation measures may be considered.

- The overflow from the chiller may be screened and reused as makeup water to the scald tank or the feather flowaway flume.
- Feather flume-water may be screened for feather recovery and recycled for flume usage or, if health authority approval is secured, it may be used in the offal flume.
- A vacuum system for feather removal, to replace the feather flume, can significantly reduce water consumption.
- Water-conserving nozzles may be used in the final bird-washing operation and for employee hand-washers, to minimize water consumption.
- Hand-washers may be knee- or foot-controlled to minimize unused water flow.

- Evisceration solids (offal) may be moved by timed flushing spray rather than by continuous flow.
- The side pan washers may be time-cycled to reduce water use.
- An automatic shutoff valve on the final carcass washer, controlled by the production line, can limit washing flow to when birds are actually being processed. Optimized spray nozzles are also desirable.
- The offal flume-water may be screened and reused in the feather flume.

A water conservation programme was carried out at one large poultry packing plant (EPA, 1973a); the water use reductions achieved are summarized in Table 3.5.

Table 3.5 Reduction of water use in a poultry products plant (after EPA, 1973a)

Process segment	Activity	Reduction in fresh water use $(m^3 h^{-1})$ from	to
Evisceration	Use of improved nozzles:		
	final bird-washers	123	74
	hand-washers	698	245
	Cycling of side-pan wash	221	74
	Rearrangement of giblet handling	880	784
Scalding and defeathering	Use of improved nozzles in whole bird-washers	110	74
	New design of feather flume for reuse of offal flume-water	230	0
	Use of chiller-water in scalder to replace fresh water	100	0
Clean-up	New high-pressure design system, with foam	$(m^3 d^{-1})$ 424	$(m^3 d^{-1})$ 174
Overall water use reduction		30%	
Overall waste load reduction		65%	

Waste segregation

Modern poultry processing plants have two main separate wastewater streams, the offal flume and the feather flume, each of which must be separately screened. Recycle options for the relatively clean feather flume-water should be considered. Other than cleanup waters, offal flume-water and scald tank drainage, most water flows are relatively clean and should be segregated for possible recycle to non-edible processing areas. As in red meat products plants, sanitary sewage should be segregated from process waste streams until after byproduct recovery.

3.3.4 Independent rendering

Process upgrading
Raw material delivery. The raw materials receiving area should be designed to contain drainage from raw materials and prevent its uncontrolled discharge to sewer. Where low-volume high-strength drainage occurs, it should be collected and sent to the cooking process. Where higher-volume lower-strength drainage occurs, as with combined feather and blood drainage, the wastewater should be steam sparged and screened to collect coagulated organics.

Where special blood recovery facilities are not available, prior to blood unloading or transfer, receiving area drains should be plugged and the operation should be supervised by responsible plant personnel.

Cooking. The cooking process should be closely monitored to prevent overfilling of cookers or use of excessive cooking rates. Internal agitation should also be controlled to prevent carryover of material to the vapour drawoff. The vapour lines should be equipped with traps to collect accidental overflow from the cookers. These traps should be inspected and cleaned regularly. Cookers for hydrolyzing hog hair and/or feathers should be provided with bypass controls to control pressure reduction after cooking cycles. Blood should be processed dry, rather than by steam sparging.

General materials processing. A regular maintenance schedule should be established to discover and repair materials leakage as soon as possible. Material processing equipment (grinders, cookers, centrifuges, filters and blenders) should be protected by floor dykes to contain spillage of materials resulting from accidents or equipment failure. Containment of spills facilitates dry cleanup.

General cleanup. All areas should be cleaned dry with scrapers, shovels, squeegees, etc., before washing with hot water. One daily cleanup should be adequate if processing equipment is properly maintained to avoid leakage and spillage. Personnel traffic lanes should be designed with gratings, handholds, etc., to provide maximum traction for operators without excessive wet cleaning.

Byproduct recovery. Byproduct recovery equipment (screens, clarifiers, etc.) should be equipped for regular or continuous removal of grease and solids. Condensed vapours from the cooking process should be cooled by heat exchange or air exposure to the greatest extent possible and preferably below 50°C to allow grease recovery by gravity separation.

Water use reduction
The greatest reduction in water consumption may be achieved by using air condensers in place of barometric leg condensers. Where the latter are in use,

water should be supplied by treated effluent rather than unnecessary use of potable water.

Water use for air pollution scrubbers designed for odour control may be reduced by the use of chemical oxidants and recycle. Makeup can be provided by treated effluent rather than potable water.

The health criteria on water reuse in an inedible rendering plant are not as restrictive as for meat and poultry plants. Thus, extensive wastewater reuse should be feasible for any given plant. For instance, well treated effluent (very low in FOG content) could conceivably be used for low-grade wash-water.

Waste segregation

Inedible rendering operations produce a wide range of wastewaters varying from relatively clean boiler and cooling tower blowdown to concentrated raw material drainage. The sewer system must be designed to segregate the various streams in order to allow efficient reuse of clean wastewater streams and to avoid hydraulic overloading of treatment equipment on the contaminated waste streams. The preferred sewer segregation scheme will largely depend upon the particular processing equipment and process layout used. However, some general observations are possible.

Relatively clean waters from boiler-water treatment, cooling and air pollution scrubbers should be segregated from more contaminated wastewaters. Condensed cooking vapours may be relatively high in FOG and soluble organic matter, but are low in solids and, therefore, can be segregated to bypass preliminary solids removal equipment (screens). The high temperature of the condensate may necessitate FOG removal by means other than gravity separation, thereby providing another reason for segregation of this stream.

Plant wash-water is high in solids and FOG levels and requires screening and primary FOG removal. This waste stream should not be diluted with cleaner wastewaters which only increase the hydraulic loading on the treatment equipment. The fluctuating nature of this waste stream suggests the need for separate equalization facilities to ensure smoothing of peak flows to the primary treatment processes.

As with meat and poultry products plants, it is good practice to segregate sanitary sewage to beyond the byproduct recovery stage. Separate sanitary waste treatment by septic tank and tile field is often feasible.

3.4 External treatment

This section will deal with treatment processes applied beyond the confines of individual process areas. As such, this classification differs from the US EPA definition (EPA, 1974a, 1974b, 1975a) wherein raw waste load was defined following byproduct recovery (e.g. FOG and solids recovery for rendering). Such recovery processes will be considered herein as part of the external treatment processes. By so doing, it is not intended to detract from the economic importance of byproduct recovery, but rather to group all treatment processes together in order to promote a better understanding of the physical, chemical and biological factors that control their performance.

Discussion will be directed towards the efficient design and operation of treatment processes in logical sequence, whether final treatment is achieved on-site, or off-site in a municipal treatment facility. Although there is a vast array of processes which could conceivably be applicable to meat, poultry and rendering industry wastewaters, a complete description of all is beyond the scope of this chapter. Accordingly, the focus will be directed towards those treatment processes which have been successfully used and documented specifically for the meat, poultry and/or rendering industries.

3.4.1 Equalization

Equalization of wastewaters both in volume of flow and in strength can contribute significantly to optimal use of treatment facilities and/or minimization of problems with discharges to municipal systems. Flow equalization allows treatment operations to be sized according to average daily flow rather than requiring oversizing to handle peak flows. This is particularly important with plant operations where waste production is often concentrated into 8 to 12 hours of the day. Equalization of such waste flows can allow significantly smaller, but more reliable, treatment facilities.

Flow equalization requires the provision of a tank with a specified retention volume and a constant flow outlet. The required volume can be determined by monitoring normal flow over the full period of flow cycling. Normally, 24 hours represents the full flow cycle, but monitoring must be performed over several days, including non-processing days, to provide an adequate design basis. From adequate flow data, the cumulative volume may be calculated and plotted against time (Fig. 3.5). Average flow from the basin is represented by a line connecting the origin to the total cumulative volume at the end of the flow period. Construction of maximum and minimum lines parallel to the average flow line and tangent to the cumulative flow curve gives the minimum storage volume required to equalize the flows over the specified time period.

Provision of flow equalization does not, in itself, ensure strength equalization because flow and strength may vary independently in some circumstances. In practice, however, flow equalization will significantly smooth out fluctuations in waste strength. It is possible to specify a given degree of strength equalization in addition to any flow equalization. This can be achieved by providing a complete mix tank with a finite retention volume, but with inflow matching outflow. For the particular case where flow variations are small (as after flow equalization), a mass balance may be performed on the equalization basin (Novotny and Englande, 1974) as

$$dC/dt \; V = C_i Q - CQ$$

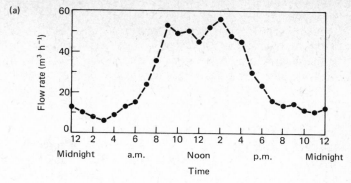

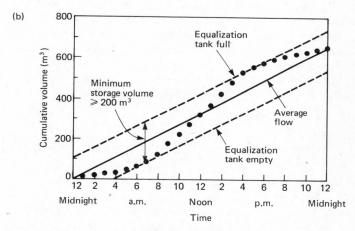

Fig. 3.5 Flow from processing plant: (a) flow hydrograph (b) cumulative flow

where C = concentration of component in basin and effluent
C_i = concentration of component in influent
V = volume of basin
Q = flow into and out of basin
t = time

Considering this system as an input–output device and applying control systems analysis techniques one can express the system response as

$$\frac{\sigma_y}{\sigma_x} = \left(\frac{\Delta}{2\tau}\right)^{0.5}$$

where σ_y is the standard deviation of the output, determined by specifying a maximum acceptable effluent concentration at a given level of probability;

σ_x is the standard deviation of the input at frequency Δ, determined by monitoring data on the input;
Δ is the time between samples;
τ is the detention time in the basin.

This expression allows calculation of the required detention time, τ.

3.4.2 Coarse solids removal

Coarse solids are usually removed by using some form of screening (Steffen, 1973; EPA, 1973a) to strain them out of the liquid flow. Thus, the controlling criterion for solids removal is particle size, which determines the required size of screen openings to ensure retention of the solids. The screening operation should therefore be sited as close as possible to the source of solids discharge in order to minimize solids disintegration during flow in the sewer. Such disintegration seriously reduces the quantity of gross solids that can be recovered by screening.

Various configurations of screening device have been employed for the pretreatment of meat and poultry plant wastewater.

Stationary screens

Conventional stationary screens have received limited application in meat and poultry plants because they tend quickly to become blocked under typical applied solids and FOG loadings. However, banks of three or four stationary screens in series may be used for feather recovery from a segregated feather flume line. Since this stream is relatively clean except for the discrete feather loading, a sequential arrangement of screens allows continued operation with intermittent removal and cleaning of individual screens. Screens for this purpose are typically of steel mesh or perforated sheet with 6–13 mm openings.

A less conventional type of fixed screen which has found wide application in these industries is the tangential or sidehill screen which has been adapted from mineral processing technology (Fig. 3.6). This device is primarily suited to dewatering slurries and sludges and can be successfully applied to highly solids-laden streams. Waste flows down a sloped face constructed of curved V-shaped wires, oriented perpendicular to the downward flow. The solids-bearing flow is found to stratify, with liquid attaching to the transverse wires and passing through the screen openings. The solids move down the slope and concentrate at the bottom of the screen. Most applications of this device in the meat and poultry industry employ wire openings of 0.25–1.5 mm.

In general, stationary screens are not well suited to heavy FOG loadings without frequent cleaning and maintenance, although steam sprays may be used to reduce clogging. Tangential screens offer relatively small space requirements, negligible energy consumption, limited operational skill requirements, relatively rugged construction and good adaptability to flow variation.

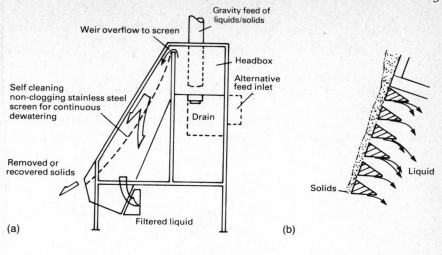

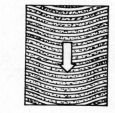

Fig. 3.6 Tangential screen (after EPA, 1973a): (a) overall cross-section (b) longitudinal cross-section through V-shaped wires (c) direction of wastewater flow over screen face

Vibrating screens

The problems associated with screen blinding for meat and poultry waste streams have been met in many cases by the use of vibrating screens. Various configurations of screen and vibration source have been used, including box screens on vibrating supports and disc screens driven in three dimensions by eccentric drives. Vibrations at 1000 to 2000 rpm with amplitudes of 0.8–13 mm are commonly used.

The screens themselves are usually made of stainless steel mesh and are designed to allow removal for periodic cleaning. During operation, the screens are prevented from blinding too rapidly by the vibrations, which dislodge particles. In addition, some designs employ scrapers to remove solids from the screen surface while others are designed to propel solids to the screen periphery by means of their vibratory motion. Success of vibrating screens is very much dependent upon the selection of the correct gauge for the solids to be removed. Heavy or abrasive solids require screens of thicker diameter wire; for light or sticky solids, wire thickness is relatively unimportant, but percentage open area is all-important.

Rotary or drum screens

Another approach to screening highly solids-laden waste streams is the use of rotary or drum screens. In these devices, a screen mesh is stretched over a drum frame (Fig. 3.7). Typically, stainless steel cloths with 0.4 mm openings at 40 mesh (40 to the inch) are employed for meat and poultry applications.

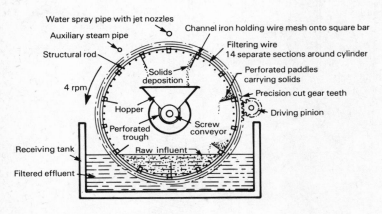

Fig. 3.7 Rotary or drum screen (after EPA, 1973a)

External-feed rotary screens provide a flow discharge onto the outside of the revolving drum, with water collection and removal from the inside of the drum. These devices are usually equipped with a knife scraper and water spray to remove collected solids. This type of rotary screen is reported as very successful for solids removal and dewatering.

One type of internal-feed rotary screen, often used for offal removal, provides a revolving horizontal drum with a tilt from the inlet end downwards. Water passes through the screen mesh, while collected solids discharge out the opposite end of the drum. Another type of internal-feed rotary screen provides wastewater inlet at one end and discharge out through the screen, as in the previous design, but solids are collected by perforated paddles on the circumference of the drum. The solids are carried up to the top of the drum and are dropped into a central collection trough which discharges them by screw conveyor. Internal-feed rotary screens are not particularly efficient at dewatering the collected solids although their performance on solids removal from the wastewater stream may be quite adequate. FOG blinding of rotary screens may necessitate steam spraying or coating of the steel mesh with Teflon.

3.4.3 Settleable solids and FOG removal

Gravity clarification
Clarifiers are essentially flow-through tanks with sufficient average detention time to allow settleable solids to reach the bottom of the tank and floatable FOG to reach the top. Since clarifiers installed in older plants were often designed strictly as catch basins for the recovery of floatable grease, many such installations were not equipped for the removal of settled sludge. In such cases, effective solids removal cannot be achieved. Moreover, in some plants the 'catch basin' was simply the wet well or sump that received wastewater drainage prior to sewer discharge. Such tanks, with their inherent turbulence, cannot effectively remove settleable or floatable matter.

The factors to be considered in clarifier design will be discussed in turn. In general, primary clarifiers tend to be rectangular in shape, with a preferred length-to-width ratio greater than 3 to 1, recommended maximum width of 3.3 m and maximum depth of about 1.8 m. It is generally good practice to provide two or more clarifiers in parallel to provide maintenance flexibility. Clarifiers may be constructed of steel, concrete or a combination of both. A good combination is the use of concrete floors with steel walls. Proper design requires knowledge of influent flow, which should be equalized, of wastewater temperature, which will affect the rate of FOG separation, and of the settling or rising velocities of particles to be removed. These velocities determine the surface overflow rate which should be less than the limiting particle settling or rising velocity. Recommended surface loading rates range from 30 to 60 $m^3 m^{-2} d^{-1}$.

Rising and settling velocities are determined by the average particle size and density as well as by water temperature. While it is possible to estimate such values, it is prudent to determine actual wastewater particle settling velocities in pilot studies to ensure adequate design. Overall detention times are often specified and a minimum of 40 min detention at the maximum hourly flow should be provided. Detention of 1 to 2 hours at the average hourly flow should ensure good separation. Efficient separation requires that turbulence be minimized and multiple inlets with baffles are recommended. Temperature fluctuation can lead to short circuiting and, therefore, should be avoided by adequate flow equalization and mixing prior to the clarifier inlet.

The outlet weir should be effective across the complete width of the basin and the overflow per hour should be limited to less than 18 m^3 per m of weir length. Weir loadings can be minimized by using one or more double-sided weir troughs. It is essential to ensure that final weir installations are level in order to avoid excessive flow over low spots.

Scum can be collected with a slotted pipe collector, but this method is inefficient for heavy FOG loads. A better system is to provide surface scrapers which draw the scum along the tank length to one end. The collection end is provided with an inclined plane up which the scrapers travel, carrying the collected scum into a collection trough. This system provides superior scum dewatering.

Sludge scraping should preferably be powered by an independent drive mechanism in order to maintain operational and maintenance flexibility. Sludge may be collected into a submerged hopper at one end of the basin, or an inclined plane rising above the water level may be used to improve sludge dewatering.

Poor clarifier performance can normally be attributed to poor initial physical design, general or intermittent hydraulic overloading or inadequate sludge and/or scum removal.

Air flotation

Air flotation as applied to meat, poultry and rendering wastewaters is an extension of the gravity separation of contaminants achieved by catch basins. Particles with a specific gravity close to that of water cannot be removed in a reasonable period of time by the action of gravity alone. Although various air flotation systems are in use in a variety of industries, dissolved air flotation is most widely used in the meat and poultry industries. In this process, the entire wastewater stream, a portion of it or recycled effluent is pressurized and injected with compressed air to produce a supersaturated condition. The supersaturated wastewater is allowed to come to equilibrium with atmospheric pressure in the flotation cell. As a result, tiny air bubbles are formed in the water which may adhere to individual particles or become trapped in flocculent particles. The effect of the air bubble attachment is to decrease the effective specific gravity of the particle and float it to the surface. The scum formed at the surface is skimmed off and may, in most cases, be recovered for rendering. Flotation processes offer the advantage of creating a sludge of lower water content than is normally obtained by gravity settling.

Flotation surface loading rates generally range from 30 to $60 \, m^3 \, m^{-2} \, d^{-1}$. These processes are also evaluated in terms of the air/solids ratio, that is, the mass of air released per unit mass of solids applied. Since operating parameters such as air/solids ratio and per cent recycle cannot be set theoretically, it is necessary to perform pilot studies prior to full-scale design. Such studies conducted at increasing values of air/solids ratio will show increasing per cent solids in the float and corresponding reduction in effluent suspended solids. A design value is chosen at which further increases produce only marginal improvement in performance.

Additional improvement in pollutant removal can be achieved by the use of chemical coagulants which promote floc formation. However, when chemical additives are used, the skimmings often contain a higher percentage of water and the tallow produced by rendering such materials is generally of a lower quality than that obtained without chemical use.

In a survey for the US EPA (EPA, 1974a), a consultant observed that many dissolved air flotation units were producing poor results largely because they were being poorly operated. Process performance depends on maintaining the optimum degree of air pressurization as well as providing good influent flow equalization and efficient skimmings and sludge removal. If air flotation units are left to run themselves, very poor performance can be expected.

Data on several dissolved air flotation units operating at meat, poultry and rendering plants are summarized in Table 3.6.

A case study of a dissolved air flotation pretreatment unit at a hog processing plant was reported by Murray (1975). Design parameters and process performance for this treatment facility, reported in 1974 (April) US dollars, may be summarized as follows. Amortization of capital has been excluded because of the variability of this factor from one financial jurisdiction to another. Clearly, however, capital amortization is a major real cost which must be calculated for any given specific application.

Capital cost			$150 000 (approx.)	
Annual operating costs:	chemicals	$12 125		
	manpower	7 500		
	maintenance	3 000		
	power	3 250		
	miscellaneous	1 500	$ 27 375*	
Cost per m³ treated	$0.012*			
Cost per kg BOD₅ removed	$0.070*			

$Cost\ per\ m^3\ treated$ — $0.012*
$Cost\ per\ kg\ BOD_5\ removed$ — $0.070*

* Excluding capital amortization.

Electrolytic coagulation and flotation

This process encompasses a wide variety of alternatives which have found application for many different industrial wastewaters. Applications to meat and poultry industry wastewater have centred on proprietary technology developed by Swifts (Beck *et al.*, 1974; Ramirez *et al.*, 1976). Rather than review all possible variations for the process, this discussion will focus on the approaches followed by Swifts.

Two distinct phases of treatment are recognized in the Swifts process, namely electrocoagulation and electroflotation. In the first stage, a reaction vessel acts as the cathode and is traversed by a series of closely spaced (2.5 cm) anodes. A DC potential maintained at less than 15 V produces cations which have the ability to promote coagulation of colloidal organic matter which tends to be negatively charged. Electrolysis of water also usually occurs to yield microbubbles of hydrogen and oxygen according to

$$2\ H_2O \rightarrow O_{2(g)} + 4\ H^+ + 4\ e^- \quad \text{anode}$$

$$2\ e^- + 2\ H_2O \rightarrow H_{2(g)} + 2\ OH^- \quad \text{cathode}$$

Finally, the electrocoagulation cell is designed to promote collisions between the destabilized particles and microbubbles through laminar vortex flow. The

Table 3.6 Dissolved air flotation performance

Plant type	Wastewater hydraulic loading ($m^3 m^{-2} d^{-1}$)	Recycle (% of influent)	Chemical dosage ($mg l^{-1}$)	BOD$_5$ Removal (%)	BOD$_5$ Effluent ($mg l^{-1}$)	Suspended solids Removal (%)	Suspended solids Effluent ($mg l^{-1}$)	FOG Removal (%)	FOG Effluent ($mg l^{-1}$)	Reference
Red meat products	60	50	N.R.	N.R.	N.R.	N.R.	N.R.	87	150	Steffen, 1973
Red meat products	46	50	N.R.	55	760	93	410	90	213	Steffen, 1973
Red meat products	24	50	ferric sulphate (N.R.)	85	200	95	60	N.R.	N.R.	Steffen, 1973
Red meat (hogs)	63	N.R.	N.R.	33	1762	32	1507	62	560	Baker and White, 1971
Red meat products	N.R.	N.R.	alum (20)	44–58	618–871	N.R.	N.R.	54–76	116–155	Hirlinger and Gross, 1957
Red meat (hogs)	88	50	ferric sulphate (100) polymer (10)	41	2400	47	690	95	31	Murray, 1975
Red meat products	50	70	aluminum chloride (1000–1500) polymer (10–20)	40	N.R.	90	N.R.	98	N.R.	Green et al., 1980
Edible oil refinery	115	N.R.	alum (450) polymer (2.6)	75–87	740	84–89	400	87–92	360	Seng, 1971
Poultry products	160	20	alum (75) soda ash (75) polymer (2)	74–98	18–294	87–99	4–94	97–99+	2–51	Woodard et al., 1972
Poultry products	106	20	alum (75) soda ash (75) polymer (2)	85	92	N.R.	55	N.R.	37	Woodard et al., 1977

N.R. = not reported.

entire electrocoagulation process is very rapid with design retention times of only 1 to 3 min. If successful, this stage is intended to produce particle–bubble aggregates with high buoyancy (specific gravity of approximately 0.8).

In the second stage, electroflotation, the wastewater flow is released to a larger flotation cell with a residence time of 20 to 30 min. This is also equipped with electrodes to promote further gas bubble formation. Apart from the mechanism of bubble formation, this stage operates similarly to the float tank of a dissolved air flotation process. In fact, electroflotation alone may be successfully applied to many wastewaters with performance comparable to dissolved air flotation.

The two-step process is claimed to be effective for high-strength wastewaters, with additional applications where surface tension is low because of the presence of surfactants. Either the two-stage process or the electroflotation process alone is claimed to produce a float of higher solids content than simple air flotation (9–12% solids generally, or up to 50% solids for high fat skimmings (Beck *et al.*, 1974)).

Additionally, where chloride concentrations are relatively high, chloride can be oxidized at the anode to yield chlorine gas. In solution, the chlorine promotes disinfection of wastewater.

Performance reported for these processes is summarized in Table 3.7: pollutant removals are seen to be very good, but coagulant dosages are high. Information on operating costs is limited. However, contrary to common expectation, power costs are relatively low. The major cost component for the process is chemical cost which will generally be a function of pollutant load and degree of emulsification. Operating costs reported by Ramirez *et al.* (1977) for a two-stage process employed at a tannery have been converted to SI units and may be summarized as follows (1977 US \$ per m^3):

Power costs at \$0.0094 per MJ
    Electrocoagulation power (0.67 MJ m^{-3})   \$0.0063
    Electroflotation power (0.29 MJ m^{-3})   0.0027
    Other power requirements (0.86 MJ m^{-3})   0.0081   \$0.017

Chemical costs
    Alum   (600 mg l^{-1} at \$0.121 kg^{-1})   0.0724
    Lime   (600 mg l^{-1} at \$0.0396 kg^{-1})   0.0237
    Anionic polymer (8 mg l^{-1} at \$5.17 kg^{-1})   0.0413   \$0.137

Total operating costs (excluding capital amortization)   \$0.154 per m^3

3.4.4 Soluble organic carbon removal

Although solids and FOG removal processes remove significant quantities of organic matter from meat and poultry industry wastewaters, they are largely

Table 3.7 Electrocoagulation–electroflotation performance

Plant type	Energy consumption ($kJ\,m^{-1}$)	Chemical dosage ($mg\,l^{-1}$) and operating parameters	BOD$_5$ Removal (%)	BOD$_5$ Effluent ($mg\,l^{-1}$)	Suspended solids Removal (%)	Suspended solids Effluent ($mg\,l^{-1}$)	FOG Removal (%)	FOG Effluent ($mg\,l^{-1}$)	Reference
Red meat slaughterhouse	N.R.	ferric sulphate (200) lime (50) polyelectrolyte (2) 15V D.C.	91	458	98	65	98	15	Beck et al., 1974
Red meat slaughterhouse	N.R.	as above plus pH control	98	86	99+	29	99	13	Beck et al., 1974
Red meat slaughterhouse	280	ferric sulphate (350–400) anionic polyelectrolyte (2) electroflotation only	91	86 (±32)	94	33 (±78)	94	16 (±22)	Ramirez et al., 1976
Red meat slaughterhouse	650	ferric sulphate (700–750) anionic polyelectrolyte (2) electrocoagulation plus electroflotation	90	205	91	96	96	23	Ramirez et al., 1976
Tannery	950	alum (600) lime (600) anionic polyelectrolyte (8) electrocoagulation plus electroflotation	62	200 (177–280)	87	85 (48–220)	95	14 (3–41)	Ramirez et al., 1977

N.R. = not reported.

ineffective for the removal of soluble organic matter. Exceptions are where specific chemical addition removes some colloidal organic matter which would otherwise not be removed by physical processes. However, the processes already described leave meat and poultry wastewaters containing relatively high concentrations of soluble and colloidal organic matter. Most of this is biodegradable and would be reflected as soluble BOD_5. Given the relatively biodegradable nature of the remaining organics, further treatment of these wastewaters is usually achieved with biological processes. As mentioned earlier, although there are many possible processes which could conceivably be applied to treating these wastewaters, this discussion will focus on those processes which have received the most widespread applications or which offer the most promise based on recent process research.

Protein recovery

Realization that a major portion of the organic matter present in meat and poultry industry wastewaters is protein has led to process development for a variety of schemes to recover protein.

Four major types of system have been reported in the literature. One incorporates adjustment to low pH and protein precipitation with the aid of chemical additives. A second approach utilizes a recently formulated cellulose ion-exchange resin to recover protein. A third involves maximization of biomass yield in activated sludge followed by protein recovery in the form of the generated biomass. The fourth approach uses ultrafiltration to separate the large-molecular-weight proteins from wastewater.

Precipitation methods may be further subdivided according to the chemical agent used. Two which have been documented are lignosulphonate (Hopwood and Rosen, 1972; Folz *et al.*, 1974; Hopwood, 1980) and sodium hexametaphosphate (Cooper and Denmead, 1979).

The lignosulphonate method functions on the basis of the interaction of negatively charged sulphonate groups on the lignosulphonate molecule and the positively charged amino groups on the proteins. Neutralization of protein charge produces a gelatinous suspension which may be efficiently removed by standard dissolved air flotation. Optimum precipitation of protein requires a low pH, generally pH 2 to 3, as well as close control to provide adequate lignosulphonate supply for the protein present. Adjustment of pH is normally achieved with sulphuric acid. Various forms of lignosulphonate are available since it is obtained from spent sulphite pulping liquor. Purified sodium lignosulphonate doses of 50 to 400 mg l^{-1}, depending on wastewater composition, have been used. Process performance on a variety of wastewaters is summarized in Table 3.8.

The sodium hexametaphosphate (HMP) precipitation method (Cooper and Denmead, 1979) also uses adjustment of pH to 3.5 with sulphuric acid. This is followed by HMP dosage proportional to blood concentration (20 mg l^{-1} for influent blood at 0.015%). Treatment performance, summarized in Table 3.8, was found to be dependent upon influent concentrations.

Cellulose ion-exchange media capable of removing proteins from solution have been known for some time. However, the cellulose media were only available as fibres which were not hydraulically amenable to large-scale wastewater treatment. Advances in the regeneration of ion-exchange media allowed their manufacture as beads, sheets or virtually any formulation desired (Grant, 1974; Jones, 1974). This improvement has been developed into a protein recovery process for food-processing wastewaters. The process is recommended as a follow-up to FOG removal by dissolved air flotation. Flotation effluent is fed to a stirred tank reactor and is mixed with the ion-exchange media. Resin with extracted protein is allowed to settle from the wastewater. The protein-loaded ion-exchange medium is fed to a second reactor where sodium salts or low pH are used to regenerate the media by displacing protein. The medium is again separated, washed and returned to the original reactor while the protein concentrate is removed for further processing.

Considerations in the design of the system include: the rate of protein uptake by the ion-exchange media, which governs the retention time in the first reactor; the protein adsorption isotherm, which dictates the required ratio of ion-exchange media to protein; and the wastewater pH and ionic strength which affect the protein adsorption. Pilot-scale tests (Jones, 1974) on slaughterhouse effluents indicate that protein recoveries in excess of 80% are possible and are coincident with BOD_5 reductions in excess of 95%.

Heddle (1979) reports research to maximize biomass production when using the activated sludge process for treatment of meat industry wastes. In this mode of operation, recovered biomass was found to contain 45 to 60% protein by weight. It was proposed that operating activated sludge with higher organic loadings to achieve maximum biomass growth would provide a viable means of protein recovery from meat industry wastewaters.

Shih and Kozink (1980) report an evaluation of membrane ultrafiltration to recover protein from poultry-processing wastewaters. This process utilizes a 0.1 to 5 μm membrane supported on a porous substructure to pass water and low-molecular-weight substances under a pressure gradient, while rejecting high-molecular-weight proteins and complexed fats to a concentrate. Operating performance on a pilot plant treating poultry industry wastewaters is summarized in Table 3.8.

Anaerobic processes
Anaerobic processes utilize anaerobic or facultative bacteria to degrade organic wastes, preferably at high temperatures (20°–35°C). The anaerobic degradation process involves essentially two stages: in the first phase, carbohydrates, fats and proteins are converted to organic acids and alcohols by acid-forming bacteria; in the second phase, methane-forming bacteria convert the intermediate organics into methane and carbon dioxide. If left uncontrolled, the

Table 3.8 Protein recovery process performance

Plant type	Protein recovery process	BOD$_5$		Suspended solids		FOG		Total nitrogen		Reference
		Removal (%)	Effluent (mg l^{-1})	Removal (%)	Effluent (mg l^{-1})	Removal (%)	Effluent (mg l^{-1})	Removal (%)	Effluent (mg l^{-1})	
Cattle slaughter (USA)	'Alprecin' lignosulphonate	83.5	208	95.7	41	95.2	21	87.3	24	Hopwood, 1980
Red meat slaughter (USA)	'Alprecin' lignosulphonate	81.8	161	90.5	71	N.R.	N.R.	71.6	23	Hopwood, 1980
Hog slaughter (UK)	'Alprecin' lignosulphonate	73.5	217	85.5	68	70.7	34	71.6	33	Hopwood, 1980
Cattle slaughter (UK)	'Alprecin' lignosulphonate	80.0	420	89.1	65	77.5	96	76.9	18	Hopwood, 1980
Poultry packing (UK)	'Alprecin' lignosulphonate	74.8	325	81.9	80	N.R.	N.R.	41.0	85	Hopwood, 1980
Poultry packing (UK)	'Alprecin' lignosulphonate	86.9	333	87.2	113	96.5	68	70.4	56	Hopwood, 1980
Cattle slaughter (USA)	lignosulphonate	79.0	506	87.0	210	95.0	15	N.R.	N.R.	Hopwood, 1980
Red meat slaughter (New Zealand)	sodium hexametaphosphate	COD 74.5 ± 7.1 to 79.1 ± 7.9	COD 361 ± 68 to 833 ± 289	85–90	60.5 ± 14.9 to 128.5 ± 45.0	91 ± 5	43.7 ± 23.8	48.8 (90.7% for protein N)	72.9 ± 24.2 (8.2 ± 8.7)	Cooper and Denmead, 1979
Poultry packing (USA)	ultrafiltration	COD 95.0	COD 131 ± 14	Total solids 85.0	Total solids 240 ± 42	N.R.	N.R.	86.0 (94.0 for protein N)	14 ± 2	Shih and Kozink, 1980

N.R. = not reported.

acid-formers may sometimes lower the pH to less than 6.5 which will inhibit methane-forming bacteria. This results in poor pollutant removal efficiency and in odour production. If the pH rises above 8.5, acid-formers will be suppressed and treatment efficiency will suffer. Thus, optimal conditions for anaerobic degradation are believed to require a pH between 7.0 and 8.5 and temperatures between 25° and 35°C. Treatment performance drops dramatically at lower temperatures. Two anaerobic processes will be considered, anaerobic lagoons and anaerobic contact processes.

Anaerobic lagoons. The most common application of anaerobic processes to meat, poultry and rendering wastewater treatment has been the anaerobic lagoon. Such systems consist of a sealed earthen basin, usually about 3 to 5 m deep, which receives the wastewater at a relatively high organic loading rate, thereby promoting anaerobic conditions throughout the lagoon depth. Given the temperature dependence of this process, the ratio of surface area to volume must be as low as possible to reduce heat loss. Detention times of 5 to 7 days are commonly employed with recommended organic loadings per unit of lagoon volume between 0.16 and 0.32 kg $BOD_5\,m^{-3}\,d^{-1}$. Shorter detention times can lead to system failure due to cell washout, while longer detention times result in an average temperature decrease for the cell which can reduce treatment efficiency. Such systems commonly achieve 85% BOD_5 removal, with some values greater than 95% reported. Operating data for several anaerobic lagoon installations described in the literature are summarized in Table 3.9.

The main problem associated with anaerobic lagoons is odour production, largely due to hydrogen sulphide and the high ammonia levels in the effluents. Odour can be controlled to some extent by ensuring adequate cover over the lagoon. Natural cover may develop where high FOG loads are applied, or may be encouraged by an initial discharge of paunch manure to the lagoon. Such covers also promote heat conservation, which is important to maintain wastewater temperatures during winter months. Artificial covers of synthetic rubber, PVC or styrofoam have been used and they may also allow the methane gas produced to be collected.

If the wastewater contains a high concentration of sulphates ($>100\,mg\,l^{-1}$), extreme problems from hydrogen sulphide may be expected and anaerobic lagoons will require additional odour control measures. Odour control by lime dosing has been reported as successful (Given *et al.*, 1974) since higher pH shifts the equilibrium from gaseous H_2S to HS^- and S^{2-}. Such pH control will also improve treatment efficiencies if the pH is outside the optimum range. Control of pH to within 7.5 to 8.2 has been found to improve treatment efficiencies as follows: BOD_5, 37.5% to 89.6%; suspended solids, 57% to 92.5%; FOG, 57% to 96.5%. Chlorination of raw wastewater, use of submerged inlets and outlets and maintenance of a complete lagoon cover have also been found useful for odour control.

Ammonia concentrations in effluents from anaerobic lagoons result from the

Table 3.9 Anaerobic lagoon performance

Plant type	Organic loading (kg BOD$_5$ m^{-3} d^{-1})	Hydraulic detention time (d)	Lagoon temperature (°C)	pH	BOD$_5$ Removal (%)	BOD$_5$ Effluent (mg l^{-1})	Suspended solids Removal (%)	Suspended solids Effluent (mg l^{-1})	FOG Removal (%)	FOG Effluent (mg l^{-1})	Ammonia in effluent (mg l^{-1})	Reference
Red meat	0.26	6.6	>22	6.5–7.0	58	460	77	130	N.R.	N.R.	64	Rollag and Dornbush, 1966
Red meat	0.18	4.6	N.R.	N.R.	65	280	N.R.	N.R.	N.R.	N.R.	N.R.	Sollo, 1960
Red meat	0.19	11	>10	N.R.	92	100	84	100	97	16	65–85	Witherow, 1973
Red meat	0.16	6.7–8.3	24	N.R.	93	122	N.R.	N.R.	N.R.	N.R.	N.R.	Hester and McLurg, 1970
Red meat	0.10	N.R.	N.R.	N.R.	86	190–260	71	220–280	88	60–150	N.R.	Saucier, 1971
Beef slaughter	0.50	3.5	21–29	6.0–7.8	87 (51–90)	300–470	67 (40–90)	300–600	N.R.	N.R.	N.R.	Enders et al., 1967
Hog packing	0.47	5	20	7	82	480	59	580	78	106	122	Baker and White, 1971
Hog slaughter	0.29	4.4	>23	N.R.	78	280	90	220	N.R.	N.R.	N.R.	Niles and Gordon, 1970
Hog slaughter	0.17–0.23	11–17	N.R.	7–7.3	56–72	640–720	44–78	440–600	42–71	220–300	156	Wymore and White, 1968
Hog processing	0.27	7.3	N.R.	N.R.	60	600	50	400	N.R.	N.R.	N.R.	Stanley Assoc., 1979
Hog processing	0.32	9.7	N.R.	N.R.	80	600	65	1100	N.R.	N.R.	N.R.	Stanley Assoc., 1979

N.R. = not reported.

decomposition of protein and are typically above $100\,\text{mg}\,l^{-1}$. Where the effluent is ultimately to be discharged to natural waters, ammonia must be removed and follow-up aerobic treatment by aerated lagoons or extended aeration is necessary. Further BOD_5 reductions following anaerobic lagoons are commonly required in such cases. Anaerobic lagoons have the advantage of relatively low capital cost, easy operation and maintenance, and the ability to handle large FOG and shock organic loads.

Anaerobic contact process. One version of an anaerobic contact process is essentially an anaerobic parallel of the aerobic activated sludge system (Rands and Cooper, 1966). This system differs from anaerobic lagoons in that it is a high-rate system requiring more equipment and process control.

Normally, an anaerobic contact system requires equalization tanks followed by digester tanks with mixers, air or vacuum gas-strippers, sedimentation tanks and sludge-return lines and pumps. Equalized wastewater flow is fed to the anaerobic digesters where it is mixed with recycled sludge flow. Digesters are maintained at 33° to 35°C and are loaded at 2.4 to 3.2 kg $BOD_5\,m^{-3}\,d^{-1}$ with an average detention time of 3 to 12 hours. The digester effluent passes through the vacuum or air gas-strippers. These remove the produced methane and carbon dioxide from the wastewater which flows into the clarifiers. Sludge is settled from the wastewater and recycled to the digester at a rate normally about 33% of the influent flow. Because of the low net synthesis of cell mass by anaerobic bacteria, sludge production is relatively low and typically only 2% of the total flow is wasted sludge.

This process, like anaerobic lagoons, has the disadvantage of ammonia production. However, odour production is generally not as great a problem because of the confined tanks used for digestion. Occasional odour problems are experienced with the clarifiers. The system has higher capital and operating costs than equivalent anaerobic lagoons, but smaller space requirements. It has the advantages that high organic waste load reduction is achieved with relatively short detention times; the methane produced can be collected and used to maintain higher digester temperatures; it is suitable for cold climate waste treatment; and it is reasonably stable under FOG and organic shock loads. Some operational data on anaerobic contact process installations are summarized in Table 3.10.

Despite several apparent advantages, this process has not been widely adopted by the industry. Another approach, the anaerobic filter (Mueller and Mancini, 1975), has been receiving increasing attention for other waste treatment applications and is undoubtedly applicable to the meat and poultry industry. This variation is a fixed-film contact process and it offers major advantages in terms of process stability.

Aerobic processes

Aerobic biological treatment processes are designed to operate with sufficient artificially supplied aeration to exceed the oxygen demand requirements of the

Table 3.10 Anaerobic contact process performance

Plant type	Organic loading (kg BOD$_5$m^{-3}d^{-1})	Sludge recycle (%)	Hydraulic detention time (h)	Temperature (°C)	BOD$_5$ Removal (%)	BOD$_5$ Effluent (mg l^{-1})	Suspended solids Removal (%)	Suspended solids Effluent (mg l^{-1})	FOG Removal (%)	FOG Effluent (mg l^{-1})	Ammonia in effluent (mg l^{-1})	Reference
Red meat	1.1–1.3	70–140	8.4–22	N.R.	94–96	43–140	0–80	39–571	N.R.	N.R.	105	Fullen, 1953
Red meat	2.1	250	3.6	22	91	100	59	160	N.R.	N.R.	N.R.	Schroepfer et al., 1955
Red meat	2.6	285	3.4	35	95	70	92	50	N.R.	N.R.	N.R.	Schroepfer et al., 1955
Red meat	0.18	N.R.	N.R.	10	97	24	82	40	81	24	N.R.	Rands and Cooper, 1966
Red meat	0.94	N.R.	N.R.	15	96	100	92	60	93	42	N.R.	Rands and Cooper, 1966

N.R. = not reported.

process organic loading. Consequently, the treatment reactor is maintained in a strictly aerobic condition. These processes may be further subcategorized into fixed-film and dispersed-growth systems; applicable processes in each category will be considered.

Fixed-film aerobic processes. Film processes involve the provision of an inert surface upon which a biological film or slime can grow. This slime is capable of absorbing organic matter upon contact with organically polluted waters and aeration is achieved by direct gas transfer from air to the exposed surface of the film. The organics are metabolized to carbon dioxide and other inorganic waste products, as well as being incorporated in new cell synthesis, resulting in overall film growth. As the film becomes thicker, the efficiency of oxygen transfer to the inner portions of the film is reduced. Eventually, anaerobic conditions may develop on the inner surface of the film where facultative anaerobic bacteria predominate. Increased film thickness also hampers the transfer of organics to inner layers, resulting in self-metabolism by the micro-organisms on the inner layers. This process weakens the bonding of the film to the inert media and the film will eventually slough off and be carried away with the wastewater. Sloughing is also promoted by shear stress from applied hydraulic loading. The two most common applications of biological film processes are the trickling filter and the rotating biological contactor.

(1) *Trickling filters.* Trickling filters were developed for the treatment of domestic sewage, originally, as beds of stone or slag media over which sewage is distributed and allowed to drain. This type of trickling filter is in wide use around the world for domestic sewage, but is not well suited to the treatment of high-strength organic wastes such as those from meat, poultry and rendering plants. More recently, plastic filter media, constructed of convoluted plastic sheets which assemble to form a honeycomb structure, have been developed. This lightweight medium allows the construction of very deep (6–12 m) compact trickling filters which are well suited to higher-strength organic wastes. These generally receive relatively high organic loading rates and are described as high-rate filters.

Trickling filters must be followed by clarifiers to settle the pieces of biological film which are continuously sloughing off. Effective pretreatment of wastewaters is essential for this process, to avoid solids-clogging of influent distributors and the filter medium itself.

Design criteria for synthetic media high-rate filters generally call for daily organic loading rates of 1 to 2 kg BOD_5 m^{-3} of filter volume. Hydraulic loading rates must generally be increased, with higher organic loading rates to encourage film attrition. Design surface loading rates range from 10 to 40 $m^3 m^{-2} d^{-1}$. Control of hydraulic loading independent of organic loading is achieved by recirculation of effluent. This generally ranges from 100 to 400% of the influent flow.

In general, trickling filters are relatively resistant to shock organic loads. However, very high FOG loads or solids loadings on filters should be avoided to prevent clogging. Tolerance of FOG levels fluctuating from 50 to 600 mg l^{-1}, pH varying from 6.6 to 9.8, and intermittent freezing was demonstrated by a plastic media roughing filter treating meat-packing wastes (Bisset and Drynan, 1973). In general, FOG loads should be kept below 300 to 500 mg l^{-1}.

Trickling filters receiving high organic loadings will produce relatively large quantities of biological solids for disposal. From 0.3 to 0.6 kg of solids are likely to be generated per kg of BOD_5 removed. BOD_5 removal is generally less than 75% for high-rate treatment. Higher overall BOD_5 removals (in excess of 90%) may be achieved by placing several filter units in series. Such staged treatment units are not commonly used for meat, poultry and rendering wastes because of the larger area requirements for latter stages of the filter if adequate final BOD_5 levels are to be achieved.

Trickling filters are suited to cold weather operation provided that measures are taken to prevent pipe freezing. Housing of the filter is recommended to ensure consistent operation in cold climates. Filters receiving light hydraulic loadings may develop problems with filter flies, but heavier hydraulic loading, which may be achieved by effluent recycle, will wash fly larvae off with the sloughed film before they can emerge as adult flies. Odours are not likely to be a serious problem provided that the filter is not organically overloaded, the biological film is wetted intermittently, and the sewage is treated before going septic.

Trickling filters receiving light organic loadings or partially treated wastes readily establish nitrifying bacteria in the lower stages of the filter. This phenomenon allows the use of trickling filters to nitrify ammonia present in effluents from anaerobic or activated sludge processes. One installation (Stracey, 1975) at a meat packing plant treats activated sludge effluent on a slag trickling filter loaded at 0.04 kg BOD_5 m^{-3} d^{-1} and reduces ammonia concentrations from 19 to 3.8 mg l^{-1}. Operating performance for a variety of trickling filter installations is summarized in Table 3.11.

Cost data for a hog processing and rendering operation which installed an anaerobic lagoon–trickling filter process are presented by Stanley Assoc. (1979). The plant in question slaughtered approximately 5000 hogs per day on an 8-hour day, 5-day week basis for an annual average production of 1.4×10^6 hogs. The treatment process consisted of dissolved air flotation, two anaerobic lagoons in parallel, followed by two-stage trickling filtration and chlorine disinfection. Process design and operating performance for the anaerobic logoons are summarized in Table 3.9 and for the trickling filters in Table 3.11.

The capital cost data for this facility were adjusted to 1977 Canadian dollars using the *Engineering News Record* construction cost index for November 1977 (ENR-cci = 2660): current capital cost may be estimated by using the correct

Table 3.11 Trickling filter performance

Plant type	Comments	Filter medium	Organic loading (kg BOD$_5$ m^{-3} d^{-1})	Hydraulic loading (m^3 m^{-2} d^{-1})	BOD$_5$ Removal (%)	BOD$_5$ Effluent (mg l^{-1})	Suspended solids Removal (%)	Suspended solids Effluent (mg l^{-1})	FOG Removal (%)	FOG Effluent (mg l^{-1})	Reference
Red meat	pilot scale recycle at 2:1	polystyrene	11	60	71	720	N.R.	N.R.	N.R.	N.R.	Garrison and Geppert, 1960
Red meat	recycle at 5:1	clay tile	2.1	19	37–74	17–269	N.R.	N.R.	39–100	0–53	Hirlinger and Gross, 1957
Hog packing	2-stage	PVC	1.2	37	74	124	80	108	60	33	Baker and White, 1971
Poultry	various treatment modes	plastic	1.8–3.2	N.R.	50–85	120–300	N.R.	N.R.	N.R.	N.R.	Summers, 1972
Hog processing	2-stage	PVC	4.2$^{(a)}$ 0.69$^{(b)}$	50–60$^{(a)}$ 25–30$^{(b)}$	>85	<70	>92	<80	N.R.	N.R.	Stanley Assoc., 1979

(a) First stage.
(b) Second stage.
N.R. = not reported.

ENR-cci. Cost data for the anaerobic lagoon/trickling filter facility may be summarized as follows (1977 Canadian dollars):

Capital cost (adjusted to 1977 ENR-cci = 2660) $1 100 000

Annual operating costs (excluding capital amortization)

Manpower at $10 h^{-1}	$20 800	
Electrical power at $0.0069 MJ^{-1}	3 700	
Maintenance allowance	15 000	$39 500

Cost per m^3 treated (excluding capital amortization)	$0.057
Cost per kg BOD$_5$ removed (excluding capital amortization)	$0.019

(2) *Rotating biological contactor.* The rotating biological contactor (RBC) is a relatively new process in North America, but this device and its predecessors have been successfully used for decades in Europe. Like the trickling filter, this process uses a biological film to absorb and metabolize organic matter from wastewaters. For an inert medium the RBC commonly uses rotating discs of lightweight plastic which are submerged up to 45% of their diameter. The biological film which develops on these discs absorbs and metabolizes organics while submerged and then is able freely to absorb oxygen from the atmosphere when exposed to the air. The basic concept of the system, that of moving the microbial biomass to the food and to the air supply, saves energy because the biomass weight is small in comparison with the total weight of wastewater treated. Moreover, the power requirement for aeration by bringing the biomass in contact with the air is limited to the torque required to turn the shaft supporting the lightweight plastic discs. Disc rotation performs the added function of mixing the tank contents. The RBC process is capable of achieving effluent quality comparable to that of activated sludge treatment but at approximately 50% of the energy demand.

As with the trickling filter, continued film growth eventually leads to sloughing of the film into the water, producing sludge at a rate of 0.5 to 1.2 kg per kg BOD$_5$ removed. Thus, sedimentation is required following RBC units to collect the pieces of sloughed biomass.

The modular nature of the rotating shaft and discs allows several stages of treatment to be conducted in series as required by effluent specifications. Each stage of units can develop its own particular microbial population which is suitable to the waste at that stage. Provision of stages beyond the point of heavy organic oxygen demand can be used to achieve nitrification of ammonia.

Design criteria for daily organic loading range from 0.005 to 0.04 kg BOD$_5$ per m^2 of disc area, while daily hydraulic loading is generally 0.02 to 0.04 m^3 per m^2 of disc area. Discs are mounted together on a horizontal shaft to comprise a stage and designs usually call for three to six such stages in series. Disc rotation provides film aeration as well as tank mixing and this is usually achieved with peripheral disc velocities of 10 to 25 m per min.

Experience of the RBC process for treating meat, poultry and rendering wastes is limited. However, some pilot-scale operating data on wastewaters from integrated meat products plants are summarized in Table 3.12.

In general, RBC units are relatively compact and modular in nature. They are relatively easy to operate, have low power and maintenance requirements, and are relatively resistant to upset by shock organic loads. Their simplicity makes RBC units particularly attractive to smaller operations. However, they should be protected from excessive FOG loadings by providing primary FOG removal and they require shelter to protect them from freezing during winter operation.

Dispersed growth aerobic processes. This class of aerobic processes involves those in which the microbial population performing treatment is kept suspended within the liquid medium during treatment rather than being attached to a fixed surface. The relevant processes to be considered here are aerated lagoons and variations on the activated sludge process.

(1) *Activated sludge.* The term 'activated sludge' covers a wide variety of aerobic processes with certain common features. In all cases, the wastewater is exposed to vigorous aeration which encourages rapid assimilation of organic matter by micro-organisms, resulting in overall microbial population growth. The microbial cell growth (sludge) is settled from the wastewater and a fraction is recycled to be mixed with incoming wastewater entering the aeration chamber. This forms the so-called 'mixed liquor' which undergoes the vigorous aeration described above. The name 'activated sludge' is derived from the recycled sludge, since it is 'activated' in the sense of having a large microbial population which exhibits a high capacity for the assimilation of organic matter. Mixed liquor suspended solids (MLSS) concentration or volatile MLSS is normally used to represent biomass in activated sludge systems.

Many process modifications have been developed from the original activated sludge process. Some of these modifications involve plug flow aeration basins, in which the waste is added in increments along the length of the basin (step aeration) or the aeration capacity is reduced towards the end of the basin (tapered aeration). In general, plug flow systems are not well suited to meat, poultry or rendering wastes unless flow and waste strength equalization is provided. Alternatively, complete mix systems rapidly dilute the incoming wastewater throughout the entire aeration basin volume and, thereby, provide such systems with better shock load resistance. However, even complete mix systems should be provided with waste equalization, particularly for processes using relatively short detention times.

Apart from selection of flow regime (complete mix or plug flow) or aeration distribution, the major activated sludge process variables are the sludge age (or mean cell residence time) and the ratio of food to micro-organisms (or specific utilization rate). The sludge age represents the average time a given microbial

Table 3.12 Rotating biological contactor performance

Plant type	Comments	Organic loading (kg BOD$_5$ m^{-2} d^{-1})	Hydraulic loading (m^3 m^{-2} d^{-1})	BOD$_5$ Removal (%)	BOD$_5$ Effluent (mg l^{-1})	Suspended solids Removal (%)	Suspended solids Effluent (mg l^{-1})	Ammonia Removal (%)	Ammonia Effluent (mg l^{-1})	Reference
Red meat products	3-stage pilot	0.009 (0.008–0.011)	0.06	83 (74–92)	27 (12–40)	N.R.	N.R.	N.R.	N.R.	Chittenden and Wells, 1971
Red meat products	4-stage pilot	0.010 (0.005–0.02)	0.015	95	26	97	20	20	54	Johnson and Krill, 1976
Red meat products	4-stage pilot	0.014 (0.007–0.028)	0.021	93	46	90	40	N.R.	30	Johnson and Krill, 1976
Red meat products	4-stage pilot	0.041 (0.020–0.08)	0.063	83	112	88	55	N.R.	N.R.	Johnson and Krill, 1976
Red meat products	4-stage pilot	0.082 (0.040–0.16)	0.126	64	260	74	100	N.R.	N.R.	Johnson and Krill, 1976

N.R. = not reported.

cell may be expected to remain in the system before being wasted from the system as excess sludge. This parameter may be estimated on the basis of the total microbial mass in the system divided by the rate of sludge wastage from the system. The ratio of food to micro-organisms is the total organic substrate applied to the system, divided by the total process microbial mass (kg BOD_5 per kg MLSS). Specific utilization rate is similar, but is derived as the total substrate removed divided by the total microbial mass. The range of values chosen for these parameters determines the organic loading, hydraulic detention time and/or sludge recycle rates to be used by the process. The two types of activated sludge process which have been reported in use for the treatment of meat, poultry and rendering wastewaters may be differentiated on the basis of these parameters.

(a) *Conventional activated sludge.* Conventional activated sludge may comprise either a complete mix or a plug flow system, although complete mix systems are likely to be preferable for these industries. In such systems, hydraulic detention times are typically 4 to 8 hours, sludge is recycled at 25 to 50% of the incoming flow, the sludge age is 5 to 15 days and ratios of food to micro-organisms are 0.2 to 0.4 kg BOD_5 per kg MLSS. Operation within these parameters normally maintains a microbial population in a declining growth phase where food is limiting. Under such circumstances, the rate of BOD assimilation per cell is less than would be possible at the maximum metabolic rate with unlimited food supply. However, heavy competition for food by relatively large numbers of cells ensures good overall BOD assimilation. Moreover, micro-organisms in this growth stage tend to form a settleable biological floc which allows settling in a secondary clarifier within reasonable detention times. Generally, the conventional activated sludge process does not perform a significant degree of nitrification of ammonia because of the relatively short sludge ages involved.

The conventional activated sludge process has not been widely used in the meat, poultry and rendering industries because of its relative sensitivity to upset by shock loadings, need for well trained and skilled operators, and its relatively high rate of sludge production (0.4 to 0.7 kg per kg of BOD_5 removed) requiring expensive sludge treatment and handling facilities.

Only two reported uses of conventional activated sludge were located in the literature. Both are located in England and their operating performance is summarized in Table 3.13. The second plant was designed to operate as an extended aeration plant for the purposes of nitrifying the effluent, but sludge settling problems forced operation into a more conventional activated sludge mode and nitrification is now performed by a nitrifying trickling filter (Stracey, 1975).

(b) *Extended aeration activated sludge.* The extended aeration modification of the activated sludge process is usually designed as a complete mix system.

Table 3.13 Conventional activated sludge process performance

Plant type	Process organic loading (F/M) (kg BOD$_5$ kg^{-1} MLSS d^{-1})	Volumetric organic loading (kg m^{-3} d^{-1})	Hydraulic detention time (h)	BOD$_5$ Removal (%)	BOD$_5$ Effluent (mg l^{-1})	Suspended solids Removal (%)	Suspended solids Effluent (mg l^{-1})	FOG Removal (%)	FOG Effluent (mg l^{-1})	Ammonia Effluent (mg l^{-1})	Reference
Red meat products	0.6	3.0	3.8	97	21	95	19	N.R.	N.R.	19	Stracey, 1975
Poultry products	N.R.	0.48	N.R.	99	19	94	21	N.R.	N.R.	N.R.	Dart, 1974

N.R. = not reported.

Hydraulic detention times generally range from 18 to 36 hours for large installations, sludge is recycled at 75 to 200% of the incoming flow, sludge age is maintained at 10 to 30 days and ratios of food to micro-organisms are 0.02 to 0.20 kg BOD_5 per kg MLSS. Corresponding MLSS concentrations range from 3000 to 8000 mg l^{-1}. Operation within these parameter values normally maintains the bacterial population in the endogenous growth phase where food is limiting. Under such conditions, a certain degree of endogenous respiration occurs whereby the large microbial cell mass which is present begins to feed upon itself. This has the advantage of reducing the quantities of excess sludge for disposal. The advantage is offset by the possibility of the microbial population tending to disperse rather than flocculate under these conditions. Such an occurrence may lead to difficulties in settling of sludge in the secondary clarifier.

Extended aeration is capable of performing a significant degree of ammonia nitrification. Autotrophic bacteria responsible for this transformation do not normally compete well with the heterotrophic bacteria involved in the assimilation and oxidation of organic matter and, thus, have difficulty in establishing themselves in conventional activated sludge systems. However, extended aeration with long mean cell residence times, normally greater than 10 days, can be expected to perform nitrification (EPA, 1973b). In designing the aeration system for extended aeration, allowance must be made for the oxygen demand of the nitrification process (4.3 kg O_2 per kg NH_3), in addition to the normal aeration requirements of the process (1.5 kg O_2 per kg BOD_5). Typical combined oxygen requirements are in the range from 2.0 to 2.3 kg O_2 per kg BOD_5 applied. The ability to nitrify ammonia is a major factor favouring extended aeration over other biological treatment systems, since ammonia is likely to be the major concern relative to fish toxicity for effluents from these industries.

In contrast to the conventional activated sludge process, extended aeration has been widely accepted by the meat, poultry and rendering industries. Long hydraulic and mean cell residence times, high levels of microbial mass in the system and complete mix characteristics provide extended aeration with relatively good resistance to shock loads. However, these systems can be upset by organic loadings which exceed the aeration capacity of the system or by shock loadings of FOG. In particular, emulsified FOG, such as occurs in tankwater from low-temperature wet rendering processes, can adversely affect final effluent clarity (Hrudey, 1981a). In such circumstances, heavy discharges of sludge with the effluent may occur.

Hrudey (1981a) reports the effects of emulsified lipid loading upon activated sludge performance and concludes that freedom from FOG-induced upset can be assured for daily loadings of less than 0.10 kg lipid per kg of MLSS. Problems with effluent quality were consistently experienced in bench-scale activated sludge units loaded in excess of 0.25 kg lipid per kg of MLSS. Although lipids were shown to be capable of upsetting process performance at

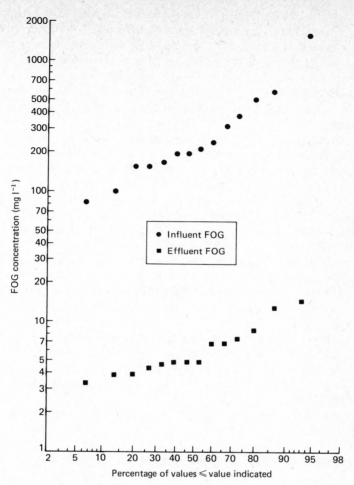

Fig. 3.8 Removal of FOG by extended aeration activated sludge plant

overload conditions, excellent FOG removal and biodegradability were confirmed. An indication of the ability of extended aeration activated sludge to remove FOG is provided in Fig. 3.8, which presents data (Hrudey, 1980) from a full-scale extended aeration activated sludge facility at a red meat products plant. It has also been found that lipid content does not adversely affect sludge settleability provided that adequate mixing and loading conditions are maintained (Hrudey, 1982). If the foregoing provisos are not satisfied, a serious lipid scum problem can arise in secondary clarification. Consequently, equalization should be provided for any emulsified FOG streams to allow them to be metered into the system gradually, rather than being discharged in slugs, for instance when batch cookers are drained.

Generally, extended aeration accepts hydraulic and organic fluctuations due to overnight and weekend process shutdown. However, in very cold climate operation, experience has shown the need for heat addition over weekends (Paulson *et al.*, 1974). This requirement has been met by discharge of condenser water to the system. Winter operation of mechanical aerators requires measures to prevent ice damage. This can usually be achieved by means of splash guards to prevent ice formation on the aerator drive mechanisms.

One variation of the extended aeration process which has found widespread application in the meat and poultry industries is the oxidation ditch. Although this process functions as extended aeration from a process viewpoint, the reactor is in fact an oval endless channel. Mixed liquor is moved around the oval and is aerated by the action of a rotor or brush aerator. A more recent adaptation is a draft tube aerator which employs a down-draft impeller aerator combined with a U-tube to maximize oxygen transfer efficiency (Gruette and Westphal, 1980). Process design considerations for the oxidation ditch are similar to those for extended aeration in general, with the additional requirement to specify a minimum horizontal velocity in the ditch of 30 cm s^{-1}.

Extended aeration offers lower capital and operating costs for sludge disposal because of the smaller quantities of sludge produced. This saving is offset by the larger aeration basin and aeration capacity requirements of extended aeration relative to conventional activated sludge. The process is somewhat easier to operate than the conventional activated sludge system, although it may require more skilled operation than biological film processes.

Operating performance data on several extended aeration plants are summarized in Table 3.14.

Cost data for the extended aeration process have been evaluated for several plants by Stanley Assoc. (1979). Two specific cases are presented for comparison.

Plant A is a poultry processing plant handling an average of 18 000 birds per day on an 8-hour day, 5-day week basis for an annual average production of 4.7×10^6 birds. The treatment facility consists of screening and extended aeration activated sludge followed by chemical coagulation, clarification and filtration. Process design and performance are summarized in Table 3.14.

Plant B is a poultry processing plant handling an average of 38 000 birds per day on an 8-hour day, 5-day week basis for an annual average production of 10×10^6 birds. The treatment facility consists of dissolved air flotation, vibrating screens and extended aeration, followed by a polishing lagoon and chlorine disinfection. Process design and performance for the extended aeration portion are summarized in Table 3.14.

The capital cost data for these facilities were adjusted to 1977 Canadian dollars using the *Engineering News Record* construction cost index for November 1977 (ENR-cci = 2660). Cost data for these two plants may be summarized as follows (1977 Canadian dollars):

Table 3.14 Extended aeration activated sludge process performance

Plant type	Process organic loading (F/M) (kg BOD$_5$ kg^{-1} MLSS d^{-1})	Sludge age (d)	Mixed liquor suspended solids conc. (mg l^{-1})	Volumetric organic loading (kg BOD$_5$ m^{-3})	Hydraulic detention time (h)	BOD$_5$ Removal (%)	BOD$_5$ Effluent (mg l^{-1})	Suspended solids Removal (%)	Suspended solids Effluent (mg l^{-1})	FOG Removal (%)	FOG Effluent (mg l^{-1})	Ammonia Effluent (mg l^{-1})	Reference
Red meat products	N.R.	N.R.	N.R.	0.32	30	95	N.R.	84	N.R.	98	N.R.	N.R.	Willoughby and Patton, 1968
Red meat products	0.25	N.R.	3000	N.R.	N.R.	84	46	80	37	N.R.	N.R.	N.R.	Hrudey, 1980
Red meat products	0.17–0.30	8–10	1000–1600	0.26–0.45	48–84	92	75	81	120	94	18	31	Paulson et al., 1971
Red meat products	N.R.	N.R.	N.R.	0.81–1.20	33	84 (52–94)	150 (67–425)	71 (16–93)	78 (32–266)	97 (92–99)	7 (3–13)	16	Hrudey, 1980
Red meat products	0.06	N.R.	5500	N.R.	N.R.	99	<20	N.R.	<20	N.R.	N.R.	N.R.	Green et al., 1980
Poultry products (A)	0.05	N.R.	4000–5000	0.24	72	>95	<30	>90	<40	N.R.	N.R.	N.R.	Stanley Assoc., 1979
Poultry products (B)	0.02–0.05	N.R.	3800–5200	0.16–0.32	132	>96	<30	>85	<75	N.R.	N.R.	N.R.	Stanley Assoc., 1979

N.R. = not reported.

	Plant A		Plant B	
Capital cost (adjusted to 1977 ENR-cci = 2660)	$400 000		$780 000	
Annual operating costs (excluding capital amortization)				
Manpower at $10 h^{-1}	$10 400		$20 800	
Electrical power at $0.0069 MJ^{-1}	8 200	$18 600	30 000	$50 000
Cost per m^3 treated (excluding capital amortization)		$0.223		$0.098
Cost per kg BOD$_5$ removed (excluding capital amortization)		$0.392		$0.124

(c) *Two-stage activated sludge.* This process is essentially two complete activated sludge systems operated in series. Thus, each stage has its own clarifier and sludge return, but the overall system offers the additional flexibility of intermixing sludge return between the two units. The two-stage concept is intended to take advantage of the ability of activated sludge to effect rapid BOD_5 reduction, even under relatively high organic loading conditions, while using a second stage to upgrade the fluctuating effluent quality which will be produced from the heavily loaded first stage. A summary of design criteria for the two-stage process is provided in Table 3.15 (Jank and Guo, 1978).

The concept of two-stage activated sludge for wastewaters with high FOG is supported by the findings of Mulligan and Sheridan (1975) and Hrudey (1980, 1981b). These indicate a high capacity of activated sludge for lipid removal, although corresponding effluent BOD_5 and suspended solids levels may become unacceptable for effluent discharge. Given a second stage of activated sludge to upgrade effluent from the first stage, acceptable effluent quality should be readily achievable.

(d) *Aerated lagoons.* Aerated lagoons are usually earthen basins similar to those used for anaerobic lagoons, but are supplied with artificial aeration from mechanical aerators or diffused air systems.

The process is theoretically a complete mix, continuous flow biological system without sludge recycle. In practice, few aerated lagoons can satisfy the complete mix model and significant plug flow inevitably occurs. Aerated lagoons are generally 2.5 to 5 m in depth and designed to provide 2 to 10 days hydraulic detention time. Poor sludge settleability conditions are common with aerated lagoons and result in relatively high suspended solids in the final effluent despite reasonable soluble BOD_5 reduction. Additional detention time of up to 40 days has been used to improve effluent quality, but this is usually achieved by provision of aerobic (non-aerated) ponds.

Organic loading to aerated lagoons is usually specified in the range of 0.03 to 0.3 kg BOD_5 m^{-3} d^{-1}. In addition to loading requirements, aerated lagoon

Table 3.15 Design criteria for two-stage activated sludge process (after Jank and Guo, 1978)

	First stage	Second stage
Organic loading rate (kg BOD_5 kg^{-1} MLSS d^{-1})	0.6–2.0	0.1–0.5
Volumetric loading rate (kg BOD_5 m^{-3} d^{-1})	1.8–7.0	0.4–1.8
MLSS concentration ($mg l^{-1}$)	4000–8000	2000–4000
Sludge return ratio (% of process influent by volume)	100–500	50–100
Oxygen requirements (kg O_2 kg^{-1} BOD_5)	1.0–1.5	1.5–2.0
Sludge production rate (kg kg^{-1} BOD_5 removed)	0.7–1.4	0.4–0.7

design depends upon oxygen supply and mixing requirements. Since both are supplied by the aerators, their sizing must be based upon the larger of the two demands. Design criteria for these parameters must be viewed cautiously since oxygen requirements will depend upon influent BOD, endogenous respiration of the micro-organisms, and nitrification of ammonia, as well as a benthal oxygen demand from solids settling out in the lagoon. Specific air requirements will also depend upon the oxygen transfer efficiency of the aerators selected. However, in general, the mixing requirements determine the necessary aeration capacity. Recommendations here range widely from 2.8 kW per 1000 m^3 to 26 kW per 1000 m^3 (Jank and Guo, 1978). The wide range reflects variations in lagoon solids, lagoon geometry and flow regime, and the mixing efficiency of the aerators.

In some cases, aerated lagoons may be operated to provide mixing and aeration in the centre of the lagoon while allowing sludge to settle along the edges. In such cases, the lagoon functions as a facultative lagoon and anaerobic sludge digestion may proceed where the sludge is deposited. Considerably lower power inputs are required for this mode of operation, ranging from 0.5 to 5.2 kW per 1000 m^3 (Jank and Guo, 1978).

Aerated lagoons have been operated successfully on domestic sewage and pulp and paper wastes in cold climates, but surface aerators must be designed to deflect water spray to avoid icing of the aerator mechanism. Sludge recycle during winter months may help maintain treatment performance which will otherwise suffer from the lower biological reaction rates at lower temperatures.

Aerated lagoons are relatively inexpensive in initial capital investment compared with mechanical treatment systems or larger area facultative lagoons, but operating costs due to the power requirements for aeration may prove higher than for other treatment systems.

Although there are reported to be several lagoon installations for meat, poultry and rendering plants (EPA, 1974a, 1974b, 1975a), performance data for only one aerated lagoon installation have been located in the literature (Clise, 1974). This installation was at a poultry processing plant and consisted of a two cell aerated lagoon 1.8 m in depth, with 12 days detention time and organic loading of 0.025 kg BOD_5 m^{-3} d^{-1}. Average pollutant removals of 93%

for BOD_5 (31 mg l^{-1} in effluent), 88% for suspended solids (103 mg l^{-1} in effluent) and 94% for FOG (24 mg l^{-1} in effluent) are reported.

3.4.5 Disinfection

Raw wastewaters from meat, poultry and rendering operations, as noted earlier, may contain significant numbers or various pathogenic micro-organisms. *Brucella, Salmonella* and *Shigella* bacteria have been reported isolated from the catch basin effluent at a meat packing plant (Tarquin *et al.*, 1974). Sampling at several meat and poultry plants indicated the common occurrence of various *Salmonella* serotypes (Vanderpost and Bell, 1976). Studies on oxidation ditches treating municipal sewage containing poultry or hog slaughter effluents indicated the consistent presence of various *Salmonella* serotypes (Kampelmacher and van Noorle Jansen, 1973).

In addition to bacterial pathogens, parasite eggs, amoebic cysts and pathogenic viruses must be removed from wastewaters in order to accomplish disinfection. In the case of meat, poultry and rendering wastewaters, these other pathogenic organisms are not normally a concern with the possible exception of tapeworm eggs.

Disinfection of wastewaters requires that the number of pathogenic organisms be sufficiently reduced or eliminated so that an infective dosage of the organisms cannot develop under environmental conditions. This is distinct from sterilization, whereby all micro-organisms, pathogenic or otherwise, are eliminated. Bacterial disinfection efficiency is commonly measured by monitoring the class of intestinal bacteria referred to as the 'coliforms' or 'faecal coliforms' (when modified culture techniques are used).

The ultimate receiving medium for wastewaters must be considered in determining the requirements for disinfection. Where effluent is to be used to irrigate human or animal feed crops, very high standards of disinfection should be sought because of the risk of pathogen survival and concentration in the soil and on plants. Reported survival times for several pathogens in soil and on plants are provided in Table 3.16 (Bryan, 1974).

Wastewater disinfection may be achieved by physical or chemical means. Various stages of wastewater treatment processes will accomplish varying degrees of disinfection and some results are summarized in Table 3.17 (Bryan, 1974). The most common means employed for wastewater disinfection are chemical in nature, with chlorine compounds or ozone as the usual disinfecting agents.

Chlorination

Chlorine may be applied to wastewaters as chlorine gas or as hypochlorite salts (sodium or calcium hypochlorite). In water, chlorine hydrolyzes to produce hypochlorous acid (HOCl) which is the most effective disinfecting agent resulting from the hydrolysis of chlorine. Hypochlorite ion (OCl$^-$), which

Table 3.16 Survival of enteric pathogens in soil and on food (after Bryan, 1974)

Pathogen	Medium	Survival (days)
Salmonellae	various soils	35->280
	tomatoes	3
	soil and potatoes	>40
Streptococcus faecalis	various soils	26-77
Shigella sonnei	apple (skin)	8
	clams and shrimp	>60
Vibrio cholerae	vegetables	5->14
Endamoeba histolytica	lettuce and tomatoes	<3
	moist soil	6-8
Ascaris ova	lettuce and tomatoes	27-35
	soil	2190

Table 3.17 Survival of pathogens through wastewater treatment processes (after Bryan, 1974)

Treatment process	Survival (%)		
	Salmonellae	Endamoeba histolytica	Tapeworm eggs
Sedimentation	<50	>50	—
Activated sludge	<10	>50	—
Trickling filter	<50	<50	>50
Anaerobic digestion	<10	limited survival	<10
Disinfection	no survival	>50	>50

results from the ionization of hypochlorous acid, exhibits only a small fraction of the disinfecting power of hypochlorous acid. The proportion of the latter which ionizes to hypochlorite ion is pH-dependent. At higher pH values (above pH 7), an increasingly higher proportion of hypochlorite ion is produced with a resultant decrease in disinfection efficiency.

The disinfection efficiency of chlorine compounds is derived from their oxidative powers which are believed to inactivate bacterial enzymes, thereby incapacitating bacterial cells. In addition to the direct reaction of chlorine with water and micro-organisms, chlorine compounds will react with reducing compounds such as sulphides, ferrous ions, manganous ions and various organic compounds. The presence of such compounds in the effluent to be chlorinated represents an unproductive chlorine demand which will increase the chlorine dosage requirements for a given level of disinfection efficiency. Chlorine also readily reacts with ammonia to produce chloramines. Monochloramine and dichloramine retain a fraction of the disinfecting power of hypochlorous acid, but these compounds are somewhat more stable than hypochlorous acid and comprise a so-called 'combined chlorine residual'. As with reducing agents, the presence of ammonia in effluents will require a large chlorine dosage to maintain a given level of disinfection efficiency.

Apart from pH effects and interfering side-reactions, chlorine disinfection efficiency is primarily a function of chlorine dosage and contact time. Chlorine contact basins should normally be designed as baffled, plug flow basins to ensure adequate contact time. Total coliform reductions from tens of millions per 100 ml down to hundreds per 100 ml (10^{-5} reduction) have been reported for meat packing effluents (Baker and White, 1971) using a chlorine dosage of approximately 10 mg l^{-1} and 50 min contact time.

Dearborn (1978) performed laboratory and pilot-scale disinfection of poultry processing wastewaters. They observed completely successful control of *Salmonella* with an applied chlorine dosage of 10 mg l^{-1} and 45 min contact time.

Unfortunately, chlorine compounds and chloramines are extremely toxic to fish and other aquatic organisms. Chlorine residuals as low as 0.006 mg l^{-1} have been reported as lethal to trout fry (Brungs, 1973). As a result, chlorination of well treated effluent which has been detoxified is likely to make the effluent very toxic. Thus, dechlorination is normally required to make chlorinated effluents acceptable for discharge to natural receiving waters.

Dechlorination may be achieved by effluent detention, chemical reduction or physical adsorption. In detention, the effluent is simply stored in a holding lagoon for a period of days in order to allow a natural reduction in the chlorine residual. Chemical procedures require effluent treatment with reducing agents such as sulphur dioxide, sodium thiosulphate, sodium bisulphite or sodium sulphite. Sulphur dioxide is relatively inexpensive and very effective at removing residual chlorine, but close control is required to avoid overdosage which may depress the effluent pH. Carbon adsorption may also be used to remove chlorine residual. This method achieves a degree of effluent polishing at the same time, by removing trace organics.

Dearborn (1978) calculated capital and operating costs for a 4560 m^3 per day chlorination/dechlorination plant as follows (1977 Canadian dollars):

Capital cost
Baffled contact chamber, 142 m^3	$33 000	
Chlorine dosage system	5 000	
Sulphur dioxide dosage system	5 000	$43 000

Operating costs (excluding capital amortization)
Chlorine, 45 kg d^{-1} at 0.44 kg^{-1}	$20.00 per day
Sulphur dioxide, 23 kg d^{-1} at 0.73 kg^{-1}	$16.50 per day
Chemical costs	$36.50 per day
	$ 0.008 per m^3

Ozonation
Ozone, O_3, is a very powerful oxidizing agent which disinfects probably by enzyme oxidation mechanisms similar to those achieved by chlorine. The main difference is that ozone does not apparently form residual compounds such as the chloramines. Ozone decomposes to only oxygen as a final product, although

it creates partially oxidized organic reaction products. Ozone itself is very toxic, but residual ozone is very unstable and toxic concentrations in ozone-disinfected organic effluents are unlikely to occur.

Unlike chlorine, ozone cannot be manufactured as a relatively pure gas and bottled. It must be generated on-site by means of corona discharge. This incurs high operating costs, relative to chlorine disinfection, because of the continuing energy requirements of the ozone generator.

Dearborn (1978) applied ozone at a dosage of 30 mg l^{-1} with a contact time of 60 min and achieved only 30% frequency of *Salmonella* control in pilot-scale evaluations at a poultry processing plant. They calculated capital and operating costs for a 4560 m^3 per day ozonation plant as follows (1977 Canadian dollars):

Capital cost
Air preparation unit	$ 70 000	
Ozone generator up to max 136 kg O$_3$ d^{-1}	140 000	
Contact system total vol. 190 m^3	50 000	$260 000

Operating costs (excluding capital amortization)
Power, 11 000 MJ d^{-1} at $0.006 MJ^{-1}	$60.00 per day
	$ 0.01 per m^3

These figures are considerably higher than those for chlorination/dechlorination and, combined with the reported superior disinfection performance for chlorination, they tend to favour that process over ozonation for disinfection of poultry plant wastewaters.

3.4.6 Land treatment

Application of meat, poultry and rendering industry wastewaters to land is considered a major category of treatment process distinct from those described above in that in some cases it combines their achievements with the provision of final effluent disposal. Specifically, land treatment processes are capable of achieving suspended matter, nutrient and soluble organic carbon removal as well as providing crop irrigation and fertilization, groundwater recharge and/or wastewater reclamation for reuse.

Wastewaters from the meat, poultry and rendering industries are particularly amenable to land treatment options for a variety of reasons including the following.

- Many plants are located in rural/agricultural settings.
- Wastewaters from these industries are known to be nontoxic and readily biodegradable.
- Land treatment systems can tolerate fluctuations in applied loading more readily than conventional processes.

Detailed design manuals have been prepared for land treatment processes in general (EPA, 1977) and specifically for food processing wastewaters (Stanley

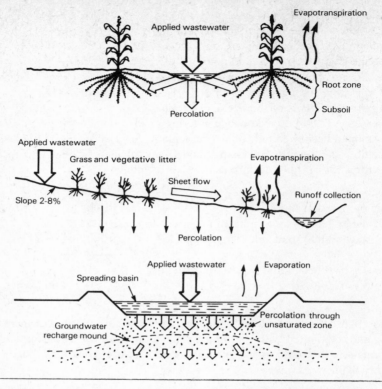

Fig. 3.9 Land treatment alternatives (after EPA, 1977; Stanley Assoc., 1978):
(a) slow-rate irrigation (b) overland flow–runoff collection (c) high rate percolation–infiltration

Assoc., 1978). Given these excellent references for process design, this discussion will simply focus on some of the major considerations concerning adoption of land treatment.

The major classes of land treatment system which are applicable to the industries under consideration are:

1. slow rate irrigation;
2. overland flow–runoff collection; and
3. high rate percolation–infiltration.

These process alternatives are schematically depicted in Fig. 3.9 (EPA, 1977; Stanley Assoc., 1978). The first process involves daily application rates of 2 to 15 mm and is intended to provide the irrigation water requirements for crop production. The second involves daily application rates of 6 to 18 mm and is intended to utilize the adsorption–biodegradation capability of the surface plant cover and soil layer to upgrade the quality of the collected runoff. The

Table 3.18 Capability of land treatment performance alternatives (after EPA, 1977; Stanley Assoc., 1978)

	Application method		
Land application objective	Slow rate[a] irrigation	Overland flow[b] runoff collection	High rate percolation[c] – infiltration
Recovery of renovated water[d]	0–70% recovery	50–80% recovery	≤90% recovery
Advanced treatment (typical average effluent concentration)			
BOD_5 removal	98 + % ($<2\,mg\,l^{-1}$)	92 + % ($10\,mg\,l^{-1}$)	85–99% ($2\,mg\,l^{-1}$)
Suspended solids removal	98 + % ($<1\,mg\,l^{-1}$)	92 + % ($10\,mg\,l^{-1}$)	85–99% ($2\,mg\,l^{-1}$)
Nitrogen removal	85 + %[e] ($3\,mg\,l^{-1}$)	70–90% ($3\,mg\,l^{-1}$)	0–50% ($10\,mg\,l^{-1}$)
Phosphorus removal	80–99% ($<0.1\,mg\,l^{-1}$)	40–80% ($4\,mg\,l^{-1}$)	60–95% ($1\,mg\,l^{-1}$)
To grow crops for sale	Excellent	Fair	Poor
Direct recycle to the land	Complete	Partial	Complete
Groundwater recharge	0–70%	0–10%	≤97%
Use in cold climates	Fair[f]	—[g]	Excellent

[a] Percolation of secondary effluent through 1.5 m of soil.
[b] Runoff of comminuted municipal wastewater over about 45 m of slope.
[c] Percolation of secondary effluent through 4.5 m of soil.
[d] Percentage of applied water recovered depends upon recovery technique and climate.
[e] Dependent upon crop uptake.
[f] Conflicting data—woods irrigation acceptable, cropland irrigation marginal.
[g] Insufficient data.

third process involves the above capabilities combined with that of the underlying soil column to upgrade the wastewater prior to groundwater recharge. Performance capabilities of these systems are summarized in Table 3.18 (EPA, 1977; Stanley Assoc., 1978).

Implementation of land treatment systems requires consideration of the characteristics of the local climate, the soil–plant system and the wastewater itself.

Climatic considerations include precipitation, temperature and wind. Precipitation characteristics will affect hydraulic loading rates, storage requirements and drainage system requirements. Temperature will affect organic loading rates, seasonal application patterns and type of application systems employed. Wind will be relevant predominantly for spray irrigation systems as well as for odour transmission from any of the systems. If wastewaters other than well treated secondary effluents are to be discharged to irrigation, buffer zones between the irrigation plots and residential areas must be provided to prevent possible nuisance conditions.

Most irrigation systems in cold climates must provide for winter storage of wastewater since irrigation is normally limited to the growing season. As a result, the combined land requirements for wastewater storage and irrigation plots can be significant.

Soil and geological characteristics which must be evaluated include permeability, infiltration rate, available water capacity, cation exchange capacity, location of groundwater table and groundwater flow regime. Relevant crop characteristics include water requirements, nutrient requirements, economic value and specific crop management requirements. Detailed analysis of these requirements is provided in the process design manuals already mentioned (EPA, 1977; Stanley Assoc., 1978).

A variety of effluent characteristics relevant to meat, poultry and rendering industry wastewaters must be considered in designing land application systems. These will be considered in turn.

Organic carbon content (BOD, COD)

The controlling factor concerning the organic content of wastewaters applied to land is their rate of degradation and oxygen demand in the soil. If organically overloaded with highly degradable materials, soil can become clogged and will become anaerobic. Decomposition under anaerobic conditions is much less effective than aerobic processes and anaerobic soil conditions are not suitable for crop production. Site-specific application rates must be determined experimentally, but wastewaters with BOD_5 in the range from 200 to 400 mg l^{-1} have been successfully applied. Overland flow systems have been loaded at daily rates of 40 to 100 kg BOD_5 ha^{-1} and infiltration–percolation systems at daily rates of up to 660 kg BOD_5 ha^{-1} (EPA, 1975b). Operation for one day followed by a few days of layoff has been successfully used to regulate application rates.

Suspended matter

The primary problems encountered with suspended matter in meat, poultry and rendering wastewaters are associated with clogging of distribution equipment. Clogging of soil pores can become a problem where there is overloading, but suspended solids concentrations in the range of 200 to 3000 mg l^{-1} have been successfully applied. Solids loadings at the higher end of the range may require rest and reaeration periods as well as periodic tillage to restore soil permeability.

Fats, oils and greases (FOG)

In the specific case of FOG, distribution equipment and soil structure clogging also represent the major concerns. Unfortunately, particles of lipid degrade very slowly because it is difficult for bacteria to attack hydrophobic globules. Consequently, rest periods for FOG degradation will be less effective than they are for decomposition of more readily degradable organic solids.

Total dissolved solids (TDS) and sodium absorption ratio (SAR)

For meat, poultry and rendering wastewaters, the TDS are predominantly inorganic in nature and, because of their ionic character, may be represented by specific conductance. Their relevance to land application systems lies mainly in possible salinity damage to crops and/or soil structure and degradation of groundwater. In general, TDS levels lower than 500 mg l^{-1} (specific conductance <75 mS m^{-1}) pose no problem, while values in the range of 500 to 2000 mg l^{-1} (specific conductance 75–300 mS m^{-1}) mark the onset of problems which become increasingly severe (Stanley Assoc., 1978).

The sodium absorption ratio is defined as

$$\text{SAR} = (\text{Na}^+) \bigg/ \left(\frac{(\text{Ca}^{2+}) + (\text{Mg}^{2+})}{2} \right)^{0.5}$$

where the concentration of ionic species is expressed in milliequivalents per litre. This parameter measures the ability of sodium to replace calcium and magnesium in a clay structure. Such ion replacement will deflocculate clay particles with resultant reduction in soil permeability. In general, no problems should be experienced for SAR < 6, but problems of increasing severity will be noted in the range of SAR from 6 to 9 and beyond. Excessive contributions of sodium from brine curing of hides and/or processed meats may severely limit land treatment applicability to meat industry wastewaters.

Nitrogenous constituents

Meat, poultry and rendering wastewaters contain nitrogen in a variety of forms ranging from organically bound nitrogen (primarily as proteins and amino acids) through ammonium ion (NH_4^+), free ammonia (NH_3) to nitrite (NO_2^-) and nitrate (NO_3^-) ions. A series of processes including decomposition (of organic nitrogen), nitrification (of ammonia to nitrite and nitrate), uptake by crops (of soluble forms), soil adsorption (of ammonium ion), denitrification (of nitrate to nitrogen gas), and volatilization (of ammonia gas) control the soil budget of nitrogen. Nitrogen is an important fertilizer for intensively cropped soil systems. Fertilization represents one of the potential advantages of land application of wastewaters from these industries. However, application rates must be controlled because nitrification to produce nitrates makes applied nitrogen soluble to the extent that a significant quantity of applied nitrogen can contaminate groundwater. Nitrate pollution of groundwater is a problem of increasing concern in many countries which rely heavily upon groundwater sources for domestic water supply. Moreover, nitrogen levels in crops can reach levels which are harmful to livestock or are directly toxic to the crops themselves.

Phosphorus constituents

Phosphorus also represents an essential plant nutrient contribution available through land applications of meat, poultry and rendering industry wastewaters.

However, unlike nitrogen, the various forms of phosphorus are readily fixed by soil adsorption and plant uptake so that contamination of groundwater is not a significant problem. Soils have been estimated to offer a phosphorus adsorption capacity ranging from 235 to 2720 kg per ha-m of soil profile (Stanley Assoc., 1978).

Pathogens

As previously noted, pathogenic bacteria, viruses and parasitic protozoa and helminths may be present in meat and poultry industry wastewaters.

The bacterial pathogens of the genus *Salmonella* and tape worm eggs represent the main risk. Soils have been found to be relatively efficient at removing pathogens from wastewater, but the possibility of subsequent crop contamination and closure of the cycle of disease transmission is the major concern. Application of wastewater to crops for human consumption is not recommended or practised in developed countries, but irrigation of forage crops may provide an indirect route to human infection via livestock. Such problems remain an area of active study, although disinfection of wastewaters prior to land application offers one solution.

Other factors

A variety of other wastewater characteristics must generally be considered in designing land application systems. These include wastewater temperature, colour, pH and concentration of trace toxic compounds and metals. Such factors are not generally a problem for meat and poultry industry wastewaters. Rendering industry wastewaters may be more prone to temperature and/or pH extremes. In such cases, it should be noted that temperatures of less than 65°C and pH in the range of 6 to 9 should be maintained to avoid problems with land application systems.

Performance of land treatment systems is difficult to summarize and case studies are more difficult to generalize than for other treatment options. However, several case studies of meat and poultry industry land application systems are available in the literature for the interested reader (Dencker *et al.*, 1977; Tarquin *et al.*, 1974; Tarquin and Bautista, 1976; Witherow, 1973; Witherow, 1975). Furthermore, a detailed manual on costing land application systems has been provided by Pound *et al.* (1975).

3.5 Summary

Water pollution control in the meat, poultry and rendering industries has evolved relatively slowly as a secondary or lesser priority compared with product safety considerations. The predominance of the latter must, of course, continue, but increasing concern in recent years has led to notable advances in water pollution control. Evaluation of the industrial processes themselves and of available technology for wastewater treatment clearly indicates that water pollution problems from this industry are manageable, given an appropriate degree of commitment.

References

APHA (1980) *Standard Methods for the Examination of Water and Wastewater*, 15th edn, American Public Health Association, American Waterworks Association and Water Pollution Control Federation, Washington, DC.

Akers, J.M. (1973) Utilization of blood, *Food Manuf.*, **48**(4), 31–32.

Baker, D.A. and White, J.E. (1971) Treatment of meat packing wastes using PVC trickling filters, *Proc. 2nd Nat. Symp. Food Processing Wastes*, US Environmental Protection Agency, NTIS No. PB215217, Denver, Colorado, 287–312.

Beck, E.C., Giannini, A.P. and Ramirez, E.R. (1974) Electrocoagulation clarifies food wastewater, *Food Technol. (Chicago)*, **28**(2), 18–22.

Bissett, D.W. and Drynan, W.R. (1973) *The Application of Trickling Filters to the Treatment of Meat Packing Wastes*, Dept. Civil Engr., University of Waterloo, Canada.

Blair, R.H. (1974) Utilization of wastes and byproducts in animal feeds, *Feedstuffs*, **46**(39), 19–24.

Brungs, W.A. (1973) Effects of residual chlorine on aquatic life, *J. Water Pollut. Control Fed.*, **45**, 2180–2193.

Bryan, F.L. (1974) Diseases transmitted by foods contaminated by wastewater, *Proc. Symp. Wastewater Use in the Production of Food and Fibre*, Report No. 660/2-74-041, US Environmental Protection Agency, Washington, DC, pp. 16–45.

Chittenden, J.A. and Wells, W.J. (1971) Rotating biological contactors following anaerobic lagoons, *J. Water Pollut. Control Fed.*, **43**, 746–754.

Clise, J.D. (1974) *Poultry Processing Wastewater Treatment and Reuse*, Report No. 660/2-74-060, US Environmental Protection Agency, Washington, DC.

Cooper, R.N. and Denmead, C.F. (1979) Chemical treatment of slaughterhouse wastes with protein recovery, *J. Water Pollut. Control Fed.*, **51**, 1017–1023.

Dart, M.C. (1974) Treatment of waste waters from the meat industry, *Process Biochem.*, **9**(5), 11–14.

Dearborn Environmental Consulting Services (1978) *Disinfection of Poultry Packing Effluents Containing Salmonella*, Econ. Tech. Rev. Report EPS 3-WP-78-9, Environmental Protection Service, Environment Canada, Ottawa.

Dencker, D.O., Landwehr, P.J. and Sallwasser, G.M. (1977) Land disposal of meat packing wastewater — case history, *Proc. 32nd Purdue Ind. Waste Conf.*, Ann Arbor Science, Ann Arbor, Michigan, 944–952.

Dept H.E.W. (1965) *An Industrial Waste Guide to the Meat Industry*, Publ. No. 386, US Dept Health, Education and Welfare, Washington, DC.

Drynan, W.R. (1974) Water and wastewater management at a rendering plant, paper presented at *47th Annu. Conf. Water Pollut. Control Fed.*, Denver, Colorado.

Enders, K.E., Hammer, M.J. and Weber, C.L. (1967) Field studies on an anaerobic lagoon treating slaughterhouse waste, *Proc. 22nd Ind. Waste Conf.*, Purdue University, Indiana, 126–137.

EPA (1973a) *In-process Pollution Abatement: Upgrading Poultry Processing Facilities to Reduce Pollution.* Vol. I, *Technology Transfer*, US Environmental Protection Agency, Washington, DC.

EPA (1973b) *Waste Treatment — Upgrading Meat Packing Facilities to Reduce Pollution*, Technology Transfer, US Environmental Protection Agency, Washington, DC.

EPA (1974a) *Development Document for Proposed Effluent Limitation Guidelines and New Source Performance Standards for the Red Meat Processing Segment of the Meat Products and Rendering Processing Point Source Category*, Report No. 440/1-73/012, US Environmental Protection Agency, Washington, DC.

EPA (1974b) *Development Document for Proposed Effluent Limitations Guidelines and New Source Performance Standards for the Renderer Segment of the Meat Products Point Source Category*, Report No. 440/1-74/031-a, US Environmental Protection Agency, Washington, DC.

EPA (1975a) *Development Document for Proposed Effluent Limitation Guidelines and New Source Performance Standards for the Poultry Segment of the Meat Product and Rendering Process Point Source Category*, Report No. 440/1-75/031-b, US Environmental Protection Agency, Washington, DC.

EPA (1975b) *Evaluation of Land Application Systems*, Tech. Bull. 430/9-75-001, US Environmental Protection Agency, Washington, DC.

EPA (1977) *Process Design Manual for Land Treatment of Municipal Wastewater*, Technology Transfer Rep. No. 625/1-77-000, US Environmental Protection Agency, Washington, DC.

Folz, T.R., Ries, K.M. and Lee, J.W. (1974) Removal of protein and fat from meat slaughtering and packing wastes using lignosulphonic acid, *Proc. 5th Nat. Symp. on Food Processing Wastes*, US Environmental Protection Agency, NTIS No. PB 237520, 85–106, Corvallis, Oregon.

FWPCA (1967) The cost of clean water, in Vol. III *Industrial Waste Profile* No. 8 — *Meat Products*, US Federal Water Pollution Control Administration, Washington, DC.

Fullen, M.J. (1953) Anaerobic digestion of packing plant wastes, *Sew. Indust. Wastes*, **25**, 576–585.

Garrison, K.M. and Geppert, R.J. (1960) Packinghouse waste processing applied improvement of conventional methods, *Proc. 15th Ind. Waste Conf.*, Purdue University, Indiana, 207–218.

Given, P.W., Hadzieyev, D., Bouthillier, P.H. and Coutts, R.R. (1974) Odour control and optimum operating conditions for the Edmonton Industrial Lagoon System, Water — 1974: I Industrial Waste Treatment, *AIChE. Symp. Ser.*, **144**(70), 219–226.

Grant, R.A. (1974) Protein recovery from process effluents using ion exchange resins, *Process. Biochem.*, **9**(2), 11–14.

Green, T., Shell, G. and Witmayer, G. (1980) Case history of nitrification of a rendering–meat packing wastewater, *Proc. 35th Purdue Ind. Waste Conf.*, Ann Arbor Science, Ann Arbor, Michigan, 653–664.

Gruette, J.L. and Westphal, G.M. (1980) Draft tube channel approach to poultry processing wastewater treatment, *Proc. 35th Purdue Ind. Waste Conf.*, Ann Arbor Science, Ann Arbor, Michigan, 577–585.

Heddle, J.F. (1979) Activated sludge treatment of slaughterhouse waste with protein recovery, *Water Res.*, **13**, 581–584.

Hester, B.L. and McLurg, P.T. (1970) Operation of a packing plant waste treatment plant at Cherokee Iowa, *Proc. 25th Ind. Waste Conf.*, Purdue University, Indiana, 436–449.

Hirlinger, K.A. and Gross, C.E. (1957) Packinghouse waste trickling filter efficiency following air flotation, *Sew. Indust. Wastes*, **29**(2), 165–169.

Hobbs, B.C. (1974) Microbiological hazards of meat production, *Food Manuf.*, **49**(10), 29–30, 33–34, 54.

Hopwood, A.P. and Rosen, G.D. (1972) Protein and fat recovery from effluents, *Process Biochem.*, **7**(3), 15–17.

Hopwood, A.P. (1980) Recovery of protein and fat from food industry wastewaters, *Water Pollut. Control*, **79**, 225–235.

Hrudey, S.E. (1980) Personal file data.

Hrudey, S.E. (1981a) Activated sludge response to emulsified lipid loading, *Water Res.*, **15**, 361–373.

Hrudey, S.E. (1981b) Evaluation of procedures for measuring fats, oils and greases (FOG) in wastewaters, *Environ. Technol. Lett.*, **2**, 503–510.

Hrudey, S.E. (1981c) Biochemistry of activated sludge: role of lipids, *Water Industry 81*, Conf. Proc. Brighton, 191–196.

Hrudey, S.E. (1982) Factors limiting emulsified lipid treatment capacity of activated sludge, *J. Water Pollut. Cont. Fed.*, **54**, 1207–1214.

Jank, B.E. and Guo, P.H.M. (1978) Biological treatment of meat and poultry wastewater, *Proc. Technology Transfer Seminar on Meat and Poultry Industry Regulations and Guidelines*, Environment Canada, Ottawa.

Johnson, D.B. and Krill, W.P. (1976) RBC pilot plant treatment of pretreated meat slaughtering/processing waste, *Proc. 31st Purdue Ind. Waste Conf.*, Ann Arbor Science, Ann Arbor, Michigan, 733–742.

Jones, D.T. (1974) Protein recovery by ion exchange, *Process Biochem.*, **9**(10), 17–19.

Kampelmacher, E.H. and van Noorle Jansen, L.M. (1973) Occurrence of *Salmonellae* in oxidation ditches, *J. Water Pollut. Cont. Fed.*, **45**, 348–352.

Mueller, J.A. and Mancini, J.L. (1975) Anaerobic filter — kinetics and applications, *Proc. 30th Purdue Ind. Waste Conf.*, Ann Arbor Science, Ann Arbor, Michigan, 423–447.

Mulligan, T.J. and Sheridan, R.P. (1975) Treatment of high strength fatty acid derivative wastewaters, *Proc. 30th Purdue Ind. Wastes Conf.*, Ann Arbor Science, Ann Arbor, Michigan, 997–1004.

Murray, D.S. (1975) Preliminary operating results of a slaughterhouse pretreatment system, *Proc. 30th Purdue Ind. Wastes Conf.*, Ann Arbor Science, Ann Arbor, Michigan, 365–376.

Novotny, V. and Englande, A.J. (1974) Equalization design techniques for conservative substances in wastewater treatment systems, *Water Res.*, **8**, 325–332.

Niles, C.F. and Gordon, H.P. (1970) Operation of an anaerobic pond on hog abattoir wastewaters, *Proc. 25th Ind. Waste Conf.*, Purdue University, Indiana, 612–616.

Paulson, W.L., Kueck, D.R. and Kromlich, W.E. (1971) Oxidation ditch treatment of meat packing wastes, *Proc. 2nd Nat. Symp. on Food Processing Wastes*, US Environmental Protection Agency, Denver, Colorado, 617–635.

Paulson, W.L., Lively, L.D. and Witherow, J.L. (1974) Analyses of wastewater treatment systems for a meat processing plant, *Proc. 27th Ind. Waste Conf.*, Purdue University, Indiana, 879–893.

Pilney, J.P., Halvorson, H.O. and Erickson, E.E. (1968) *Industrial Waste Study of the Meat Products Industry*, Contract No. 68-01-0031, US Environmental Protection Agency, Washington, DC.

Porges, R. and Struzeski, E.J. (1962) Characteristics and treatment of poultry processing wastes, *Proc. 17th Ind. Waste Conf.*, Purdue University, Indiana, 583–601.

Pound, C.E., Crites, R.W. and Griffes, D.A. (1975) *Costs of Wastewater Treatment by Land Application*, Report No. 430/9-75-003, US Environmental Protection Agency, Washington, DC.

Ramirez, E.R., Johnson, D.L. and Clemens, O.A. (1976) Direct comparison in physicochemical treatment of packinghouse wastewater between dissolved air and electroflotation, *Proc. 31st Purdue Ind. Waste Conf.*, Ann Arbor Science, Ann Arbor, Michigan, 563–573.

Ramirez, E.R., Barber, L.K. and Clemens, O.A. (1977) Physicochemical treatment of tannery wastewater by electrocoagulation, *Proc. 32nd Purdue Ind. Waste Conf.*, Ann Arbor Science, Ann Arbor, Michigan, 183–188.

Rands, M.B. and Cooper, D.E. (1966) Development and operation of a low cost anaerobic plant for meat wastes, *Proc. 21st Ind. Waste Conf.*, Purdue University, Indiana, 613–637.

Rollag, D.A. and Dornbush, J.N. (1966) Anaerobic stabilization pond treatment of meat packing wastes, *Proc. 21st Ind. Waste Conf.*, Purdue University, Indiana, 768–781.

Saucier, J.W. (1971) Anaerobic or aerated lagoons? *Water Wastes Eng.*, 8(5), C10–C12, C26.

Seng, W.C. (1971) Removal and recovery of fatty materials from edible fat and oil refinery effluents, *Proc. 2nd Nat. Symp. on Food Processing Wastes*, US Environmental Protection Agency, NTIS No. PB215217, Denver, Colorado, 337–366.

Schroepfer, G.J., Fullen, W.J., Johnson, A.S., Ziemke, N.R. and Anderson, J.J. (1955) The anaerobic contact process as applied to packinghouse wastes, *Sew. Indust. Wastes*, **27**, 460–486.

Shih, J.C. and Kozink, M.B. (1980) Ultrafiltration of poultry processing wastewaters and recovery of a nutritional byproduct, *Poultry Sci.*, **59**, 247–252.

Sollo, F.W. (1960) Pond treatment of meat packing plant wastes. *Proc. 15th Ind. Waste Conf.*, Purdue University, Indiana, 386–391.

Stanley Associates Engineering Ltd (1978) *Land Application of Food Processing Wastewater*, Econ. Tech. Rev. Rep. EPS 3-WP-78-5, Environmental Protection Service, Environment Canada, Ottawa.

Stanley Associates Engineering Ltd (1979) *Biological Treatment of Food Processing Wastewater, Design and Operations Manual*, Econ. Tech. Rev. Rep. EPS 3-WP-79-7, Environmental Protection Service, Environment Canada, Ottawa.

Stanley Associates Engineering Ltd (1981) *Water and Waste Management in the Canadian Meat and Poultry Processing Industry*, Econ. Tech. Rev. Rep. EPS 3-WP-81-3, Environmental Protection Service, Environment Canada, Ottawa.

Steffen, A.J. (1973) *In Plant Modifications and Pretreatment — Upgrading Meat Packing Facilities to Reduce Pollution*, Technology Transfer, US Environmental Protection Agency, Washington, DC.

Stracey, I.G. (1975) Treatment of abattoir and meat processing effluent at Haverhill Meat Products Ltd, *Water Pollut. Control*, **74**, 101–108.

Summers, T.H. (1972) Wastes from poultry processing — origins and treatment, *Effluent Water Treat. J.*, **12**, 299–305.

Tarquin, A., Applegate, H., Rizzo, F. and Jones, L. (1974) Design considerations for treatment of meat packing plant wastewater by land application, *Proc. 5th Nat. Symp. on Food Processing Wastes*, US Environmental Protection Agency, NTIS No. PB237520, Corvallis, Oregon, 107–113.

Tarquin, A. and Bautista, H. (1976) Treatment of high strength meat packing wastewater by overland flow, *Proc. 31st Purdue Ind. Waste Conf.*, Ann Arbor Science, Ann Arbor, Michigan, 479–484.

Thurston, R.V., Russo, R.C. and Vinogradov, G.A. (1981) Ammonia toxicity to fishes. Effect of pH on the toxicity of the un-ionized ammonia species, *Environ. Sci. Technol.*, **15**, 837–840.

Vanderpost, J.M. and Bell, J.B. (1976) *A Bacteriological Investigation of Meat and Poultry Packing Plant Effluents with Particular Emphasis on Salmonella*, Econ. and Tech. Rev. Rep. EPS 3-WP-76-9, Environmental Protection Service, Environment Canada, Ottawa.

Wells, W.J. (1970) How plants can cut rising waste treatment expense, *Natl. Provis.*, **163**, 82–91.

Willoughby, E. and Patton, V.D. (1968) Design of a modern meat packing waste treatment plant, *J. Water Pollut. Control Fed.*, **40**, 132–137.

Witherow, J.L. (1973) Small meat packers waste treatment systems — I, *Proc. 4th Nat. Symp. on Food Processing Wastes*, US Environmental Protection Agency, NTIS No. PB234 606, Syracuse, NY, 276–313.

Witherow, J.L. (1975) Small meat packers waste treatment system — II, *Proc. 30th Purdue Ind. Waste Conf.*, Ann Arbor Science, Ann Arbor, Michigan, 942–955.

Witherow, J.L., Yin, S.C. and Farmer, D.M. (1973) *National Meat Packing Waste Management Research and Development Program*, Robert S. Kerr Environmental Research Lab., US Environmental Protection Agency, Ada, Oklahoma.

Woodard, F.E., Sproul, O.J., Hall, M.W. and Ghosh, M.M. (1972) Abatement of pollution from a poultry processing plant, *J. Water Pollut. Control Fed.*, **44**, 1909–1915.

Woodard, F.E., Hall, M.W., Sproul, O.J. and Ghosh, M.M. (1977) New concepts in treatment of poultry processing wastes, *Water Res.*, **11**, 873–877.

Wymore, A.H. and White, J.E. (1968) Treatment of slaughterhouse waste using anaerobic and aerated lagoons, *Proc. 23rd Ind. Waste Conf.*, Purdue University, Indiana, 601–618.

4 The treatment of waste from the fruit and vegetable processing industries

J R Harrison, L A Licht and R R Peterson,
CH2M Hill, Corvallis, Oregon

4.1 Introduction

Pollution control is becoming an increasing problem for fruit and vegetable processing plants. The reasons for this growing problem vary widely, depending on the circumstances of the individual processor. For example, rising costs for energy, the application of new byproduct recovery processes, increased needs for better wastewater effluent quality and recent changes in equipment technologies for wastewater treatment are causing many fruit and vegetable processors to re-evaluate their existing wastewater treatment systems. For processors constructing new plants, there is an increased awareness of the impact of wastewater treatment on the overall cost of processing food. Also, modifications to reduce product losses, such as changes in peeling systems, new processing methods, water conservation and reuse systems, have all combined to change waste characteristics. These changes in wastewater characteristics have sometimes reduced the efficiency of previously effective wastewater treatment hardware.

In view of these technological changes and the implications of the increased costs and environmental requirements, there is ample reason to examine closely the methods currently in use throughout the industry. For the food processor and the consulting engineer facing challenges in pollution control this chapter brings together and reviews, from the viewpoint of engineering experience in the field, much previously reported information and data.

This chapter describes the general wastewater characteristics associated with the production of fruit and vegetable products. It includes the application of current technologies now available for treatment of fruit and vegetable wastes, and their advantages and disadvantages. Methods that, until recently, have been considered suitable for treatment of fruit and vegetable wastes are reconsidered in the light of today's rising electrical power costs, strict treatment standards and limited suitable sites for treatment. Current and probable trends in solids disposal and byproduct recovery systems are also presented.

4.2 Waste characteristics

The composition and source of wastewater from the fruit and vegetable processing industry is largely determined both by the industrial end-product and by the manufacturing processes used. Major wastewater loads are usually generated by washing, fluming, trimming, peeling, blanching, can washing, cooling and plant cleanup. Although manufacturing and product differences can cause considerable variation in waste characteristics, all food processing plants share some common characteristics and associated wastewater treatment problems.

4.2.1 Pollutants

In general, wastewater from fruit and vegetable processors contain compounds that are mainly organic, primarily soluble (dissolved) and highly reactive when contacted with bacteria. Wastes from fruit and vegetable processing plants are amenable to biological treatment by conventional methods used for treating domestic wastewater. However, a combination of factors unique to fruit and vegetable wastes has resulted in major problems in their treatment. Such factors include the rapid demand for oxygen necessary for aerobic treatment, seasonal load variations, high variations in waste strength and volume, low pH, lack of nutrients, and changes in waste characteristics with product changes. Major pollutants of concern include biochemical oxygen demand (BOD), total suspended solids (TSS), settleable solids, wastewater pH, nitrogen, oil and grease.

Wastewater flows from fruit and vegetable plants are often intermittent and vary greatly, depending on in-plant operations. Because of these variations, flow equalization or other special scheduling of plant discharge may be required.

Waste from most fruit and vegetable processing plants has a considerably higher BOD per unit of flow than domestic and other wastewaters. The failure to recognize this difference caused the failure of many early designs for treatment facilities at food processing plants. This was especially true of staged or series oxidation reactors such as rotating biological contactors or plug flow activated sludge systems. Underestimation of the rate of BOD oxidation also resulted in problems because aeration systems were undersized for peak oxygen demands. Low dissolved oxygen concentrations in aeration basins and the inability of oxygen to penetrate the biological floc resulted in the growth of filamentous bacteria that are especially troublesome in the treatment of fruit and vegetable wastes.

Wastes from fruit and vegetable industries also differ from domestic wastewater in the availability of BOD substrate or rate of BOD available for biological breakdown. With fruit and vegetable wastes the availability of BOD is almost immediate. Wastes from domestic sewage and from many other industries contain more complex compounds that must be hydrolyzed or broken down before the BOD becomes available for biological oxidation. Soluble sugars, for example, contained in fruit and vegetable wastes have a very rapid rate of oxidation compared with most other wastes that require a sequential breakdown to simpler substances before becoming available for bacterial synthesis.

Another major pollutant in fruit and vegetable waste is the solids content associated with the flow. This is usually measured by the settleable solids (SS) test or the total suspended solids (TSS) test. The SS test is often the only test used by waste treatment plant staff at fruit and vegetable plants. This is because it is easy to perform and is an effective indicator of clarifier performance. Drawbacks of the SS test are that it does not give an accurate measurement of the mass of solids in the wastewater. Variations in solids settling rate, solids compaction, temperature and density currents result in a wide variation in the actual mass of solids associated with the settled volume measured in the SS test. Because of this, the TSS rather than the SS test should be used to determine accurately the mass of solids or milligrams of solids per litre of wastewater.

An example of the different results of SS and TSS tests is given by the test work of a french fry (potato chip) manufacturer, in which the SS test indicated that a screening device was removing 84% of the incoming solids. On this basis, the manufacturer considered the screen a good replacement for his existing primary clarifier, which removed only about 77% of the total suspended solids from the incoming flow. However, additional test work on the screening device for TSS removal indicated only a 45% solids removal. The discrepancy between the two test results, as shown by this example, is due to the inherent variability of the SS test. Had these differences not been specifically evaluated, the manufacturer might have replaced the clarifier with a less efficient screening device.

Wastewater pH is a common problem at fruit and vegetable plants. Where plants use caustic or lye peeling methods, high pH values can exceed standard limits for sewer discharge and result in poor sludge settleability or biological inhibition. Low pH or acidic wastewater is perhaps the most common problem with many fruit and vegetable wastes. Although waste containing acid brines is the most usual cause of trouble, even wastes with near neutral or slightly high pH levels can develop an eventual low pH problem; this is because of the rapid biological breakdown (often under anaerobic conditions) of fruit and vegetable wastes. The discharge of low pH wastewater can result in excessive corrosion of concrete and sewers, toxic levels of hydrogen sulphide, poor settling of primary sludges and upset of secondary biological processes.

The nitrogen content of fruit and vegetable wastes occurs in various forms; however, the majority is usually present as complex organic nitrogen. Unless the fruit and vegetable wastes undergo anaerobic digestion to break or hydrolyze complex organic nitrogen, most such wastes are deficient in the ammonia nitrogen necessary for the biological growth associated with conventional aerobic treatment. The lack of available nutrients and especially nitrogen has historically been a problem in treating fruit and vegetable wastes.

4.2.2 Waste survey

Because the quantity and characteristics of waste discharged from fruit and vegetable processing plants are highly variable, there can be no substitute for a plant survey to determine the actual waste characteristics and loads. Factors causing variation in waste strength often include the commodity being produced, the product mix, storage requirements for materials, the field conditions and time from harvest, the method of harvest and transportation, the fruit or vegetable maturity, processing methods and the product condition. Only crude generalizations can be made concerning food processing waste characteristics; however, Table 4.1 lists pollutant loads commonly found, on the basis of quantity of raw product used by the typical fruit or vegetable processor (EPA, 1977).

The gathering of accurate flow and waste characterization data is probably the most important step the fruit and vegetable processor can take to ensure that waste treatment facilities are not undersized, but unfortunately it is often overlooked until after an equipment or treatment system failure occurs.

The use of tests for pollutants to determine design criteria for waste treatment facilities is shown by an example from a frozen foods manufacturer processing broccoli and cauliflower. Tests at the plant yielded the following pollutant discharge per amount of end-product:

average flow = 26.7 litres kg^{-1} product with a ratio of peak to average day flow = 1.3:1;
BOD average = 5.1 with a range of 3.3 to 9.2 kg per tonne product;
TSS average = 2.2 with a range of 1.2 to 3.2 kg per tonne product.

Using these pollutant loads for production rates of 176 tonne per average day and 227 tonne per peak day, the design criteria for wastewater loadings were calculated (Table 4.2).

Materials balance

The first step in determining the origin and extent of waste products is to draw a simple diagram of waste sources from their point of discard to their final point of discharge. From this, a materials or mass balance should be made. First, all of the incoming raw materials are listed, including water used in the manufacturing process. The most important commodity to identify for a material balance is the amount of raw fruit or vegetable being processed during the test period. This amount is used as a common denominator for later analysis of pollutant content as a percentage of product.

Characterization of all of the individual waste streams is necessary to

Table 4.1 Average waste characteristics (EPA, 1977)

Crop	Flow (l tonne^{-1} raw product)	BOD (kg tonne^{-1} raw product)	TSS (kg tonne^{-1} raw product)
Apples	10	9.0	2.2
Apricots	23	20.0	4.9
Asparagus	35	2.5	3.8
Dry beans	37	30.0	21.0
Lima beans	32	24.0	19.0
Snap beans	18	7.5	3.0
Beets	11	26.5	11.0
Broccoli	38	10.0	8.5
Brussels sprouts	34	3.8	7.5
Berries	15	9.5	3.5
Carrots	14	15.0	8.5
Cauliflower	71	8.0	3.0
Cherries	16	19.0	1.0
Citrus	13	4.8	1.8
Corn	8	13.5	5.0
Grapes	6	4.5	0.8
Mushrooms	33	7.0	3.6
Olives	34	13.5	13.0
Onions	23	28.5	8.5
Peaches	13	17.5	4.3
Pears	15	25.0	6.0
Peas	23	19.0	5.5
Peppers	19	16.0	29.0
Pickles	15	21.0	4.1
Pimentos	29	27.5	2.9
Pineapples	11	12.5	4.6
Plums	10	5.0	1.0
Potato chips	7	12.5	16.0
Potatoes, sweet	9	46.5	28.0
Potatoes, white	1.5	42.0	64.0
Pumpkin	1.2	16.0	38.0
Sauerkraut	0.4	2.8	0.5
Spinach	3.7	7.0	3.0
Squash	2.5	10.0	7.0
Tomatoes, peeled	0.9	4.7	6.0

Table 4.2 Wastewater design criteria

Time description	Period (days)	Load			Concentration	
		Flow (m^3 d^{-1})	BOD (kg d^{-1})	TSS (kg d^{-1})	BOD (mg l^{-1})	TSS (mg l^{-1})
Average	172	4700	898	387	191	82
Maximum daily	22	—	1620	563	345	120
Peak	5	6110	2090	726	445	154

complete the material balance. If done correctly, the balance should closely correlate the amount of waste lost during production (or gained, in some cases) with the amount of waste measured in the individual sewers. More important than obtaining an exact balance, the individual sources of waste streams or points of product loss will have been characterized. With this information available, waste treatment efforts can be concentrated on those sources where the greatest benefits can be achieved.

Building, staffing and maintaining a wastewater analysis laboratory to perform an accurate mass balance is costly. Unless the in-plant measurements are to continue over a long period, the use of an outside laboratory or consulting service may be less costly and may result in better data, rather than burdening the processing staff with the work.

The need for material balances is emphasized by the example of the discrepancy of waste loads sometimes found in the potato processing industry. Here, it is common practice to discharge recycle streams from solids dewatering equipment and combine them with raw wastewaters before pumping to a primary clarifier or screen for solids removal. Samples of the combined recycle and raw wastewater streams are often reported as 'raw waste discharge'. Since many vacuum filters have only 50% solids capture efficiency, the presence of recycle in the 'raw wastewater' sample can lead the processor to overestimate the waste solids load in the raw wastewater at 1.5 times its actual value. If a material balance has been completed for the primary clarifier, vacuum filter and influent pump station, the food processor can determine whether loads being reported as raw waste are being influenced by recycle streams. Another cross-check in the material balance is to compare the amount of filter solids cake (dewatered primary solids) being hauled from the processing plant with the amount of solids in the raw wastewater. A large discrepancy between the incoming solids and those solids being hauled from the plant indicates either very poor solids capture in the primary treatment unit or a problem in waste measurement or testing.

Flow measurement and sampling
Perhaps the single most important wastewater characteristic to determine during the waste survey is flow. This is because flow is the limiting design factor used to size many treatment units. Types of wastewater flow often measured for efficient treatment design are: (1) average flow, (2) maximum daily flow and (3) peak instantaneous flow.

Flow measurement and sampling equipment should be capable of providing the basic information necessary for performing the material balance. The information usually required is that necessary (a) to assess the economics of in-plant waste reduction versus end-of-pipe treatment and (b) to select and size treatment equipment.

Selection of measurement and sampling points is important since the mass balance is only useful if a representative sample location has been chosen. Flow measurement and sampling equipment should be reliable, nonclogging, accurate, accessible for maintenance and as inexpensive as possible. As minimum criteria, automatic sampling should include metering and recording of flows both entering and discharging from the production plant. In-plant flow measurements are often taken by measuring the time required to fill channels, gutters or storage vessels. Weirs or Parshall flumes can also be used to obtain accurate flow measurements.

Care must be taken to obtain representative samples. The samples taken should be proportionate to the flow being discharged whenever possible. It should be recognized that the waste strength may vary in a slow-flowing open channel or gutter where adequate mixing is not possible. Inadequate mixing may result in solids separation, so that the waste strength varies from the top of the water surface to the bottom of the channel. Sample points should be situated where the waste is thoroughly mixed. If a grab or instantaneous sample, rather than an automatic sample, is taken, more frequent sampling may be required to obtain samples representative of the entire day's activity.

A complete waste evaluation is based on flow, the amount of solids present and the percentage of pollutants associated with solids. To obtain this information, wastewater tests on both the total (as sampled) and filtered (soluble) samples of the waste stream are necessary. By analyzing both the total and soluble pollutants, then dividing the difference by the wastewater TSS, the amount of pollutant associated with the suspended solids can be determined. Table 4.3 shows the results of tests on waste from the processing of frozen

Table 4.3 Wastewater particulate pollutants, frozen french fries processing

Pollutant	Concentration		Particulate fraction (kg pollutant kg^{-1} TSS)
	Total (mg l^{-1})	Soluble (mg l^{-1})	
TSS	2600	—	1.0
COD	5400	2540	1.1
BOD	3600	1780	0.7
TKN	230	100	0.05
NH$_3$	30	30	—

french fries, to calculate the amount of pollutants associated with the suspended solids. Stated simply, the particulate fraction values allow the wastewater manager to calculate the mass of chemical oxygen demand (COD), BOD, total Kjeldahl nitrogen (TKN), and ammonia (NH$_3$) removed for each unit mass of suspended solids removed from the wastewater stream. The use of the information in Table 4.3 for the sizing of wastewater treatment equipment is demonstrated in Section 4.5.

The extent of waste flows or pollutant strengths cannot safely be assumed. Making assumptions without checking or confirming the information is perhaps the most common and costly error made in a waste survey. Plant personnel may be experts in processing their food product, but to make assumptions about waste strength or loadings without actual tests can lead to disaster. Only checking and rechecking through the use of on-site measurements and laboratory test work will produce an accurate waste balance.

4.3 Aims of waste treatment

There are two major reasons for the treatment, recycle or reuse of wastes: either for economic advantage or to comply with wastewater discharge standards. Being a good neighbour, reducing shock loads to the city sewer or eliminating odours is sometimes important but these are generally secondary reasons for action. Although in-plant water and wastewater conservation are good alternatives to waste treatment, in most cases such in-plant practices are not sufficient to eliminate completely the need for some form of treatment.

4.3.1 Economic considerations

The high capital costs of wastewater pollution control equipment have been a major factor in locating food processing plants in urban areas where discharge to city sewers is feasible. A survey in the USA indicates that approximately 55% of the fruit and vegetable processing plants discharge wastes to city sewers. About 30% discharge directly to land disposal systems, and only about 15% have their own wastewater treatment systems that discharge directly to receiving waters (Schmidt *et al.*, 1975). Food processors who have experience with wastewater treatment realize that treatment usually means not only removing BOD and suspended solids from the wastewater, but also concentrating and disposing of primary or biological solids and screenings that may be produced by or removed as part of the treatment system.

The decision to own and operate a wastewater treatment system can be costly, not only in the initial removal of pollutants, but also in the subsequent steps necessary to handle solids generated from the initial treatment process. However, the cost of discharge to city sewers means that many processors must install some degree of treatment. The exact extent of that treatment varies from screening — the cheapest method — to more costly methods that produce an effluent of near drinking-water quality. The economics of deciding how much to treat must be carefully calculated, accounting for trade-offs in annual versus capital costs (see Section 4.12).

4.3.2 Meeting discharge standards

Regardless of the location of a fruit and vegetable processing plant, discharge requirements of one or more governmental agencies must be met; usually some combination of national, state or province, and local regulations needs to be observed.

National regulations are generally less stringent than local, state or province restrictions. This is because the specific control required to protect the water quality of a receiving stream or to ensure the protection of a city's wastewater treatment system is often more limiting than the broad-based national standards. National regulations in the USA provide common guidelines for all fruit and vegetable processing plants that discharge into national waterways. These restrictions are based on the application of conventional wastewater treatment equipment for removal of BOD and suspended solids. Allocations for direct discharge to a receiving stream are often expressed as kg per tonne (or lb per ton) of either BOD or TSS pollutants from the final food product. Waste discharge allocations based on production rates are usually relatively liberal so that an unnecessary burden is not placed on industry but, at the same time, national continuity of wastewater goals using conventional wastewater treatment technology is achieved.

In the mid-1970s national standards were established in the USA for canned fruit and vegetable processors. These standards required that by July 1977 industry was to achieve a reduction of pollutants equivalent to secondary treatment or levels referred to as 'Best Practical Technology Currently Available' (BPT). The same national regulations established a target date of 1983 to achieve further reductions in pollutants, referred to as 'Best Available Technology Economically Achievable' (BAT). The BAT technology represented what the government believed to be the very best control and treatment measures, including in-plant modifications and process changes, that could be developed within each industry category or subcategory. A partial list of both BPT and BAT limitations for the canned fruit and vegetable category is given in Table 4.4 (Bureau of National Affairs, 1979). Limits for BPT and BAT were established for both the peak day and 30-day average allowable discharge loads.

In the late 1970s a National Commission on Water Quality reported to the United States Congress that for many industries reduction of pollutant standards beyond the BAT limits would not greatly improve the water quality of most receiving waters. The commission's findings further emphasized the need to remove 'priority' pollutants (129 specific compounds and heavy metals which EPA identified as environmentally harmful) rather than the 'conventional' (BOD, TSS and pH) pollutants which had been previously targeted by the BPT regulations. The commission's findings caused the BAT regulations to be temporarily withdrawn, while BPT regulations remained in effect and at their previous levels.

At present, the US government is in the process of re-evaluating the BAT limitations for the fruit and vegetable industry. The government has proposed to establish what have been described as limits more stringent than the present enforceable BPT levels and possibly less stringent that the recently withdrawn BAT levels. The proposed regulations are referred to as 'Best Conventional Pollutant Control Technology' (BCT) and are, at the time of this writing, yet to be issued.

Table 4.4 US national discharge standards

Category	BPT (kg tonne^{-1})				BAT (kg tonne^{-1})			
	BOD		TSS		BOD		TSS	
	Peak day	30-day average	Peak day	30-day average	Peak day	30-day average	Peak day	30-day average
Apple juice	0.60	0.30	0.80	0.40	0.20	0.10	0.20	0.10
Apple products	1.10	0.55	1.40	0.70	0.20	0.10	0.20	0.10
Citrus products	0.80	0.40	1.70	0.85	0.14	0.07	0.20	0.10
Frozen potato products	2.80	1.40	2.80	1.40	0.34	0.17	1.10	0.55
Dehydrated potato products	2.40	1.20	2.80	1.40	0.34	0.17	1.10	0.55
CANNED AND PRESERVED FRUITS								
Apricots	3.00	1.81	5.36	3.74	1.26	0.94	1.26	0.94
Caneberries	0.77	0.46	1.88	6.95	0.18	0.13	0.18	0.13
Cherries:								
brined	2.87	1.78	5.18	3.68	0.76	0.62	0.76	0.62
sour	1.77	1.11	3.20	2.30	1.10	0.84	1.10	0.84
sweet	1.12	0.60	2.01	1.43	0.45	0.34	0.45	0.34
Cranberries	1.71	1.03	3.06	2.14	0.62	0.47	0.62	0.47
Dried fruit	1.86	1.13	3.34	2.34	0.73	0.56	0.73	0.56
Grape juice:								
canning	1.10	0.60	1.99	1.44	0.77	0.58	0.77	0.58
pressing	0.22	0.14	0.40	0.29	0.11	0.09	0.11	0.09
Olives	5.44	3.34	9.79	6.92	2.29	1.61	2.29	1.61
Peaches	1.51	0.93	2.72	1.03	0.77	0.58	0.77	0.58
Pears	1.77	1.12	3.21	2.32	0.86	0.66	0.86	0.66
Pickles:								
fresh pack	1.22	0.75	2.19	1.54	0.64	0.46	0.64	0.46
process pack	1.45	0.92	2.63	1.91	0.65	0.51	0.65	0.51
Pineapples	2.13	1.33	3.86	2.76	1.48	1.11	1.48	1.11
Plums	0.69	0.42	1.24	0.87	0.28	0.20	0.28	0.20
Raisins	0.43	0.28	0.78	0.57	0.20	0.16	0.20	0.16
Strawberries	1.79	1.06	3.19	2.20	0.62	0.45	0.62	0.45
Tomatoes	1.21	0.71	2.16	1.48	0.52	0.38	0.52	0.38

In addition to national limits based on kilogram per day, more stringent standards required by local agencies usually include limits on the concentration of pollutants.

Another common limitation placed on industrial discharges in the USA is that of dilution. As an example, a common requirement for discharge to some streams is that the mixture of treated industrial waste and receiving stream water shall not result in a mixed water stream BOD concentration of greater than $1.0\,\text{mg}\,\text{l}^{-1}$.

If local conditions allow, food processing industries have sometimes been successful in obtaining a double set of discharge limits: that is, one set of limits is enforced during periods when the receiving stream flow may be small and another, more liberal, set of limits when the receiving stream has a larger flow, such as during periods when water runoff occurs, which will increase the dilution of the industrial discharge.

A food processor discharging into a public sewer must often comply with local sewer ordinances. The ordinances are intended to prevent: (1) blockage and damage to the collection system; (2) hazards to workers in the sewers and at the treatment plant; and (3) interference with the operation of the city's wastewater treatment plant. Ordinances usually contain specific limits on heavy metals, toxic compounds, oil and grease, temperature, pH and other waste characteristics. The ordinances may also include special user charges for the ongoing expenses of operation and upgrade of the publicly owned treatment works.

Depending on whether local, regional or national funds were used for financing the wastewater facility, each industry discharging to a city sewer may also be required to repay its proportionate share of the original cost of constructing the publicly owned treatment works. A charge to cover the cost of constructing publicly owned treatment works is commonly described as an industrial cost recovery, or ICR, charge. In the USA, industrial cost recovery applies only to the federal grant portion of the original capital cost (without interest) of public treatment plant construction. An industry's share of the public treatment works costs is usually based on the quantity of major pollutants actually being treated: flow, TSS and BOD. As an example of billing using this method of cost recovery, a sewer rate ordinance usually establishes charges such as the following:

$$34¢ \text{ per } m^3 \text{ of wastewater}$$
$$13¢ \text{ per kg BOD}$$
$$9¢ \text{ per kg TSS.}$$

For publicly owned treatment works that have tertiary treatment requirements including the removal of nutrients such as nitrogen and phosphorus, additional charges based on the dry weight of nutrients discharged to the city sewer may apply. The exact basis and amount of the sewer charge for discharge to a city sewer vary greatly from one locale to another, and a processor should investigate local sewer ordinances and governmental regulations that apply to the particular location of the production plant.

4.4 Pretreatment

Most food processing plants that discharge into a public sewer are required to provide some method of pretreatment. Pretreatment usually refers to gross solids removal, silt removal, screening of coarse or large solids and any other steps necessary to prevent sewers from plugging and to prevent upset at the public treatment facilities. Pretreatment is also required when discharging wastewater to land application systems, to prevent both plugging of spray nozzles and crusting of untreated food processing solids on the ground surface.

In some cases, pretreatment before discharging wastes to a city sewer or land application system may also include further removal steps, such as both primary and secondary treatment. However, for many processors, only the following pretreatment steps are required.

4.4.1 Flow measurement and sampling

Sampling of pollutants and measurement of flows discharged to a city sewer are usually the minimum pretreatment processes required by municipalities. The use of open channel flow-measuring devices is often preferred in the food processing industry because automatic flow-measuring equipment can be easily checked and calibrated by manual measurements. Also, open channel structures can be designed for easy access for sampling and cleaning.

Since flows often vary greatly, food processors sometimes install parallel flow-measuring devices for extremely low or high flows. This allows accurate flow measurement over a greater range of flow variations than when a single measuring device is used. A combination of rectangular, triangular and Cipolletti weirs may also be used, depending on the flow rate and accuracy of the flow measurement required. A wide range of flow measurements can be achieved by using removable weir plates arranged so that a weir of appropriate size can be selected should flow change drastically as a result of production, maintenance or other operational changes at the plant. Details of flow measurement and waste sampling are available in established handbooks of hydraulics (King and Brater, 1963) and trade association publications (Katsuyama, 1979).

4.4.2 Screening

Most screens used in pretreatment are not intended to remove a significant amount of suspended solids, but only to remove coarse solids that might otherwise interfere with normal sewer flow or further treatment. For coarse solids removal, screen sizes of 20 to 40 mesh are commonly used. Although finer screens (200 to 400 mesh) can remove smaller suspended solids, they also produce greater amounts of wet screenings and frequently become clogged with fine solids or oil and grease. Three types of screens are typically used: vibrating, static and rotating.

Vibrating screens

A vibrating or shaker screen is often used when a large volume or high rate of solids must be screened. The use of a vibrating screen for separation of fruit and vegetable solids is illustrated in Fig. 4.1(a,b). Vibrating screens generally have a high capacity for handling large amounts of flow and solids through a

THE FRUIT AND VEGETABLE PROCESSING INDUSTRIES 221

Fig. 4.1 Screening (courtesy CH$_2$M Hill)
(a) Shaker screen

given screen area. The disadvantages of vibrating screens are high maintenance requirements and the frequency of mechanical failures directly related to the dynamic motion exerted on the screen parts. Vibrating screens have been used extensively where a high volume of coarse or large screenings must be handled. This occurs with the production of frozen french fries or canning of other large root crops that require considerable solids loss to obtain a suitable end-product.

Static screens
Solids can also be screened through the use of a tangential or static screen (also called a side hill screen). This screen is shown in Fig. 4.1(d), where wastewater

Fig. 4.1(b) Vibrating screen

from raisin manufacture passes through the screen while coarse solids and denser material slide tangentially into a solids hopper. An advantage of static screens is that no moving parts are required, thus maintenance is minimal; however, care must be taken in screen selection, otherwise the raw wastewater may also shoot tangentially into the solids hopper. Changes in wastewater characteristics or periodic dumps of oil and grease or other solids that blind screens can also cause problems with static screens.

Rotating screens
Rotating centrifugal screens are especially useful when a high solids capture

Fig. 4.1(c) Rotating screen

rate is required or fine mesh screens are used. Fruit and vegetable wastes are discharged to the outside of a rotating screen drum (Fig. 4.1(c)). The solids are usually scraped from the drum by a doctor blade, while the screened effluent passes through the drum and is discharged from the drum's centre. Advantages of the rotating screen are that it achieves a high rate of solids removal and is self-cleaning; its disadvantages can be related to potential limitations on the screen's hydraulic loading or solids handling capacity because of the screen size selected.

4.4.3 Silt removal

Large amounts of soil and silt are brought in with raw products, especially with root crops such as potatoes, carrots and beets. Dirt and silt that adhere to the root crops are removed in initial washing of the product or during the first

224 INDUSTRIAL WASTEWATER TREATMENT

Fig. 4.1(d) Static screen

hydraulic fluming for product transfer. Since municipal wastewater treatment plants are not designed to deal with large quantities of silt, and because of the abrasive nature of this material, the food processor is usually required to handle and remove this silt as a pretreatment step. Often, after removal of the silt, flume-water is recycled as product transfer water.

The quantity of silt to be handled varies greatly with harvesting methods, soil conditions, crop type, soil moisture and other factors. One potato processor indicated that up to 4% of the incoming raw potato weight consisted of silt, soil and other debris that required removal in a silt system. For a potato processor

THE FRUIT AND VEGETABLE PROCESSING INDUSTRIES 225

Fig. 4.2 Silt removal
(a) Pond

handling daily 91 tonnes of potatoes with only a 2% silt content, from 1.9 to 3.8 m³ d⁻¹ of silt and other debris will require disposal. This assumes that the solids content of separated silt and removed sludge will be between 35 and 65%.

Silt pond

In the past, settling ponds or lagoons (silt ponds) have often been used for silt removal. Figure 4.2(a) shows a silt pond used for the separation of dirt carried in from the fields with carrots and other root crops. One major advantage of the use of a silt pond is that solids are allowed to settle over a long period of time,

Fig. 4.2(b) Thickener

resulting in good solids removal and a relatively low volume of sludge as compared with other systems. Disadvantages of the silt pond are the large areas required for storage and the odours formed during the long detention times, the large exposed water surface areas, and difficulties with pond cleaning and removal of solids.

Thickener
To minimize the space required for silt removal and to allow silt to be removed and disposed of on a continuous basis, circular solids thickeners or clarifiers

Fig. 4.2(c) Channels

can be used to remove silt (Fig. 4.2(b)). One of the major disadvantages reported with thickeners is the high maintenance requirement, caused by mud and silt solids plugging pipes and valves and blocking rake mechanisms.

Channels

A series of concrete channels with removable baffles for draining and solids removal has also been used for silt removal (Fig. 4.2(c)). The channels have no moving parts and therefore avoid the mechanical disadvantages of silt thickeners. Channels also allow periodic draining and cleaning, in contrast to the annual chore of cleaning a single silt pond. Disadvantages of the channel system are that a large land area is required and that silt must be removed frequently, even during bad weather.

Cyclones

Grit-removing cyclones (Fig. 4.2(d)) have also been used in some plants for silt removal. Cyclones are usually preceded by coarse screening devices to prevent plugging. Cyclones are relatively inexpensive to operate and are often used in recycle systems where only coarse solids removal and water reduction are required. Their disadvantages are poor solids capture, high maintenance requirement, high equipment wear due to abrasion, and a large amount of dilute mud that may still require further settling or solids separation.

Fig. 4.2(d) Cyclone

4.4.4 Neutralization

To protect municipal sewerlines and to prevent pH shocks to the public treatment works, most municipalities limit the pH of wastewater discharged into the sewer to a range from pH 6 to pH 9. Such a range may be established arbitrarily, and if the food processor's waste flow is small compared with the other sewage flows, a variance (exception) to the sewer ordinance may be granted by the local sewer authority. Also, the problem of discharging highly alkaline or acidic wastewater may be overcome by discharging the food processing waste at a constant flow rate, since the adverse effects of extreme pH often result not from the actual pH value itself, but from sudden shock loading. A constant or paced discharge may allow discharge of wastewater with a pH level that would cause problems in sudden loads.

If a variance to a sewer ordinance cannot be acquired, or if upsets to the municipal treatment works and problems with the city sewer exist, neutralization of the extreme pH wastewater may be necessary. A reliable neutralization system usually includes mixing tanks, chemical storage areas and equipment, and pH monitoring and recording equipment.

4.4.5 Oil and grease removal

Many sewer ordinances require that wastewater discharges contain less than $100\,\text{mg}\,l^{-1}$ of oil and grease. This standard is based largely on interferences from oil and grease of petrochemical origin. Some municipalities allow variances in oil and grease limitations if the oil and grease do not interfere with the sewer or treatment works and they originate from an animal or vegetable source. Regardless of their origin, it may be beneficial for the food processor to remove and salvage spent oil and grease and sell it to rendering and recovery firms.

A grease trap can be used to recover oil and grease from small wastewater flows. The grease trap usually consists of a device similar to a septic tank, which detains the floating oil and grease by use of baffles at the entrance and exit points to the tank. Where quantities of oil and grease are large or the volume of wastewater is great, grease may be skimmed from a gravity clarifier, floated in a flotation clarifier as described in Section 4.5, or removed in collection channels with a belt oil-skimming device.

4.4.6 Flow equalization

Control of surges is not usually a pretreatment requirement. However, when the food processing wastes have a significant effect on the municipal treatment plant, the smoothing-out of flow variations, accidental spills and surges from plant cleanup can sometimes allow the public facility to accommodate industrial waste loads that would otherwise upset the plant.

One method of using flow equalization to the food processor's advantage is to schedule the industrial discharge during periods when waste loads to the municipal plant are otherwise low and therefore the treatment plant has unused capacity. For example, discharging a majority of the food processing waste between midnight and 6 a.m., when domestic flows have typically diminished, may lessen the overall impact on the municipal plant. However, the advantage of such scheduling may be offset by the need to dilute the high-strength food waste with domestic waste — also to reduce the impact on the municipal system.

Because each production plant must be evaluated individually, the need for flow equalization varies greatly. For example, significant settling of solids in a flow equalization basin may necessitate solids removal prior to equalization, or mechanical mixing during the equalization step.

4.5 Primary treatment

The success of primary treatment in removing pollutants depends largely on the type of fruit or vegetables being processed. For example, primary treatment can significantly reduce suspended solids with raw products like tomatoes, potatoes and beets, and in some types of corn processing. On the other hand, the wastes from canning peas, peaches, pears, or the wastes from processing juice, produce few settleable solids; as a result, fewer benefits are achieved by primary treatment of wastes from these products.

The effect of primary treatment can be illustrated by considering the particulate pollutants in the wastewater described in Table 4.3 (page 215). Assuming that 77% of the raw wastewater TSS (2600 mg l^{-1}) is settleable, then the equivalent of 2000 mg l^{-1} TSS (0.77 × 2600 mg l^{-1} TSS) will be removed. The corresponding effect on other pollutants would be the removal of:

$$2200 \text{ mg l}^{-1} \text{ COD } (2000 \text{ mg l}^{-1} \text{ TSS} \times 1.1 \text{ kg COD kg}^{-1} \text{ TSS})$$
$$1400 \text{ mg l}^{-1} \text{ BOD } (2000 \text{ mg l}^{-1} \text{ TSS} \times 0.7 \text{ kg BOD kg}^{-1} \text{ TSS})$$
$$100 \text{ mg l}^{-1} \text{ TKN } (2000 \text{ mg l}^{-1} \text{ TSS} \times 0.05 \text{ kg TKN kg}^{-1} \text{ TSS})$$

No ammonia would be removed, since ammonia is a soluble (completely dissolved) chemical.

In addition to the type of product being processed, the equipment selected, actual waste loading to primary treatment units, and the design of the system can significantly affect the success of primary treatment. Although bench-scale and pilot tests can yield information on the benefits of primary treatment, direct scaling-up from such tests can lead to gross errors in estimating full-scale equipment performance. As an example, the settling-column test on potato wastewater shown in Fig. 4.3(a) can give useful information on the thickening characteristics (indicated by the percentage of solids concentration) of a full-scale primary clarifier (Fig. 4.3(b)). However, if the clarifier design were based solely on the results of the ideal, quiescent or undisturbed settling-column test, the full-scale clarifier would probably be undersized. The effects of peak loads, hydraulic short-circuiting in the larger full-scale clarifier, temperature variation and other factors must be considered when selecting suitable design parameters based on pilot or small-scale tests.

THE FRUIT AND VEGETABLE PROCESSING INDUSTRIES 231

Fig. 4.3 Primary treatment
  (a) Settling column for waste evaluation

4.5.1 Gravity clarifier

A gravity clarifier or settling tank simply provides a quiescent space through which solids particles fall to settle on the bottom. A gravity clarifier fulfils two functions: solids separation and thickening. If the primary clarifier achieves good solids separation, a majority if not all of the settleable solids will be removed. However, unless the solids are removed in a thickened state, further solids dewatering will be necessary to reduce the sludge volumes to quantities reasonable to handle.

Gravity clarifiers (Fig. 4.3(b)) are usually sized on the basis of the hydraulic

Fig. 4.3(b) Gravity clarifier

loading to the clarifier surface area in $m^3 m^{-2} d^{-1}$. Typical values are between 12 and 41 $m^3 m^{-2} d^{-1}$, with effluent quality from the primary clarifier usually deteriorating at the higher overflow rate. The gravity clarifier water depth must be adequate to allow solids to accumulate. Many gravity clarifiers used in the treatment of fruit and vegetable wastes are equipped with oil and grease skimmers.

Gravity clarification provides good solids capture and removal. Primary clarifiers have a long-proven history of efficient use in the treatment of wastes from fruit and vegetable industries. Their disadvantages are the need for large water surface areas and the formation of acidic sludge when solids are held too long in the bottom of the clarifier. Acidic sludges can damage concrete and pumping equipment and are difficult to dewater and concentrate with some solids handling equipment.

Fig. 4.3(c) Flotation clarifier

4.5.2 Flotation clarifier

The use of a flotation clarifier for the removal of combined domestic and food processing wastes is shown in Fig. 4.3(c). The flotation clarifier has a sludge removal mechanism that operates in a fashion similar to conventional gravity clarifiers for heavy solids. In addition to gravity separation, flotation of solids occurs because fine air bubbles introduced into the unit attach to and float the solids that do not settle by gravity. Floated solids are skimmed into a trough and either augered or pumped to downstream solids dewatering or disposal equipment.

Although it is common practice in the treatment of municipal waste to equip flotation clarifiers with chemical feed systems that enhance the flotation process, most food processors whose primary solids are used for animal feed are hesitant to add chemicals to waste. This is partly because of the added cost and also because of concern for the quality of byproduct animal feeds.

The main advantage of flotation clarifiers is their ability to remove both settleable and nonsettleable solids. Also, the hydraulic flow rate is claimed by manufacturers to be higher than with conventional gravity clarifiers. However, they present the disadvantages of initial high capital costs, high power requirements of the air pressurization system, and increased requirement for mechanical maintenance.

Fig. 4.3(d) Inclined plate separator

4.5.3 Inclined plate separator

The inclined plate separator is a high-rate gravity clarifier consisting of inclined parallel plates stacked to form small channels or tubes and usually housed in a metal structure. The inclined plate separator in Fig. 4.3(d) is being used to remove primary solids from raisin processing wastewater.

The advantage of this settling device over other primary treatment methods is that the effective settling area is increased by the closely spaced plates, so that the overall area required may be only one-fifth to one-tenth that necessary with a conventional gravity clarifier (depending on the specific equipment and wastewater being considered). Inclined plate separators are also ideal for use inside buildings where space is at a premium. On-site construction is often minimal, since they usually come as shop-fabricated units. Also, with fruit and vegetable wastes that are especially prone to go septic, shorter detention times in the separator can be an advantage.

THE FRUIT AND VEGETABLE PROCESSING INDUSTRIES   235

A disadvantage of inclined plate separators is that sludge thickening may not be as effective as with either flotation or gravity clarifiers. Also, solids can bridge or blind openings between the plates. If discharge of settled solids from the bottom of the inclined plate separator is not frequent, plugging or bridging of solids can become a problem. Another disadvantage of the inclined plate separator is that limited full-scale operating data are available on its application to fruit and vegetable wastes. This means that bench or pilot tests should be performed to obtain sizing criteria.

4.5.4 Centrifugal concentrators

Centrifugal concentrators consisting of rotating screens as fine as 400 mesh have also been used in place of conventional primary treatment systems. Rotating screens operate like a basket centrifuge (see Section 4.10.3), the screen taking the place of the solid basket. Raw wastewater enters the rotating screen at the centre and a strong centrifugal force drives water through the screen, leaving heavier solids on the screen surface from which they can be collected and removed.

The advantages of centrifugal concentrators are low initial costs and short hydraulic retention times, which reduce the potential for anaerobic decomposition. Their disadvantages are poor solids capture compared with other primary treatment alternatives, a high rate of mechanical failure due to the dynamic action of the machine, and the inability to handle variable waste loadings.

4.5.5 Complete primary system

The success or failure of a primary treatment system depends on both the performance of the primary treatment device and the interaction and successful operation of solids concentration equipment, screens, and many other equipment items that make up the complete primary treatment system (Fig. 4.4).

Solids concentration is discussed in more detail in Section 4.10, but it is important at this stage to understand how the performance of concentration equipment can affect clarifier or primary influent and underflow loadings. Consider Fig. 4.4, which shows the use of either a vacuum filter or a basket centrifuge for solids concentration in a primary treatment system, and assume that the wastes treated are those from the potato processor with the characteristics given in Table 4.3. On the basis of experience with full-scale wastewater treatment systems for potato plants with either steam or dry caustic peel, the probable effects on the primary clarifier are shown in Table 4.5. Because the vacuum filter has a lower solids capture rate (50% rather than 80%) and a more moist solids cake (10% rather than 20% solids), the result is 41% more clarifier underflow, or sludge that needs to be dewatered with the vacuum filter. Also, with the vacuum filter there is 48% more solids cake to be disposed of

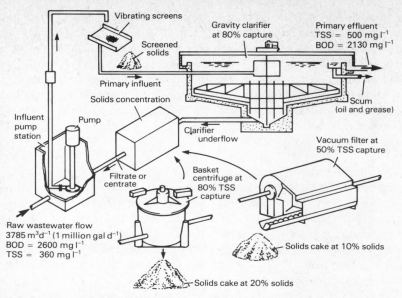

Fig. 4.4 Complete primary wastewater treatment system

than with the basket centrifuge. The added 48% of sludge from the vacuum filter consists entirely of water, since the dry weight of the solids cake (8000 kg d^{-1}) taken to cattle feeding or other disposal locations is the same whichever device is used.

The actual raw wastewater may be difficult to sample because waste gutters are often well below ground level. An influent pump station is used to transfer the raw wastewater to pretreatment or primary treatment equipment (Fig. 4.4). In most cases, recycle streams from the solids dewatering process are added to the influent pump station prior to sampling. When sampling and testing occurs after the recycle streams are added, a greater solids discharge is reported for the plant than actually occurs in the raw wastewater. For the example shown in Table 4.5, recycle streams for a vacuum filter would place an additional 22% TSS and 13% BOD load on the primary clarifier, compared with use of a basket centrifuge. In evaluating the complete primary system, these recycle streams should be accounted for both in equipment sizing and in performance evaluation.

4.6 Anaerobic treatment

Anaerobic treatment is a promising, but as yet infrequently applied method of treating fruit and vegetable wastes. Although anaerobic treatment will generally not produce the effluent quality that can be achieved using aerobic

Table 4.5 Effects of solids concentration equipment

	Vacuum filter	Basket centrifuge	Difference
RAW WASTEWATER			
Flow (m^3 d^{-1})	3 785	3 785	0
TSS (kg d^{-1})	9 800	9 800	0
BOD (kg d^{-1})	13 600	13 600	0
PRIMARY INFLUENT			
TSS (kg d^{-1})	13 800	10 800	22%
BOD (kg d^{-1})	16 700	14 500	13%
PRIMARY EFFLUENT			
TSS (mg l^{-1})	500	500	0
BOD (mg l^{-1})	2 130	2 130	0
CLARIFIER UNDERFLOW			
Flow (m^3 d^{-1})	239	140	41%
TSS (kg d^{-1})	12 000	8 900	42%
SOLIDS CONCENTRATION EQUIPMENT PERFORMANCES			
Solids capture rate (%)	50	80	38%
Cake solids concentration (% TSS)	10	20	50%
Dry cake solids (kg d^{-1})	8 000	8 000	0
Cake volume (m^3 d^{-1})	80	42	48%

treatment methods, it offers specific advantages over aerobic treatment. The reported advantages of anaerobic treatment are that it uses less electrical energy (since oxygen is not required); it produces a usable byproduct, methane; there is less biological sludge to dispose of; and nutrient requirements are lower than with aerobic treatment.

With these advantages, why have fruit and vegetable processors not installed more anaerobic treatment systems? The answer lies partially in the disadvantages associated with anaerobic treatment. Anaerobic bacteria are generally more sensitive to toxic compounds than are aerobic bacteria and, since anaerobic bacteria grow more slowly than aerobic bacteria, they require greater time for waste digestion and additional time to respond and recover from shock loads. Anaerobic systems are also sensitive to pH change, low temperature and the presence of oxygen (toxic to some anaerobic bacteria). These disadvantages, plus the fact that electrical power costs and the other advantages of anaerobic treatment have not been great enough to cause food processors to seek alternatives to better-known waste treatment methods, have resulted in comparatively little use of anaerobic treatment.

Even with these disadvantages, anaerobic treatment has been used, and sometimes successfully, for the treatment of fruit and vegetable wastes. Anaerobic ponds are perhaps the most widely applied method; a variety of anaerobic filters and anaerobic contact processes have also been used with success. However, it is doubtful that anaerobic treatment will be used to treat fruit and vegetable wastes to what are generally considered secondary treatment standards. This is because of the high strength of raw fruit and vegetable processing wastes and the problems of removing oxidizable material (as tested by the aerobic BOD or COD tests) with a treatment process that is essentially oxygen-free. High-strength wastes generally contain too many anaerobically nonbiodegradable compounds that can be oxidized and reported as BOD or COD to produce an effluent of secondary treatment quality.

Anaerobic treatment may be considered as an intermediate step prior to further aerobic treatment or effluent polishing to meet pretreatment goals, and it is also appropriate where an acceptable effluent quality can be achieved through dilution and either BOD or COD loads need to be reduced. Ideal applications of anaerobic treatment are as a pretreatment system to a municipal sewer or a land application system, or to precede and reduce what would otherwise be a much larger aerobic treatment system. As energy costs continue to rise, food processors can be expected to place a greater premium on the advantages of the anaerobic treatment process; this will probably increase its use.

4.6.1 Ponds

Processors of potatoes, corn, apples and some other products have successfully used anaerobic ponds for treatment. The ponds are usually about 4.5 to 6 m deep, to minimize the surface and land area requirements and also the odours that can result from air–water contact. Anaerobic ponds are relatively inexpensive to construct and, unless problems occur, require little maintenance. Odour emissions from anaerobic ponds limit their practical application to rural or industrialized locations where nuisance complaints are unlikely to occur.

4.6.2 Filters

In an anaerobic filter, media of either rock, plastic or some other material are housed in a closed container (Fig. 4.5(b)). The raw waste is distributed in the bottom of the container and treated effluent, as well as methane and other off-gases, is taken from the top of the filter structure. Filters such as those shown treating starch-gluten wastes in Fig. 4.5(a) have been reported to remove 1.6 to 3.2 kg COD m^{-3} d^{-1}. Although anaerobic filters are relatively inexpensive to construct, their application is limited by problems in obtaining good distribution of the raw waste and in maintaining proper pH for biological growth, and the inability to obtain a treated effluent of secondary treatment standard quality. However, since off-gases can be collected and odours scrubbed, the anaerobic filter can be made suitable for use in urban areas. This is a distinct advantage over ponds, where collection of off-gases is difficult because of the large water surface area.

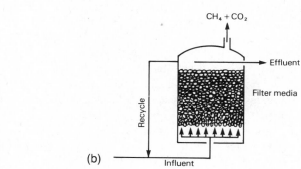

Fig. 4.5 Anaerobic filter

4.6.3 Contact process

To achieve higher waste removals per volume of reactor, several investigators have tested the anaerobic contact process. This process is similar to an activated sludge system (see Section 4.7.3) in which anaerobic conditions are maintained by mixing raw wastewater and return sludge in a closed container without air. Although removals of 3.2 to 6.4 kg COD $m^{-3}\,d^{-1}$ have been reported, practical operating and design problems with this process have limited its use. Like the other anaerobic treatment processes, anaerobic contact has been subject to upset from shock loadings and problems in separating the anaerobic bacteria from the treated effluent.

4.6.4 New anaerobic processes

New anaerobic treatment technologies, some of which are patented, are being tested on fruit and vegetable wastes. For several of these new technologies, daily removal rates of 16 to 24 kg COD m^{-3} are claimed. The new technologies include fluidized sand beds, effluent and/or solids recycle, and the use of new filter media. Whether these innovative approaches can successfully be used for the full-scale treatment of fruit and vegetable wastes remains to be proven.

4.7 Secondary treatment

Aerobic systems are the treatment processes most commonly used to achieve the BOD removals usually necessary to meet secondary effluent discharge limits. Such limits vary depending on the receiving stream and the requirements of local regulatory agencies. Common requirements in the USA are 80% BOD removal or 30 mg l^{-1} BOD and TSS, whichever achieves the better quality effluent.

Aerobic treatment may also correctly be called aerobic waste conversion because dissolved starches, sugars and other carbohydrates are utilized by micro-organisms as food, and in the process new bacteria are generated or converted to waste bacterial solids. For this conversion process the micro-organisms require oxygen that must be supplied through an aeration or treatment unit. Also, the new biological cell matter that is produced must be disposed of. If proper conditions are not supplied for bacterial growth, the settleability of the micro-organisms decreases so that solids cannot be separated and removed from the treated effluent. Poor settling of solids may result from too much BOD being available to the micro-organisms, or insufficient amounts of oxygen or nutrients for proper biological growth. The relationship of solids produced, oxygen required and sludge settleability, to the amount of waste loadings discharged to the aerobic treatment system is shown in Fig. 4.6. Waste loading is expressed as food (BOD) available in the secondary influent, divided by the amount of micro-organisms available in the aeration basin (mixed liquor volatile suspended solids); this is commonly referred to as the food-to-micro-organism ratio, or F/M, in units of kg BOD applied daily to the secondary system per kg volatile solids in the treatment basin. As shown in Fig. 4.6, at higher waste loadings less oxygen is required, more solids are produced, and sludge settleability can decrease.

Experience has shown that lack of a proper environment for the growth of bacteria and other micro-organisms results in system failure. For example, neglecting to provide a sufficient supply of oxygen, even during peak waste discharge periods, can result in septic sludge which enables undesirable, nonsettleable micro-organisms to predominate; inadequate design of sludge concentration or disposal equipment can result in a solids buildup within the treatment system until the accumulation of solids in the final settling structures

THE FRUIT AND VEGETABLE PROCESSING INDUSTRIES   241

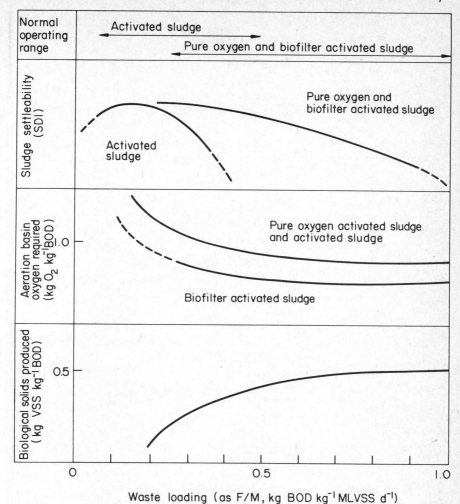

Fig. 4.6 Effects of waste load on activated sludge system performance

overflows into the treated effluent, thereby resulting in poor treated effluent quality. Attempts to treat without having proper nutrients, temperature or pH in the main biological reactor can similarly result in the predominance of micro-organisms that are difficult to settle and in poor effluent quality.

Further information about the environment necessary to maintain biological growth is available in numerous texts (Grady and Lim, 1980). The biological response to various environmental conditions is illustrated in Fig. 4.7. The biological response when treating domestic and apricot-canning wastes at an inappropriate F/M ratio of 1.4 is shown in Fig. 4.7(a). Treatment plant

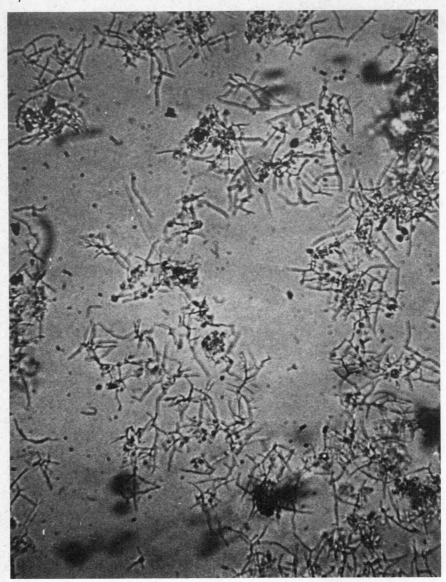

Fig. 4.7 Biological response of activated sludge
    (a) Non-settleable bacteria (unidentified), apricot-canning high F/M (1.4)

Fig. 4.7(b) Fungi, low pH wastewater

response to such loadings is illustrated in Fig. 4.8(a), showing that foam in the treatment reactor and poor effluent quality can occur with inappropriate F/M. (Foaming can also be caused by the presence of excess surfactants, a high concentration of oils and grease or the accumulation of surface microorganisms, often as a result of poor surface hydraulic characteristics.)

Similar treatment problems occur where low pH in the biological reactor results in the growth of undesirable fungi (Fig. 4.7(b)).

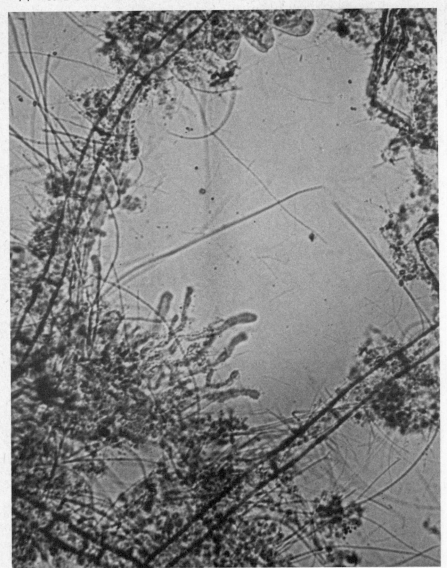

Fig. 4.7(c) Filaments, nutrient deficiency

Deficiencies in proper nutrients can also result in the shift from desirable single-cell micro-organisms to multicellular microbial growths, generally termed filamentous micro-organisms (Fig. 4.7(c)). Oxygen deficiency and other operational shortcomings can also cause filamentous bacteria to predominate.

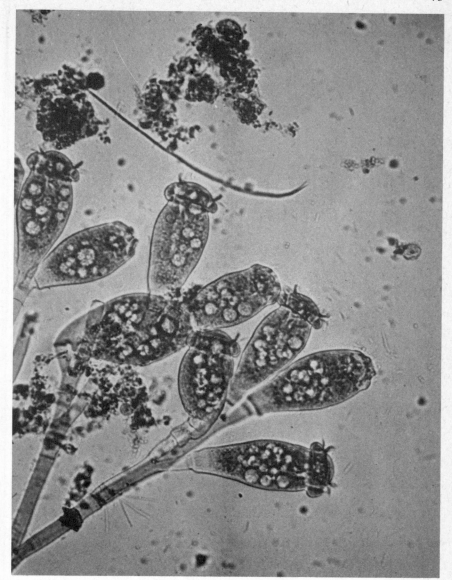

Fig. 4.7(d) Protozoa in a healthy sludge

These micro-organisms do not settle or compact well in the clarification or sedimentation steps necessary to obtain good effluent quality. The filamentous growth is commonly called 'bulking sludge' and often results in carryover of solids into the treated effluent; Fig. 4.8(d) shows this effect in the treated

Fig. 4.8 Activated sludge treatment plant response
    (a) Foam

effluent of apple juice processing mixed with domestic sewage. Solids carryover can also occur when the F/M ratio is kept too low and the bacteria become dispersed; this causes a condition commonly referred to as 'pinpoint floc' (Fig. 4.8(c)). Maintaining sludge for too long a time in the secondary clarifier can also result in floating or rising sludge (Fig. 4.8(b)). Rising sludge in the secondary clarifier is especially a problem at low F/M values where excessive nitrification has taken place.

Fig. 4.8(b) Rising solids, excessive clarifier retention time

With proper design and operation, these problems of sludge and bacterial response can be overcome. Aerobic treatment plants have been operated successfully with fruit and vegetable wastewaters. Proper design and operation result in healthy (settleable) bacteria and protozoa, as shown in Fig. 4.7(d). As a consequence, good effluent quality can be achieved.

A number of different treatment processes can be used to achieve good

Fig. 4.8(c) Pinpoint floc, low F/M ratio

effluent quality at the lowest possible capital and operating costs. The best choice of treatment varies, depending on individual site restrictions, waste characteristics, local power costs, effluent requirements and a number of other factors. In all cases, the secondary treatment units must provide an environment suitable for the growth of the desired biological organisms.

Design criteria typically used to provide a suitable environment for various secondary treatment processes are summarized in Table 4.6, which reflects recent experience in the successful application of secondary treatment technology. These criteria are, at best, rules of thumb, and the application of design factors to a specific fruit and vegetable wastewater should be judged individually. Since there is a range of design criteria that might be appropriate, depending on the particular waste treatment situation, criteria other than those listed in Table 4.6 might be given consideration.

4.7.1 Ponds and lagoons

Non-aerated ponds are usually enclosed by earthen dikes (Fig. 4.9). Because mechanical aeration is not used in ponds, oxygen must be supplied by biological means. Potential sources of oxygen include photosynthetic algae, oxygen previously entrained in the pond influent and oxygen supplied from contact of

Fig. 4.8(d) 'Bulking sludge', oxygen deficiency in apple juice processing–domestic wastewater

the pond water surface with the atmosphere. Depending on the degree of treatment desired, ponds may be used in a variety of ways, including the treatment or stabilization of untreated primary or raw wastewater. Because of the high strength of food processing wastes, BOD loadings to such stabilization ponds must be kept low to prevent odour and other nuisances — even with low loadings such problems tend to arise periodically since operator control is minimal.

Aeration is often used to supply oxygen needed for waste stabilization and to allow the size of the pond to be reduced. Ponds used in this manner are generally called aerated lagoons. Non-aerated settling ponds often follow the

Table 4.6 Secondary treatment design criteria for food processing wastes

PONDS
Depth	0.9–1.8 m
Hydraulic detention time	7–30 days
BOD loading	1.7–5.6 g m^{-2} d^{-1}

AERATED LAGOONS
Lagoons
Depth	2.1–4.6 m
Hydraulic detention time	>5 days

Aerators
Oxygen requirements	1.3–2.0 kg O$_2$ kg^{-1} BOD
Minimum mixing	2.0–3.9 W m^{-3}

TRICKLING FILTER
Filter
Media depth	0.9–2.4 m with rock media
	3.0–12.2 m with high-rate media
Configuration	— Circular with rotating distributor
	— Rectangular with fixed nozzle distributors
BOD loading	0.32–0.64 kg m^{-3} d^{-1}
Minimum wetting rate	0.11 m^{-2} s^{-1} for circular
	0.71 m^{-2} s^{-1} for rectangular

Clarifier
Overflow rate	16–24 m^3 m^{-2} d^{-1}
Solids loading	Not critical
Configuration	Circular recommended
Settled sludge concentration	0.5–0.75% TSS

ACTIVATED SLUDGE
Aeration basin
Mixed liquor suspended solids (MLSS)	2000–4000 mg l^{-1}
Per cent volatile suspended solids	50–90%
Maximum food-to-micro-organism ratio (F/M)	0.2 kg BOD kg^{-1} MLVSS d^{-1}
Water depth	3–6 m

Aerators
Oxygen requirements	1.0–1.3 kg O$_2$ kg^{-1} BOD
Minimum mixing	7.9–15.8 W m^{-3}

Clarifier
Overflow rate	16 m^3 m^{-2} d^{-1}
Solids loading	49–73 kg m^{-2} d^{-1}
Configuration	Circular recommended
Settled sludge concentration	0.5–0.75% TSS

BIOFILTER ACTIVATED SLUDGE
Filter tower
Media depth	3.0–12.2 m
Configuration	— Circular with rotating distributor
	— Rectangular with fixed nozzle distributors
BOD loading	1.6–4.8 kg m^{-3} d^{-1}
Minimum wetting rate	0.3–0.71 m^{-2} s^{-1} for circular
	1.71 m^{-2} s^{-1} for rectangular

Aeration basin
  Mixed liquor suspended solids
    (MLSS) — $2000-4000\ \text{mg l}^{-1}$
  Per cent volatile suspended solids — $50-90\%$
  Maximum food-to-micro-organism
    ratio — $0.7\ \text{kg BOD kg}^{-1}\ \text{MLVSS d}^{-1}$
  Water depth — $3-6\ \text{m}$
Aerators
  Oxygen requirements — $0.6-0.9\ \text{kg O}_2\ \text{kg}^{-1}\ \text{BOD}$
  Minimum mixing — $7.9-15.8\ \text{W m}^{-3}$
Clarifier
  Overflow rate — $16\ \text{m}^3\ \text{m}^{-2}\ \text{d}^{-1}$
  Solids loading — $49-73\ \text{kg m}^{-2}\ \text{d}^{-1}$
  Configuration — Circular recommended
  Settled sludge concentration — $0.5-1.0\%$ TSS

PURE OXYGEN ACTIVATED SLUDGE
Aeration basin
  Mixed liquor suspended solids — $3000-6000\ \text{mg l}^{-1}$
  Per cent volatile suspended solids — $50-90\%$
  Maximum food-to-micro-organism
    ratio — $0.7\ \text{kg BOD kg}^{-1}\ \text{MLVSS d}^{-1}$
  Water depth — $3-6\ \text{m}$
Aerators
  Oxygen requirements — $0.6-0.9\ \text{kg O}_2\ \text{kg}^{-1}\ \text{BOD}$
  Minimum mixing — Greater than activated sludge, as recommended by aerator manufacturer
Clarifier
  Overflow rate — $16\ \text{m}^3\ \text{m}^{-2}\ \text{d}^{-1}$
  Solids loading — $49-73\ \text{kg m}^{-2}\ \text{d}^{-1}$
  Configuration — Circular recommended
  Settled solids concentration — $0.5-1.0\%$ TSS

ROTATING BIOLOGICAL CONTACTORS
Contactors
  BOD loading (first stage only) — $29\ \text{g m}^{-2}\ \text{d}^{-1}$
    (total stages) — $10-15\ \text{g m}^{-2}\ \text{d}^{-1}$
  Disc diameter — $3.7\ \text{m}$
  Length, each shaft — $7.6\ \text{m}$
  Surface area per shaft — $9290\ \text{m}^2$
  Drive power (per mechanical
    shaft) — $3.6\ \text{kW}$
Clarifier
  Overflow rate — $16\ \text{m}^3\ \text{m}^{-2}\ \text{d}^{-1}$
  Solids loading — $49-73\ \text{kg m}^{-2}\ \text{d}^{-1}$
  Configuration — Circular recommended
  Settled solids concentration — $0.75-1.5\%$ TSS

aerated lagoon to settle and store biological solids, as illustrated in Fig. 4.9. Eventually, settling ponds must be dewatered and waste solids removed and disposed of.

Stabilization ponds have been successfully used for the treatment of fruit and vegetable wastewater. However, many processors have added aeration because

(a)

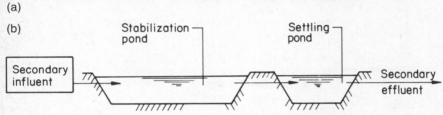
(b)

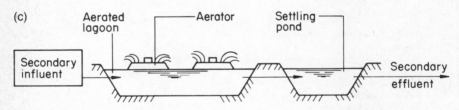
(c)

Fig. 4.9 Ponds and lagoons
    (a) Aerated lagoon and settling pond
    (b) Stabilization pond
    (c) Aerated lagoon

of complaints about odours that result from the periodic decline in algae growth; odours resulting from insufficient oxygen during warm weather or during peak processing loads are also common. The large amount of land required for stabilization ponds has also limited their use. Nonetheless, if a plant is located in a rural setting, if complaints are unlikely, if land is relatively inexpensive and if periodic poor effluent quality due to excess algae and suspended solids can be tolerated, stabilization ponds with their low capital costs may be viable.

In many cases, however, the food processing plant is in an area where stabilization ponds cannot be used. An aerated lagoon followed by solids settling and storage ponds may be the next most economical waste treatment system. Although more land and greater power requirements for mixing are necessary for aerated lagoons than for some other secondary treatment options, aerated lagoons have a distinct advantage in being relatively easy to operate and requiring no costly final clarifier.

4.7.2 Trickling filters

One of the oldest aerobic biological treatment systems is the trickling filter. In the past, filters have generally consisted of a bed of rock medium about 1.8 to 2.4 m in depth (Fig. 4.10). Air fills the void spaces between the stones and provides oxygen that is entrained into the wastewater as it trickles through the filter medium. Waste loadings to trickling filters of rock medium have been limited to 320 to 640 g BOD $m^{-3} d^{-1}$ of medium. At higher loadings, odour and insect problems can occur as voids between the stones become plugged.

Recent use of horizontal redwood or vertical plastic media has allowed daily loadings to be increased to 3.2 to 4.8 kg m^{-3} without problems of plugging or odour. These new media, commonly referred to as high-rate media, present a large surface area and also a high percentage of voids or airspaces. Figure 4.11 shows two types of high-rate media commercially available: redwood (horizontal redwood slats) and a plastic medium made of vertical corrugated PVC sheets. High-rate media allow more air to circulate and better sloughing of bacteria than rock media. Thus, trickling filter loadings are not usually limited by lack of oxygen but solely by the amount of BOD that can be removed per unit volume of medium. Reported BOD daily removal rates range from 0.8 to 1.6 kg m^{-3}, the lower value being more common.

Although numerous equations are available for predicting effluent quality from trickling filters (Gromiec and Malina, 1970; Benefield and Randall, 1980), they are dependent on numerous assumptions concerning waste treatability and coefficients of performance specific to each type of filter medium. Because of these variables, there is no substitution for pilot testing to determine the actual removal values.

The advantages of a trickling filter are its ease of operation, relatively low maintenance requirement, low power requirement, and the relatively stable effluent obtained under varying waste discharges from fruit and vegetable processors. Its disadvantages are high initial capital costs, the relatively large land area required, and uncontrolled sloughing of biological solids that can sometimes impair treated effluent quality and cause odours.

Trickling filters have been used in conjunction with activated sludge aeration basins. Here the trickling filter is used as a roughing filter to precede the aeration basin — return activated sludge is not recycled over the trickling filter.

(a)

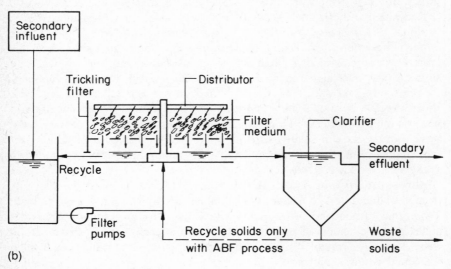

(b)

Fig. 4.10 Trickling filter plant
    (a) Rotating distributor on traditional rock medium filter
    (b) Flow diagram

As a roughing filter, a trickling filter usually has BOD loadings ranging from 3.2 to $4.8\,\mathrm{kg\,m^{-3}\,d^{-1}}$.

Other relatively new processes using trickling filters have been used for waste treatment at lower filter loads, ≤ 0.8 to $1.6\,\mathrm{kg\,BOD\,m^{-3}\,d^{-1}}$: these are the patented activated biofilter, or ABF process (Schaumburg and Lasswell, 1970),

Fig. 4.11 High-rate biofilter media
    (a) Redwood medium detail (courtesy Neptune Microfloc)

and the trickling filter solids contact, or TF/SC process (Norris *et al.*, 1982). At the time of writing, the ABF process has been successfully used on food processing wastes. The TF/SC process is relatively new and has not been used on food processing waste but has been tested at several sites on domestic wastewater.

In the ABF process (patented by Neptune Microfloc), solids from the clarifier bottom (see Fig. 4.10) are recycled over the top of the trickling filter so that a mixed liquor, similar to that in an activated sludge plant, is maintained throughout the trickling filter. In this process, the trickling filter serves as the aerator. Since rock trickling filter media would become plugged with the recycle solids, the ABF process can only be used with high-rate filter media.

The TF/SC process can be used with rock media as well as with high-rate media. This process is essentially a very lightly loaded ($<0.8-1.6$ kg BOD m^{-3} d^{-1}) roughing filter, followed by a contact or aeration basin with approximately 30 min hydraulic detention time. Since the roughing filter has a low BOD loading, there is little BOD remaining for the aeration basin to remove. This allows the size of the aeration basin and aerators to be significantly smaller than is required for conventional activated sludge.

The main advantage of both the ABF and TF/SC processes is their ease of operation. For low-strength food processing wastes they may also provide

Fig. 4.11(b) Distribution over redwood medium

significant power savings over other processes. However, for high-strength food processing wastes, the need to maintain minimum wetting rates by recycling tower effluent to the filter may minimize or eliminate these power savings. Another advantage of these processes is the improved effluent quality compared with that obtained using the trickling filter alone, so that they are particularly suitable for consideration when improving existing treatment facilities. A disadvantage of both ABF and TF/SC is that the filter volume required is two to four times that of other treatment options at higher loadings, thus capital costs are higher. There is also less process flexibility than is afforded by some other treatment processes.

4.7.3 Activated sludge

In the 1940s the activated sludge process (Fig. 4.12) began to replace the use of trickling filters to achieve secondary treatment standards. For treatment of

Fig. 4.11(c) Plastic medium (courtesy B F Goodrich)

domestic waste, activated sludge achieves good effluent quality, requires less land and is lower in capital costs than the earlier rock trickling filter plants. However, the application of activated sludge for treatment of fruit and vegetable wastes has been less successful. Field experience has shown that organic loadings to the aeration basin need to be maintained at an F/M ratio of 0.2 or less to achieve continuously high effluent quality and process stability. Attempts to use the activated sludge process at higher loadings have produced effluent with a predominance of undesirable filamentous micro-organisms (Fig. 4.7). Sludge bulking, foaming and poor effluent quality have been the result of operating at high F/M loading rates.

Activated sludge can be used successfully to treat most fruit and vegetable wastes, but usually when the F/M ratio is less than 0.2. A major advantage of the activated sludge process is the treatment plant operator's ability to control the system by varying the number of micro-organisms present in the aeration basin. This allows proper operation of the system under a variety of waste load conditions. The treatment plant operator can modify conditions to achieve good effluent quality or reduce power requirements and sludge production. A major disadvantage of activated sludge is its susceptibility to upset from shock loading and sludge bulking problems once undesirable bacteria are present.

258 INDUSTRIAL WASTEWATER TREATMENT

Fig. 4.11(d) Distribution over plastic medium

4.7.4 Biofilter activated sludge

Attempts to find treatment alternatives that would overcome the disadvantages of both the trickling filter and the activated sludge processes have resulted in numerous variations of combined trickling and activated sludge plants. Besides those systems already mentioned in Section 4.7.2, a combination of technologies for the treatment of fruit and vegetable wastes, the biofilter activated sludge process, has been used successfully. As illustrated in Fig. 4.13, this process includes a trickling filter tower that precedes the conventional activated sludge aeration basin and clarifier. The term 'biofilter' denotes that returned

THE FRUIT AND VEGETABLE PROCESSING INDUSTRIES   259

(a)

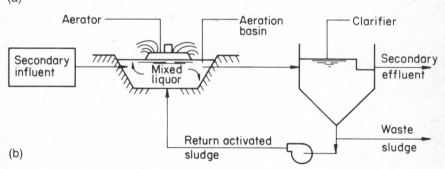

(b)

Fig. 4.12 Activated sludge plant
 (a) Surface aerated activated sludge
 (b) Flow diagram

sludge from the clarifier along with secondary influent is recycled over the filter. The inclusion of recycled sludge in this process differentiates it from the roughing filter activated sludge process, where return sludge is transferred directly from the clarifier to the aeration basin. Also, the biofilter activated sludge process is distinctly different from the ABF process in that the mixed liquor is both aerated and grown primarily in the aeration basin which follows the filter.

A number of different tests have compared the roughing filter activated

260 INDUSTRIAL WASTEWATER TREATMENT

(a)

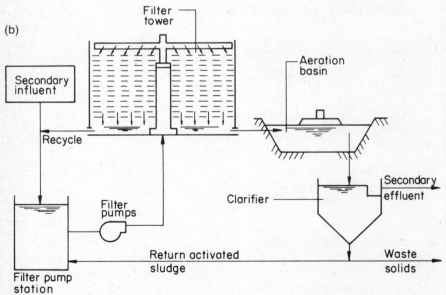

(b)

Fig. 4.13 Biofilter activated sludge
    (a) Filter tower
    (b) Flow diagram

sludge process with the biofilter activated sludge process. Results indicate that the biofilter activated sludge process produces a more stable system than does the roughing filter activated sludge process (CH2M HILL, 1972; Williams *et al.*, 1973; CH2M HILL, 1976) when treating fruit and vegetable wastes. Since only minor piping modifications are necessary to accommodate either process, it is usual to design a combined filter and activated sludge plant that allows use of either process.

There are over fifty biofilter activated sludge plants in the USA; most treat industrial waste, often food processing waste. While some have failed to achieve design expectations (Harrison, 1980), most have been very successful. For example, a french fry (potato chip) producer uses a trickling filter at a loading of $10\,kg\,BOD\,m^{-3}\,d^{-1}$ followed by an activated sludge basin loaded at an F/M ratio of $0.56\,kg\,kg^{-1}\,d^{-1}$; this treatment process has a history of over 9 years of successful operation.

The main advantage of biofilter activated sludge is that the activated sludge portion of the plant can be operated at organic loadings several times higher than those possible with the conventional activated sludge process. High effluent quality and savings in operating costs are also achieved with the biofilter activated sludge process. The main disadvantage of the process is its relatively high capital costs compared with other secondary treatment processes. This is especially true in locations where earthen basins would make conventional activated sludge or aerated lagoons more cost-effective. The difficulty of obtaining good distribution of wastewater to the filter and icing problems during cold weather are further disadvantages.

4.7.5 Pure oxygen activated sludge

The pure oxygen activated sludge process, sometimes used to overcome the shortcomings of the better-known activated sludge process, uses a covered aeration basin divided into stages for mixing and aerating the wastewater. With pure oxygen activated sludge, high purity (approximately 90%) oxygen rather than air (which contains only 21% oxygen), is contacted with the mixed liquor in the aeration basins. A reported advantage of using pure oxygen, rather than air, is that a larger biomass concentration can be supported per volume of basin, so that smaller aeration basins can be used. Special oxygen generation equipment and covered basins are required for pure oxygen activated sludge (Fig. 4.14).

In designing pure oxygen activated sludge plants, professionals differ as to whether the size of the secondary clarifier can be reduced from that necessary with conventional activated sludge. Moreover, several studies have concluded that pure oxygen activated sludge treatment is more expensive than use of the conventional activated sludge. Despite the claims of equipment manufacturers, the high capital costs of this process have undoubtedly been one reason for limiting its use in the treatment of fruit and vegetable wastes. A further

262   INDUSTRIAL WASTEWATER TREATMENT

(a)

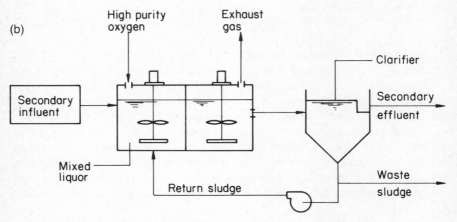

(b)

Fig. 4.14 Pure oxygen activated sludge
  (a) Full-scale plant (courtesy Letepro)
  (b) Flow diagram

disadvantage is the sophisticated operation necessary for maintenance of the pure oxygen generation equipment and the specialized aerators required. A recognized benefit of the pure oxygen activated sludge process is that smaller, although more expensive, aeration basins are required.

4.7.6 Rotating biological contactors

Rotating biological contactors (RBCs) are devices that use plastic filter media which rotate on a horizontal shaft to allow contact between bacteria and raw wastewater (Fig. 4.15). The flow pattern is similar to that of a conventional trickling filter plant, but recycle, although possible, is probably the exception rather than the rule. Mixing and oxygen transfer are accomplished by the rotation of the disc. Like the trickling filter, biological solids periodically slough from the filter medium and are removed but not recycled in a secondary clarifier. Unlike the trickling filter, which requires that water be pumped to trickle through the medium, no pumping is required with the use of RBCs since the medium itself is rotated through the wastewater.

When RBC equipment was first introduced into the USA in the early 1970s, manufacturers suggested that these systems be designed on the basis of daily hydraulic loading ($m^3 m^{-2}$) commonly used for low-strength domestic wastewater. But for high-strength waste from fruit and vegetable producers it is the organic rather than the hydraulic load that limits the amount of waste discharge that can be applied to an RBC. The initial loadings suggested by the equipment vendors had to be reduced drastically before successful RBC operation could be achieved (Cochrane and Dostal, 1972; Thomas and Sanborn, 1973). BOD removal rates with the RBC process can vary from 0.21 to 0.82 $kg m^{-2} d^{-1}$ of filter media. Slight improvements in RBC performance have been reported with the use of diffused air in conjunction with biodiscs, effluent recycle and other process modifications. However, these modifications have also added to the cost and complexity of RBC installations.

Advantages of the RBC method are low power requirements and ease of operation. Disadvantages are the lack of operator control, inferior filter media offered by some vendors, and limited ability to cope with shock or fluctuating waste loads. A feature that should be noted in the design of such treatment processes is that, in scaling up from a small unit, rotation speed and related variables must be adjusted if the performance of the full-scale unit is to be predicted reliably.

4.7.7 Pilot testing

Design criteria for secondary wastewater treatment systems, as well as other waste treatment technologies, vary greatly from plant to plant with different products, maintenance practices and process selections. The principles embodied in such commonly used phrases as 'Based on our present

Fig. 4.15 Rotating biological contactor
(a) Media detail

THE FRUIT AND VEGETABLE PROCESSING INDUSTRIES 265

(b) Full scale plant

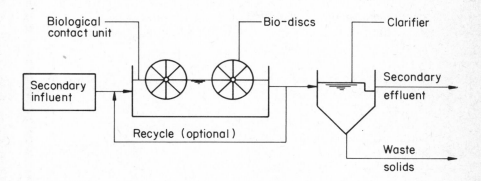

(c) Flow diagram

knowledge...', 'From experience with similar wastes...', 'Other investigators have shown...' or 'The equipment manufacturer recommends...' are not usually adequate to ensure success of the final treatment system. Certainly, when thousands or sometimes millions of dollars are being invested in waste treatment, further assurance of successful treatment is often necessary: a properly operated and evaluated pilot plant is a logical method of obtaining design criteria specific to the food processor's waste. Through the course of the pilot study and evaluation, a better characterization of the waste loads at the plant is often obtained, and design criteria for the full-scale wastewater treatment plant can be established more accurately. The feasibility of obtaining the desired effluent quality is also determined during pilot study, as is useful information on operational requirements of the full-scale system.

Pilot plants should be designed to model, as closely as possible, the operation of the full-scale treatment plant. Figure 4.16 shows just a few of the different reactors used in pilot-testing secondary wastewater treatment facilities. Aerated lagoons or ponds can be adequately tested by using common household swimming pools as reactors (Fig. 4.16(a)). The basins of pilot activated sludge plants may be constructed of a number of different materials and include steel tanks (Fig. 4.16(b)). The pilot-testing of more sophisticated facilities, such as those used with the pure oxygen activated sludge process, often requires specialized equipment that is supplied ready for use as trailer- or skid-mounted units (Fig. 4.16(d)). Other equipment manufacturers may supply their hardware in modules or pieces that require on-site construction, such as the activated biofilter and roughing filter pilot units shown in Fig. 4.16(c).

A successful pilot plant study starts with obtaining a representative source of raw wastewater for use as feed to the pilot reactors. This includes obtaining wastewater from sewers or gutters that will actually need treatment. Cooling-water or other noncontaminated wastewaters that will not be treated in the full-scale system should be separated out prior to pilot testing.

Design criteria that need to be established during the pilot programme for biological treatment include the following:

- Waste treatability
- Oxygen requirements
- Sludge production rate
- Solids settleability
- Nutrient requirements
- Sludge dewaterability
- Ability to achieve desired effluent quality

To determine these criteria the pilot plant must be operated under a number of varying conditions. The operation usually includes a test period when the waste loading nearly equals the estimated design load of the facility. To determine reactions to peak loads and to ensure that the design load originally estimated

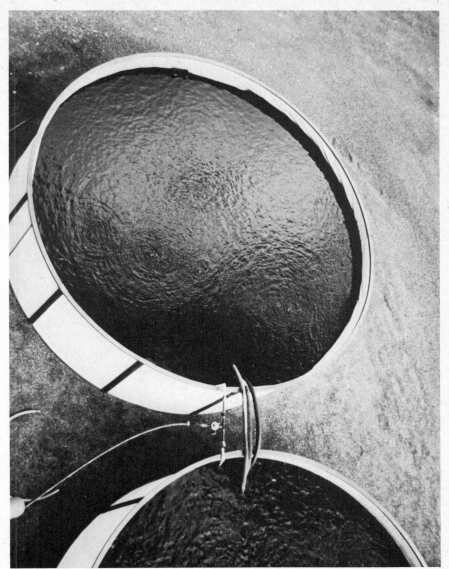

Fig. 4.16 Pilot plants for process
      (a) Aerated lagoon evaluation (above-ground swimming pools)

was not too conservative, there is a further period of testing at loadings above those anticipated to be suitable for average design conditions.

A pilot study does not guarantee a successful secondary wastewater treatment facility: pilot plant studies and evaluations conducted by inexperienced personnel can lead to inappropriate conclusions and the use of improper design

Fig. 4.16(b) Activated sludge

Fig. 4.16(c) Biofilter activated sludge

Fig. 4.16 (d) Pure oxygen activated sludge

criteria. For example, a pilot plant operated for only a short period of time, or one that does not exactly model all of the load conditions of the full-scale plant, may lead to wrong conclusions. Moreover, full-scale operation may involve different ventilation requirements, sidewall-to-treatment-volume ratios, effects from sunlight, short-circuiting or wave action, and a number of other inter-related physical and environmental factors that may or may not be present in the pilot-scale unit. Only through a carefully conducted pilot study, careful interpretation of data, and proper scale-up of information can the risk of failure of a full-scale wastewater treatment plant be minimized.

4.8 Tertiary treatment

Tertiary or advanced wastewater treatment encompasses a large number of individual processes that can be employed to remove the organic, solid or inorganic pollutants that remain after secondary treatment. Depending on the characteristics of the fruit and vegetable wastes being treated and the secondary treatment process used, tertiary treatment may be required if the final effluent desired is to contain less than 30 to 100 mg l^{-1} BOD or 50 to 100 mg l^{-1} TSS. Tertiary or additional secondary treatment may also be required if removal of nitrogen or other nutrients is required.

Treatment processes that are theoretically applicable to secondary effluents include nitrification, chemical precipitation and sedimentation, filtration, carbon adsorption, ion exchange, reverse osmosis, ammonia stripping and others. Of these, only the first three — nitrification, filtration, and chemical precipitation and sedimentation — are likely to be applied to waste from fruit and vegetable processing plants. For a discussion of other tertiary processes, the reader is referred to guides on treatment plant design (Sanks, 1979; Culp and Culp, 1971).

4.8.1 Nitrification

Nitrification can often be achieved in the aeration basin, biofilter or other BOD removal device used in a secondary wastewater treatment plant. Other methods use special basins solely for nitrification (Sutton and Jank, 1977). Regardless of the treatment method used, the key to obtaining biological nitrification is the provision of a suitable environment for the nitrifying organisms that oxidize ammonia to the nitrate form of nitrogen. Since nitrifying organisms (*Nitrosomonas* and *Nitrobacter* bacteria) are more sensitive to environmental conditions than the bacteria that normally remove BOD and other organic compounds, in order to establish nitrification the waste manager must identify and observe critical design parameters such as temperature, pH, dissolved oxygen concentration, possible shock or surge waste loads, and bacterial solids retention time.

An example of special requirements for nitrification is given by a french fry producer who manufactures identical products at two different plants. Waste from Plant 1 directly enters an aerobic biological treatment system, so that organic nitrogen compounds do not have sufficient time or appropriate conditions to be converted into ammonia. At Plant 2, wastewater first enters an anaerobic pond, then is treated in an aerobic activated sludge basin; excessive amounts of organic nitrogen are anaerobically broken down into ammonia, resulting in the need for nitrification prior to discharge to the local receiving stream. To achieve nitrification, existing aeration equipment has had to be replaced by equipment that does not cool the aeration basin contents excessively, so that nitrifying bacteria can be maintained during winter months.

The waste manager should be aware that, as in the preceding example, special process changes, equipment modifications or different operational procedures are likely to be required in the conventional secondary treatment process if nitrification is to occur.

4.8.2 Chemical precipitation and sedimentation

This process involves the addition of a chemical or polymeric agent to the secondary effluent. The coagulant helps to form a precipitate that is sufficiently large and dense to settle out by gravity. Coagulants commonly used are lime, alum, ferric chloride and certain polymers. A tube or plate gravity settling device is adequate to remove the coagulated solids, because of the relatively small solids load and low volume requiring treatment. The use of a specially designed chemical clarifier with flocculating influent well is another possible method of chemical precipitation and sedimentation. Effluent from the settling or sedimentation step can be either discharged or transferred to an effluent filtration process for final removal of any nonsettleable solids.

4.8.3 Filtration

Filtration to remove suspended solids particles that might otherwise be carried out in the effluent from a secondary wastewater treatment plant reduces not only the solids content but also the associated BOD, nitrogen and other compounds.

In the past, a single sand medium has been used for what is known as slow-sand filtration. The filter is usually loaded at 0.03 to $0.091 \, m^{-2} \, s^{-1}$ and is left in service until the head loss is near the top of the filter wall. At that time, the filter is drained and allowed partially to dry; then sludge that may have accumulated on the filter surface is removed. Although slow-sand filters are easy to operate and construct, they require a relatively large amount of land area and cost more than filters with faster flow rates.

Rapid-sand filtration now allows loading rates of 0.7 to $3.41 \, m^{-2} \, s^{-1}$, and dual- or tri-media filters can be used. These filters can either be pressurized, as in the system illustrated in Fig. 4.17, or operated with gravity flow in open tanks

THE FRUIT AND VEGETABLE PROCESSING INDUSTRIES 273

(a)

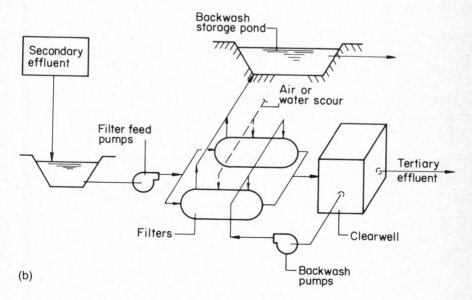
(b)

Fig. 4.17 Filtration system for tertiary treatment
    (a) Full-scale plant (courtesy Neptune Microfloc)
    (b) Flow diagram

or containers. Dual- or tri-media filters are sometimes preferred over single-media filters because more time can elapse before backwashing is required.

One disadvantage of rapid-sand filtration is the amount of operator attention necessary to backwash or reverse the flow of treated effluent through the filters. However, frequent backwashing is necessary to remove suspended solids that have been deposited on the bed. Backwash systems necessitate a significant provision of additional pumps, clear wells and storage ponds. Cleaning by hydraulic backwash is generally accomplished at pumping rates of $10 \, l \, m^{-2} \, s^{-1}$. Backwash may be preceded by air-scouring of the filter media.

As a rule, filters cannot be used economically when the suspended solids concentration of the secondary effluent exceeds 100 to $250 \, mg \, l^{-1}$ suspended solids, because of the frequency and volume of backwashing needed. Where high concentrations of suspended solids are anticipated, the secondary effluent should pass through effluent polishing ponds or be given further treatment by chemical precipitation and sedimentation prior to filtration.

4.9 Land treatment

Land disposal or treatment of fruit and vegetable waste is often the preferred method of pollution control, especially when the processing plant is in a rural area where land is cheap, where no city sewer is available for pretreatment and municipal discharge or where there is no receiving stream nearby for discharge of treated effluent. Since discharge of treated effluent to a receiving stream can often be avoided with land application, the regulatory requirements for waste monitoring are also minimized. Another often-cited benefit of land application is the production of a usable crop on the disposal site, whose sale can sometimes help to offset the cost of wastewater treatment and disposal.

Although land treatment has advantages, the waste manager should realize that treatment by this method can appear deceptively simple. As shown in Fig. 4.18, application of wastewater during winter months can present operational challenges. There are many successful land application systems, but there are also many that have failed as a result of inappropriate application procedures causing contamination of groundwater, surface runoff, loss of soil infiltration capacity, decrease in plant growth and other environmental damage.

System failures generally are not apparent at first but are likely to be seen after five or more years because, with most land application systems, the effects of poor land management are cumulative. Failure can only be avoided or corrected by means of good design and system management.

4.9.1 Process variations
Land treatment can be accomplished in several different ways. The two basic methods generally used by food processors are high-rate irrigation and overland flow (Fig. 4.18).

THE FRUIT AND VEGETABLE PROCESSING INDUSTRIES   275

(a)

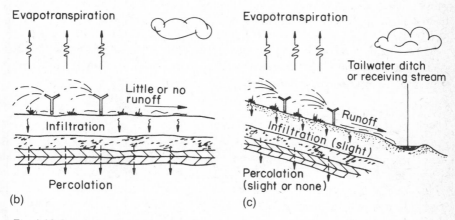

(b) (c)

Fig. 4.18 Land treatment
(a) Application in winter
(b) High-rate irrigation
(c) Overland flow

In high-rate irrigation, which is used by most food processors, conventional sprinklers apply the wastewater. Hydraulic loading rates to the irrigation field exceed the water requirements of the crop (this differentiates high-rate from standard irrigation where only the water needed by the plants is added) and much of the water percolates below the root zone. Irrigation is stopped when the infiltration–percolation capacity of the soil is reached, to avoid runoff. Another constraint may be groundwater contamination with nitrogen, organics or other compounds.

Overland flow utilizes runoff from the land application system. The land surface must have a uniform slope, generally ranging from 2 to 8%, so that runoff will move in the desired direction. A series of mounds parallels the ground slope to discourage the applied water from short-circuiting the full width of the field. Examples of such mounds can be seen in the freshly cropped overland flow site to the left of the ponds in Fig. 4.9. Overland flow is used where soil permeability is low or the groundwater table is high, so that the majority of water does not move into the soil profile. Runoff from the site is collected at the base of the slope and given additional treatment or discharged.

High-rate irrigation presents relatively few nuisance problems if the system is properly designed and operated. Treated effluent is disposed of without discharge to a receiving stream, and power requirements for treatment are relatively low. The disadvantages are: the large amount of land required; the need for a suitable soil, which restricts the location of the disposal site; the need to store wastewater during rainy periods or when irrigation cannot take place without runoff; and operational difficulties in freezing weather.

Overland flow application has the advantages of reduced land area requirements compared with conventional irrigation methods, reduced wastewater sprinkler or application equipment costs, and the ability to use soils that might otherwise be unsuitable for conventional irrigation methods. Its disadvantages are the lack of suitable design criteria for high-strength food processing waste, the greater possibility of nuisance problems with increased waste loading, and the need to discharge treated runoff.

4.9.2 Constraints

Normally, it is necessary to pretreat wastewater before land application. Even when suspended solids concentrations are low, the wastewater is usually screened to ensure that sprinkler nozzles and other equipment items are not blocked. Silt and other suspended solids particles that cause rapid wearing of equipment should also be removed. Adjustment of wastewater pH, removal of oil and grease, and the removal of sodium or other specific ions may also be necessary to maintain the infiltration capacity of the soil or to avoid poisoning of plants.

Hydraulic constraints
In addition to pretreatment requirements, the hydraulic or infiltration capacity of the soil may limit the amount of wastewater that can be applied. Previous erosion, rough topography, or lack of a suitable cover crop will reduce the hydraulic application rate possible before runoff occurs. The infiltration rate

may be limited either by the immediate daily and hourly rate or by the accumulative total water that can be applied in a growing season or year. Table 4.7 gives the maximum hydraulic load applicable under ideal conditions to different soil types (evaporation and precipitation not included). Maximum hydraulic loadings actually obtainable under field conditions are often less than those shown in Table 4.7 because of sloping ground, groundwater table, poor crop cover or frozen or crusted ground.

Table 4.7 Estimated maximum hydraulic loading for irrigation — ideal conditions

	Infiltration rate		Percolation rate $(m^3\,yr^{-1})$
	Hourly $(m^3\,ha^{-1})$	Daily $(m^3\,ha^{-1})$	
Fine sand	254+	3810	7.58
Sand loam	127–254	1900	4.57
Silt loam	76–178	890	2.29
Clay loam	51–102	380	1.02
Clay	25–51	130	0.25

BOD

BOD associated with both suspended solids and dissolved organic material can also limit the amount of wastewater that can be applied to an irrigation field. If the BOD is in the particulate or solid form, the pollutants tend to lie on the ground surface and putrefaction is a greater problem than when the BOD is in a dissolved form that can percolate into the ground.

Typical average annual BOD loadings with high-rate irrigation range from 3.4 to 11 $g\,m^{-2}\,d^{-1}$. Higher loadings for the peak day or peak week often exceed the yearly average values. However, should too much BOD be applied to the soil, anaerobic conditions can occur. When anaerobic bacteria predominate, not only is the cover crop likely to die because of poor conditions for growth, but the hydraulic infiltration rate of the soil may be greatly reduced. This reduction occurs because anaerobic bacteria produce a slime that normally protects the bacteria but which tends to clog the soil particles, reducing the soil's ability to transmit liquid beneath the ground surface.

Suspended solids

Suspended solids do not usually limit the amount of food processing wastewater that can be applied to land, except when large quantities, such as peelings, remain on the ground surface and putrefy, attract flies or cause other nuisance conditions. Good management of the spray irrigation field with yearly harrowing or disking will usually minimize any surface sealing or crusting from suspended solids. Treated wastewaters that contain significant concentrations of suspended solids have been used for irrigation at daily application rates of 22.4 $g\,m^{-2}$, although daily loadings of less than 11.2 $g\,m^{-2}$ are more typical.

Chemicals

Nitrogen associated with the fruit and vegetable wastewater often limits the amount of waste that can be applied to irrigated land. Although not all the nitrogen may be immediately available for plant use, the amount of organic nitrogen that mineralizes to ammonia, and subsequently to nitrate, is often used as a limiting criterion by regulatory officials in restricting the amount of wastewater that can be land-applied. The amount of available nitrogen that can be applied by a land application system depends upon the length of the growing season and the type of cover crop. Typical annual nitrogen requirements for forage or grass crops range from 11.2 to $22.4\,g\,m^{-2}$, but can be as high as $67\,g\,m^{-2}$ for some grasses. Annual nitrogen requirements for trees or field crops such as oats, wheat or potatoes, generally range from 5.6 to $16.8\,g\,m^{-2}$.

The amount of sodium in the wastewater may be another chemical limitation. This is especially true if caustic peeling is used. Sodium can reduce the permeability of some clay-bearing soils, but is of less concern in silt or sandy soil. The potential effect of sodium on the soil is measured by the sodium adsorption ratio (SAR) of the waste. The waste manager should consult the numerous texts available on the effects of sodium (McNeal, 1981).

4.9.3 Operation and management

The major tasks involved in operating a land treatment system include:

- maintaining the proper application rate and frequency;
- managing the soil and cover crop;
- monitoring the performance of the system.

Although these considerations are more critical with high-rate irrigation, even overland flow wastewater applications need to be scheduled according to weather and crop needs. Waste applications also have to be coordinated with harvest requirements. Storage ponds, sized adequately to hold wastewater when irrigation cannot take place, require aeration to prevent odours.

Experience has shown that both spray and surface irrigation can take place during the winter when fields are icy, provided that the system is designed especially to be operated in such conditions and maintenance standards are high. The amount of wastewater applied in winter must be regulated so that, when the spring thaw occurs, the release of water and BOD previously frozen on the ground does not result in temporary problems of odour, groundwater pollution and runoff.

Proper soil management is required to maintain the infiltration rate and to prevent soil erosion. To accomplish this, a healthy cover crop should be established and general soil conservation practised. Grasses and other crops help to retain the soil's infiltration rate by preventing sprinkler irrigation droplets from puddling and sealing the ground surface. A good cover crop is also necessary to remove nutrients from the ground. For successful operation, or as required by the local regulatory authority, monitoring of wastewater characteristics, soil, crop, groundwater and runoff may be necessary.

The waste manager should consult local agronomists, soil conservation personnel and farm advisory staff for expert assistance. The farming operation may be contracted to a private farmer, a share of the proceeds from the harvest going to the contractor.

4.10 Solids concentration and disposal

Whether the food processor's product loss is 10, 20 or 50%, there is a corresponding amount of waste solids from the screening, primary, secondary and tertiary treatment processes that must be concentrated and disposed of. The volume and difficulty of handling of the waste sludges generated in the treatment of fruit and vegetable waste should not be underestimated. The cost and inconvenience of concentrating and disposing of waste solids is often as great as, if not greater than, the inconvenience and expense of removing primary pollutants such as BOD or TSS.

Table 4.8 Typical waste solids characteristics

Type	Character
Screenings	20–50% solids screenings are usually dry and are handled as a solid rather than as a liquid sludge.
Silt	20–40% solids silt at the lower solids concentration is a thick mud, difficult to handle. At higher solids concentrations, silt has properties of a solid and handling is easier.
Primary solids	Solids usually range from 1 to 5%. The presence of silt or chemicals can raise the percentage to 10–30%.
Secondary solids	Claimed solids concentrations range from 0.5 to 2% for clarifier underflow. Concentrations less than 1% are more typical with food processing wastes, especially if the treatment plant is heavily loaded or nutrient-deficient.
Tertiary solids	For clarifier underflow preceded by lime precipitation, 7% solids may be possible. For solids settled from sand filter backwash, settled solids are likely to be less than 1%.

Table 4.8 lists the types and characteristics of the sludges generally encountered in the treatment of fruit and vegetable waste. Since sludges vary greatly from one production facility to another, a study is often required to determine the unique characteristics of the solids to be handled. Pilot tests of sludge thickening or concentrating equipment are often warranted prior to final sizing and designing of solids concentrating and disposal equipment.

Past failure to heed warnings of the difficulty of solids concentration and disposal is now being corrected at many processing plants due to the availability of new solids handling equipment, an increased understanding by engineers of how to design solids facilities, and the emphasis many processors are placing on converting what were formerly waste solids into useful byproducts or animal feed. Whether its final destination is conventional disposal or conversion, it is usually necessary to reduce the volume of the sludge by means of one of the following methods of concentration.

4.10.1 Conditioning

Solids conditioning is often necessary prior to further thickening, dewatering or disposal. The type and degree of conditioning varies with the type of solids produced and the chemicals used at the food processing plant. For instance, vacuum filters that dewater waste potato solids from plants using steam and noncaustic peeling processes often produce a relatively wet filter cake of 8–10% solids content and the rate of solids capture can drop dramatically. To improve the solids content it may be necessary to capture and condition the sludge before vacuum filtering it. Application of basket centrifuges to these same waste solids gives a cake of between 15 and 25% solids content with good capture, without any conditioning. However, with another type of solids the vacuum filter, rather than the centrifuge, may be the dewatering device that does not require the sludge to be conditioned.

The amount of conditioning needs to be evaluated for each sludge individually. Often, the conditioning tests are performed in conjunction with dewatering tests so that both can be properly evaluated.

Chemical conditioners

The use of chemical conditioners such as alum, ferric chloride, lime and other chemicals, although common in municipal waste treatment, is not general in food processing waste treatment systems. This is especially true of primary sludges destined for use as an animal feed or byproduct, where addition of a conditioning agent can reduce the sale value of the product or increase the food processor's liability for altering a 'natural' solids byproduct. The added expense and difficulty of chemical addition has also tended to discourage food processors from using chemical or concentrating agents. The tendency has been to search for concentrating and dewatering equipment that does not require chemical conditioning.

Digestion

There are two types of sludge digestion process: anaerobic and aerobic.

Anaerobic digestion has been practised for many years at municipal treatment plants. Although not a method of concentrating the solids that need disposal, it does destroy pathogenic bacteria and reduce the amount of volatile dry solids in the sludge. Anaerobic digestion has not been widely used for conditioning solids from food processing plants because of the dilute solids that are usually present and the seasonal variation of solids loadings. Both dilute solids and seasonal variation require relatively large and correspondingly expensive anaerobic digesters that are not economic for most food processing operations. Also, with solids from fruit and vegetable processing plants there is usually no need to kill pathogenic bacteria, since the wastes are not of human or animal origin.

Aerobic digestion is more practical than anaerobic digestion for seasonally operated plants treating only fruit and vegetable wastewater. Its disadvantage is the relatively large amount of energy required to meet oxygen demands. Aerobic digestion usually reduces the volatile solids content of the sludge and may reduce odour problems where liquid sludge is to be applied directly to farmland or stored.

Heat treatment
By heating sludges under pressure to temperatures that range from 150° to 300°C, solids are coagulated, bacterial cell structure breaks down and the sludge solids can be separated more readily. As a result, the sludge is sterilized, deodorized and often made more suitable for dewatering. A major disadvantage of heat treatment is the amount of high-strength BOD recycled in the supernatant after heat treatment, requiring further biological treatment. Also, heat treatment is expensive in terms of capital costs and requires high amounts of power or fuel.

4.10.2 Thickening
Thickening of solids consists of removing water to reduce the sludge volume. The thickened sludge remains a fluid with solids content usually between 3 and 8%. Thickening can be accomplished by means of gravity or flotation, centrifugation or sludge lagoons.

Gravity and flotation
Thickening using gravity or flotation methods can reduce the volume of sludge prior to further dewatering or disposal. The benefits of thickening must be tested on the sludges that need concentration, to determine its effectiveness: if adequate primary or secondary settling has already taken place, thickening may not significantly increase the solids concentration, since much the same processes are involved.

Centrifuges
Many of the centrifuge units discussed in Section 4.10.2 may also be used to dewater sludges. To achieve dewatering, the centrifuges are usually modified to handle a dry cake and the feed rate is significantly reduced.

There are basically three types of centrifuge commercially available: solid bowl, disc nozzle and basket centrifuges. Their requirements for prior sludge conditioning and methods of operation differ. Because of the tendency to clog at high solids concentrations, disc nozzle centrifuges often require grit removal or prior screening. Both disc nozzle and solid bowl centrifuges operate on a continuous basis, while basket centrifuges operate on a fill and draw cycle. Advantages of centrifugation are that the operator can control the degree and performance of thickening and that thickening devices, when lightly loaded, can sometimes be modified to dewater solids. Disadvantages of thickening by centrifugation are the high capital cost of equipment and the relatively high maintenance costs for machine upkeep.

Lagoons
Anaerobic and facultative sludge lagoons can be used to thicken solids. A conditioned sludge (well stabilized) should be discharged to either type of lagoon to reduce nuisance or odour problems. Sludge is usually conditioned by aerobic or anaerobic digestion (preferably the latter) prior to discharge to a thickening lagoon. The use of anaerobic sludge lagoons has not always been successful because of periodic and often uncontrollable odours, which occur when lagooning is done without regard to loading or the maintenance of a water cover over the sludge layer to minimize direct dispersion of odorous compounds into the air. These problems can be relieved by reducing solids loadings to lagoons and maintaining an aerobic surface layer of relatively solids-free water above the sludge. When this is done, the lagoon becomes a 'facultative' type, which contains an aerobic water surface layer with an anaerobic sludge layer at the bottom. Regardless of the type of lagoon loading, current design practice is to locate lagoons away from highways and buildings to minimize nuisance complaints.

Thickening lagoons are relatively low in cost and easy to operate compared with other dewatering alternatives. Additional advantages are low power costs (little required other than pumping), and the reduction of solids by means of anaerobic digestion. However, thickening lagoons present potential nuisance problems, particularly in urban areas. A further disadvantage is the large area of land required. Solids removal from lagoons can also be a problem, since the thickened sludge generally will not flow to a common point for pumping: pontoon-mounted pumps or sludge dredges must be moved into contact with the sludge so that solids can be removed. Despite these problems, for plants located in a suitable area lagoons continue to be a cost-effective method for thickening solids because of their ease and low cost of operation.

4.10.3 Dewatering
Dewatering involves the transformation of a liquid sludge to a cake of dry, or nearly dry, solids (usually >10 to 15% solids). The following methods are commonly considered.

Vacuum filters

Vacuum filters have been commonly used for dewatering food processing waste. With some waste solids, a dry cake can be peeled from the drum without prior chemical conditioning. With other waste sludges, chemicals or other additives may be required if a cake is to form.

A disadvantage of vacuum filters is their sensitivity to changes in waste sludge characteristics, which can result in poor performance or problems with cake formation. High power requirements and capital costs are sometimes additional disadvantages of vacuum filters. Vacuum filters often include a belt wash system where sometimes a significant amount of fresh water or filtrate water is pressurized and sprayed on to the belt for cleaning. This pressurization stream usually needs treatment and can add to the amount of flow to the treatment system.

Centrifuges

Centrifuges generally require less power than vacuum filters but can require more power to operate than filter presses or some other dewatering devices. Another advantage of certain centrifuge operations is that, where other dewatering devices may require sludge conditioning prior to dewatering or may be sensitive to changes in sludge pH or other characteristics, centrifuges may not be sensitive or have conditioning requirements. Disadvantages of centrifuges are the high initial capital cost and the maintenance required for proper operation.

Pressure filters

Pressure filters (commonly called plate and frame filter presses) are used where a very high solids cake concentration is desired. Sludge conditioning is usually required to form a cake and to obtain good solids capture. Pressure filters are not commonly used where large volumes of sludge require dewatering because of their limited cake-holding capacity. Pressure filters generally have a high capital cost and their batch or cyclic operation makes them unattractive for dewatering at most large food processing plants.

Filter presses

Filter presses accomplish sludge dewatering by pressing free water from the solids through a series of pressure rollers. The advantage of these units is their ability to obtain extremely high cake solids while requiring very low power. Disadvantages are that sludge conditioning is required with some food processing waste, and also that a large amount of high-pressure belt wash-water may be needed to clean the filter belt after cake discharge.

Dual-cell gravity concentrators

The dual-cell gravity concentrator has been used successfully to dewater some food processing sludges. These units require relatively little power for operation and have a low capital cost compared with other sludge dewatering devices. However, dual-cell gravity concentrators require high maintenance and operator attention and have limited cake-handling capacity; chemical conditioning is almost always required. Dual-cell gravity concentrators become most attractive where relatively small volumes of sludge need dewatering and other conventional dewatering devices require too much capital to be practical.

Drying beds and lagoons
In areas with dry weather, sludge-drying beds or lagoons have been used for dewatering solids. Aerobic digestion is commonly applied to waste solids prior to the use of drying beds or lagoons, to reduce the volatile content and subsequent odour problems of the disposed sludge.

Drying beds are often constructed with sand bottoms and often have an underdrain system to capture water that percolates down through the sludge. Sludge-drying beds are open-air basins or shallow ponds which hold the sludge until the moisture content drains out or evaporates. Drying beds may be constructed with a wedge wire layer at the bottom. The wire bottom has been found effective in reducing blinding and increasing the drying rate, which has been a problem with units having sand bottoms.

Sludge-drying lagoons (sometimes referred to as humus ponds) are similar to drying beds; however, the depth at which sludge is placed is three to four times greater. Also, the filling period of a drying lagoon is much longer (often a year or more) before solids are dried and removed so that the lagooning cycle can be repeated.

Drying beds and lagoons have the advantages of relatively low cost and ease of operation compared with other dewatering methods, but present potential nuisance problems in locations other than rural areas. A further disadvantage is the large area of land required. Sludge-drying beds also require a dry climate. However, for plants located in a suitable area, drying beds and lagoons continue to be a cost-effective method for dewatering solids because of their ease of operation.

4.10.4 Disposal
Residuals that are not converted to a byproduct for animal feed can constitute a major disposal problem. Sludges that are in a liquid form (generally ranging from 3 to 8% solids) are usually applied to land by either surface or subsurface application. When sludges have a higher solids content, application to farmland becomes difficult and disposal to landfill is usually chosen. Because of concerns for groundwater contamination, regulations on both land disposal and landfill are becoming more stringent. Pertinent local and national regulations should be carefully reviewed to avoid a problem with the final disposal of waste solids.

Land application

Many of the constraints associated with land treatment of either raw or primary treated food processing waste also apply to the disposal of waste solids. Nitrogen loading is usually the parameter that limits the amount of sludge that can be applied to the soil. The BOD content and solids loadings are generally not a problem with sludge disposal, because sludges usually have been digested or oxidized in the preceding treatment. Also, effects from BOD and solids loading of raw or untreated sludges are minimized if sludges are incorporated by disking or harrowing into the ground after application.

Because of the high solids content, special equipment is needed for both transporting and applying waste solids. For example, if sludges are to be pumped, special solids pumps are required: positive displacement, open impeller, and special manure, solids-handling or slurry pumps are used in transporting sludges. A number of different commercially manufactured pumps are available. In sizing both pumps and pipes, the properties of sludge, density, viscosity and erosion must be accounted for. Tanker trucks having a maximum hauling capacity of $11.4 \, m^3$ (3000 gal) have been used to transport sludges to land application sites that are remote from the food processing facility. These trucks are similar to those used for hauling and disposing of septic tank sludges. Since tank trucks are designed for roadway use, their use for off-road hauling is limited, especially during bad weather. Where hauling distances are long, or if large volumes of sludge must be transported, larger tankers should be considered. Liquid tank trucks up to $22.7 \, m^3$ (6000 gal) capacity are commercially available. If the waste sludges are dried to the extent that they no longer flow as a liquid, transportation by conveyors, front-end loaders, or other mechanical means may be required.

Special application equipment is also required for sludge disposal on land. Specially designed sprinklers with nozzle openings ranging from 5 mm to over 50 mm in diameter are used for applying solids. Special tank trucks with injector rakes designed for disposing of sludges beneath the ground surface can be used with liquid sludges. For dry cake sludges, trucks equipped with manure-spreading boxes that both break and broadcast the solids over the ground surface can be used. Also, if large volumes of liquid sludge need to be applied below the ground surface, special injectors fed by pumps are available for continuous sludge transportation and disposal.

The choice of both transportation and application equipment varies greatly depending on the type of sludge, sludge volume, aesthetic and neighbourhood requirements and soil conditions. Specialists in farm and land management should be consulted during design of the sludge disposal programme. If the food processor does not dispose of sludge on his own land but depends on the cooperation of farmers for the use of their land, then local agricultural practices must be given close consideration.

Landfill

To be acceptable for landfill, sludge is usually required to be at least 20% solids or to be movable by tractor or other equipment blades. Sludge dryness is required so that leachate from the landfill does not become a problem. Also, landfill operators desire dry solids so that waste sludges can be worked and incorporated with other dry rubbish at the site.

Because of the nature of solids from fruit and vegetable waste, hazardous or toxic materials are usually not a problem. However, a chemical analysis of the sludge may be required by local or regulatory agencies to ensure that significant concentrations of toxic compounds are not present. Local regulations should be reviewed to determine whether the dewatered sludges can meet local landfill standards.

One advantage of landfill is that pickup, transport and disposal of waste sludges can often easily be contracted to another party. For small volumes of sludge, landfill can be both a cost-effective and a convenient method of disposal. However, where larger volumes are involved landfill may not be cost-effective, especially where there is a charge for use of the landfill site. Also, the use of a publicly owned landfill site does not allow the food processor to control disposal prices, or to guarantee that a further disposal site will be available when the landfill becomes full.

Other methods

Other methods of sludge disposal that are sometimes used in the fruit and vegetable industry are incineration, ocean discharge, effluent discharge, composting and drying.

Incineration, or thermal oxidation, reduces the sludge volume by approximately 90% and sterilizes the remaining ash solids. The liquid content in the solids is removed by evaporation and the volatiles in the solids are burned. The final product is an ash which requires cooling or quenching and final disposal, which is usually to landfill. The most common types of sludge incineration reactor are the multiple hearth and the fluid bed furnaces. Because of the high cost, both of initial equipment and operation, incineration has not been widely used for disposal of waste solids from the fruit and vegetable industry.

The disposal of waste sludges over a wide area in the ocean is another possible alternative. Ocean disposal methods might include pumping sludge into barges, towing it far to sea and releasing it in deep water, or pumping sludge through an outfall for disposal on the ocean floor. Besides the obvious disadvantage of possibly adversely affecting the environment, ocean disposal may be more expensive than conventional land application of waste solids.

Should the food processor's discharge permit allow the wasting of solids, then it may be less expensive to waste by this method rather than disposing of solids separately. One instance of effluent disposal of sludges which may be cost-effective is where the food processor has been required to pretreat BOD to acceptable levels for discharge to a city sewer. Sometimes the levels of allowable

TSS discharge are significantly greater than the allowable BOD discharge values. Some food processors have found it beneficial to treat the BOD by conventional methods and then discharge the generated biological solids from the conventional treatment step to the city sewer along with the treated effluent.

Composting can be done prior to the marketing of waste sludges as a fertilizer or soil conditioner. The purpose of composting is to reduce simultaneously the weight and the moisture content of the sludge and, at the same time, to stabilize the sludge to increase its acceptability as an agricultural product. Composting is a simple natural process requiring only a proper environment to allow the sludge to stabilize. The composting process is considered complete when the sludge product can be stored without giving rise to nuisances such as odours or insects that would otherwise present problems in disposing of it competitively in the fertilizer market. For most sludges the addition of a bulking agent (for example, recycled compost, wood chips, waste hay or other dry products) to raw waste sludges may be required to obtain a mixture suitable for composting. High-rate sludge composting methods include windrow, deep-pile and mechanical composting. Although composting is a relatively simple process, both in operation and maintenance, it is land-intensive and involves relatively extensive materials handling. Other disadvantages are odours, insect problems, dust or mud, and generally higher operating costs than other methods of sludge disposal. Composting has not been widely used for the disposal of waste solids from fruit and vegetable processors. Operational, packaging and transportation costs of making available a composted sludge byproduct are generally such that it is not competitive in the marketplace with other commercially available fertilizers, soil conditioners and animal manures.

Sludge drying is another alternative that can produce a sludge suitable for marketing as a commercial fertilizer although it is not commonly used. The three methods of sludge drying generally considered are (1) heat drying, (2) basic extraction sludge treatment and (3) oil immersion dehydration. Major disadvantages that have limited the use of sludge drying processes for fruit and vegetable waste solids are the high electrical energy requirements, the high capital, operation and maintenance costs, and the availability of less expensive sludge disposal options.

4.11 Byproduct utilization

In utilization of byproducts, or of what may have previously been a waste material, the former waste is converted into some additional product such as livestock feed, alcohol, a human food or a commodity for manufacturing. Of these potential uses, livestock feeding is often found to be the most cost-effective method of converting a waste product into a saleable byproduct. The following discusses steps that lead to a successful market programme for byproducts as livestock feed.

4.11.1 Conversion of waste to livestock feed

Criteria for livestock feed

For body maintenance and acceptable growth of livestock, several key requirements for nutrition must be met by any feed:

- The ration must be palatable and acceptable to the animal.
- The ration must be usable by the animal's digestive tract.
- The ration must be nutritionally balanced, meeting some of the animal's nutritional requirements.

It is not essential that a nutrient, such as carbohydrate, be in the form of a conventional feed such as corn-grain or barley. Other feeds may also be acceptable such as wastes from potato, corn, or other fruit and vegetable products. It is, however, essential that the carbohydrate feed be ingested and digested, and produce the nutritional energy required by the animal.

For any carbohydrate source to work with conventional livestock feeding systems it must perform satisfactorily as judged by the following:

- The feed must be available initially in a high-quality, nonspoiled form.
- The feed must be transported and stored in a form that preserves its nutrient content and prevents putrefaction or degeneration.
- The feed must be blended with other feed ingredients to provide a balanced ration.

Processing operations

Segregation of materials that could contaminate the byproduct is the first basic requirement in producing a usable feed from fruit or vegetable wastes. Regulatory officials often require demonstrated assurances that the food processor can deliver a wholesome product.

The sewering of all food-quality waste to a central collection point makes it easier to process the byproduct stream. Operations that increase the byproduct's market value include dewatering, drying and the addition of preservatives and other ingredients. Dewatering can be accomplished by those devices described earlier in this chapter. Drying the byproduct is a final dewatering step and is usually accomplished by thermal drying using either rotary, flash, sonic, or spray dryers.

Preservatives to prevent bacterial breakdown of the carbohydrate food source can be added by the metered blending-in of organic acids (propionic and acetic), salts or bactericides. The metering and blending of other ingredients such as supplemental conventional feeds, flavouring agents, and minerals to maintain a specified nutrient level may also be desired.

Appraisal of value

It is often difficult to determine a byproduct's fair market value. A representative sample analysis will provide the basis for evaluation of the product's feed

Table 4.9 Potato byproduct centrifuge cake

Constituent	As received (%)	Dry (%)
Protein	1.18	2.64
Fat	0.04	0.09
Fibre	1.97	4.42
Moisture	55.34	0.0
Mineral matter	5.20	11.65
Carbohydrate	36.27	81.20
Digestible protein	0.91	2.03
Digestible fat	0.01	0.02
Digestible fibre	1.26	2.83
Digestible	35.54	79.58
Total digestible nutrients	37.72	84.46
Total dry matter	44.66	100.00
Calcium	0.08	0.17
Phosphorus	0.06	0.14
Magnesium	0.05	0.11

value: a potato byproduct analysis, as shown in Table 4.9, provides specific information on the product's moisture, protein, fat, fibre, ash and carbohydrate content. From these analyzed values, digestibility of the components, total digestible nutrients, net energy of maintenance, net energy of gain and digestible energy can be estimated using formulae established by the animal feed industry.

From representative analysis, two marketing tools can be generated: (1) a balanced ration using the byproduct as a source of livestock nutrition, and (2) a dollar-per-tonne feed value based on the price of competing energy and protein feeds.

The analysis presented in Table 4.9 contains all the information required to include the potato byproduct in a nutritionally balanced ration for cattle, swine, horses or poultry. A standard ration balancing programme requires data about the amount and nutritional quality of an animal's daily feed requirements, such as body size, sex and breed. Available conventional feeds and their respective prices are entered into the programme. From this information a nutritionally balanced ration with a least-cost mix can be designed.

To determine the byproduct's cash value, the nutritional content (protein, energy, minerals and fibre) is compared with that of conventional feeds at their respective prices. With this input, a value that reflects the byproduct's contribution to a feeding ration can be determined.

Marketing
The goal of the marketing effort is to provide a reliable outlet for the byproduct. Processing, pricing, transporting, storing and monitoring of the byproduct all contribute to establishing a successful long-term programme. Normally, financial incentives are necessary while system start-up problems are being solved.

A satisfactory livestock feed programme using byproducts requires delivery of a consistently wholesome byproduct to the customer. Blending the byproduct with conventional feed is necessary to produce a nutritionally balanced ration. Furthermore, the marketing programme must provide a pricing structure that will both reward the food processor for the additional capital and management costs, and provide an incentive to the livestock feeder to handle and use the product.

Livestock feed can serve as a long-term utilization option if properly planned and monitored. Also, capital costs for in-plant process modifications to manufacture the feed are relatively small compared with other byproduct options, such as alcohol fuel production.

Disadvantages of feed as a byproduct include:

- the limited information available on storage, handling and use of feed byproducts;
- the fluctuating moisture content of byproducts;
- the seasonal nature of byproducts, which makes it difficult to establish a feed production programme;
- the possible need for additional capital outlay by both customer and food processor.

4.12 Treatment costs

As the reader is aware, costs for wastewater treatment are increasing at a rapid rate. Also, costs vary greatly with local conditions, required construction methods and operating conditions. Ever-increasing factors in the cost of treatment are the local cost of electrical power and the availability and distance of acceptable sites for waste solids disposal. To complicate matters further and to render generalized cost information of little value, treatment costs vary with wastewater quantity and character. This occurs not only between different products or fruit and vegetable processing methods, but significant differences can also occur between production plants, depending on the housekeeping and in-plant recycle systems being used.

The reader being properly forewarned that waste treatment costs should be developed on a case-by-case basis, the following sections generalize both present and future treatment costs. These estimates are presented to illustrate the relative magnitude of costs for the various treatment processes and should not be used to project actual costs.

4.12.1 Present-day costs

A summary of wastewater treatment costs for the more significant treatment processes already discussed is given in Table 4.10. These costs were developed

Table 4.10 Wastewater treatment costs (US $ — 1983)

	Capital[a] Initial investment	Annual[b] Operation and maintenance	Payment[c]	US $ per 1000 gal	UK £ per m³
90-DAY OPERATION					
Flow measurement and screening	119 000	5 300	29 000	0.32	0.05
Neutralization	167 000	38 000	71 000	0.79	0.13
Aerated lagoon	863 000	34 000	206 000	2.29	0.30
High-rate irrigation	290 000	20 000	78 000	0.87	0.14
Activated sludge:					
without sludge concentration	1 075 000	292 000	506 000	5.62	0.93
with sludge concentration	1 858 000	98 000	468 000	5.20	0.86
Filtration	452 000	6 700	97 000	1.07	0.18
360-DAY OPERATION					
Flow measurement and screening	119 000	21 000	45 000	0.13	0.02
Neutralization	167 000	153 000	186 000	0.52	0.09
Aerated lagoon	863 000	137 000	309 000	0.86	0.04
High-rate irrigation	290 000	30 000	98 000	0.24	0.04
Activated sludge:					
without sludge concentration	1 075 000	1 167 000	1 381 000	3.84	0.63
with sludge concentration	1 858 000	390 000	760 000	2.11	0.35
Filtration	452 000	27 000	117 000	0.33	0.06

[a] US *Engineering News-Record* Construction Cost Index = 4000.
[b] US $1.60 = UK £1.00.
[c] 15% interest — 10 year loan with annual payments. Payment includes annual retirement of capital plus operation and maintenance costs.

for a hypothetical food processing plant producing $3785\,m^3\,d^{-1}$ (1.0 mgd), computed for 1983 using the *US Engineering News-Record* construction cost index of 4000 (*Engineering News-Record*, 1982). The annual costs are based on the operating and maintenance costs plus equal annual payments of capital expenditures based on a 10-year loan at a 15% interest rate. It has been assumed that significant site grading or preparation is not required and that good soil conditions exist. Costs do not include items such as site acquisition, fencing, yard lighting and access roads.

It has been assumed that the food processing plant discharges a peak day flow of 2 million gallons per day ($8000\,m^3\,d^{-1}$) and that average wastewater concentrations are $1000\,mg\,l^{-1}$ for both BOD and TSS. Capital costs include an additional 25% of the actual construction costs to account for the engineering, legal and contingency costs associated with any construction project. Electrical power costs are based on hydroelectric rates of 3.3¢ per kWh experienced in the Pacific Northwest section of the USA.

Until recently, world conditions have favoured, for the treatment of wastes from most fruit and vegetable processors, the approach of simply not treating beyond simple flow measurement and screening, unless required. A favoured point of discharge for screened wastewater has been the city sewer, for which the monthly charge is often more attractive than the large capital expense and inconvenience of an industry-owned treatment facility. Where an industry-owned treatment facility is to be constructed, either land treatment or aerated lagoons have been used. As shown by Table 4.10, the additional cost of activated sludge treatment over both aerated lagoons and high-rate irrigation gives little incentive to treat with activated sludge, unless these alternatives are not available or are restricted by local conditions. However, the recent rise in electrical power costs and the increasing difficulty in finding suitable sludge disposal sites have had the result that, in many places, the least-cost treatment option has changed. These changes are discussed in the following section.

4.12.2 Future trends

Wastewater treatment for fruit and vegetable wastes must change to meet present and near-future conditions. Although exact predictions cannot be made, most experts agree that the rising cost of electrical power and, to a lesser degree, the lack of suitable land will have a large influence on future trends in waste treatment. For example, in the past many fruit and vegetable processors have not found it advantageous to remove primary solids prior to discharge to a city sewer system. With the increased power costs that many city facilities will incur in treating industrial wastewater, a corresponding increase in city sewer rate charges is likely to encourage many food processors to re-evaluate the need to install primary sedimentation.

Pretreatment steps that require relatively little power, such as anaerobic treatment, will be given greater consideration by fruit and vegetable processors in the future, both where discharge is to a city sewer (to reduce sewer costs) and as a means of reducing electrical costs at an industry-owned waste treatment facility.

Interest in land application is likely to reach a new peak in future years. Although land costs have increased, they have generally not kept pace with the increased cost of electrical power. Because of this, where land is available, high-rate irrigation will be considered by many fruit and vegetable processors as a means for both treating and disposing of their wastewaters. Where little land is available and discharge to a receiving stream or city sewer is possible, pretreatment by use of overland flow methods is likely to be considered by more fruit and vegetable processors.

Where the fruit and vegetable processor requires conventional aerobic treatment, further consideration will be given to the use of combined treatment processes to minimize power requirements. For example, the use of roughing filter or biofilter activated sludge is likely to be given greater consideration than the more energy-intensive approach of conventional activated sludge. This is especially true where a food processor has an existing activated sludge plant that requires upgrading and improvement.

Regardless of the method of treatment considered, the environmental engineer working in the fruit and vegetable industry will be giving greater consideration to wastewater treatment costs. There has been and will continue to be no universal 'best' process or treatment scheme available. Each facility must be judged individually with consideration given to the following:

- Waste type
- Strength of wastewater
- Seasonal needs for treatment and processing
- Individual value of capital and annual cost
- Process flexibility requirements
- Operator attention
- Concerns for odour and public nuisances
- Effluent quality requirements
- Utilization of existing treatment facilities
- Solids and disposal requirements
- Power costs
- Site locations, restrictions and conditions.

References

Benefield, L.D. and Randall, C.W. (1980) Attached-growth biological treatment processes, in *Biological Process for Design for Wastewater Treatment*, Prentice-Hall, Englewood Cliffs, NJ, 391–410.

Bureau of National Affairs, Inc. (1979) Environmental Protection Agency Effluent Guidelines and Standards for Canned Fruits and Vegetables, *Environmental Rep.*, **135**, 0221–0228.

CH2M HILL (1972) *1971 Pilot Plant at the Willow Lake Sewage Treatment Plant*, Project Report C6519, Corvallis, Oregon.

CH2M HILL (1976) *1975 Wastewater Treatment Plant Study for the City of Turlock, California*, Project Report M1023.18, Corvallis, Oregon.

Cochrane, M.W. and Dostal, K.A. (1972) RBC treatment of simulated potato processing wastes, *Proc. 3rd Nat. Symp. on Food Processing Wastes*, EPA-R2-72-018, Corvallis, Oregon, 99–115.

Culp, R.L. and Culp, G.L. (1971) *Advanced Wastewater Treatment*, Van Nostrand Reinhold, New York, NY.

Engineering News-Record (1982) **209**(20), 106.

EPA (1977) *Pollution Abatement in the Fruit and Vegetable Industry*, EA-625/3-77-007, Volume 3, US Environmental Protection Agency, 6.

Grady, C.P. and Lim, H.C. (1980) The ecology of biological reactors, in *Biological Wastewater Treatment*, Cheremisinoff, P.N. (Ed.), Marcel Dekker, New York, NY, 197–227.

Gromiec, M.J. and Malina, J.F. (1970) *Verification of Trickling Filter Models using Surface*, Technical Report EHE-70-13 CRWR-60, Center for Research in Water Resources, Austin, Texas.

Harrison, J.R. (1980) Survey of plants operating activated biofilter/activated sludge, paper presented at the 1980 California Water Pollution Control Association Northern Regional Conference and Training School.

Katsuyama, A.M. (1979) Monitoring liquid waste flows, in *Guide for Waste Management in the Food Processing Industry*, Food Processing Institute, Washington, DC.

King, H.W. and Brater, E.F. (1963) *Handbook of Hydraulics*, 5th edn, McGraw-Hill, New York, NY.

McNeal, B.L. (1981) Evaluation and classification of water quality for irrigation, in *Salinity in Irrigation and Water Resources*, Yaron, D. (Ed.), Marcel Dekker, New York, NY, 21–45.

Norris, D.P., Parker, D.S., Daniels, M.L. and Owens, E.L. (1982) High-quality trickling filter effluent without tertiary treatment, *J. Water Pollut. Control Fed.*, **54**, 1087–1098.

Sanks, R.L. (1979) *Water Treatment Plant Design*, 2nd edn, Ann Arbor Science, Ann Arbor, Michigan.

Schaumburg, F.D. and Lasswell, S.S. (1970) Novel biological treatment process utilizes unique redwood media, *Water Wastes Eng.*, August, 34–37.

Schmidt, C.J., Clements, E.V. and LaConde, K. (1975) Treatment alternatives for the fruits and vegetables processing industry, *Proc. 30th Purdue Ind. Waste Conf.*, Ann Arbor Science, Ann Arbor, Michigan, 318–331.

Sutton, P.M. and Jank, B.E. (1977) Principles and process alternatives for biological nitrogen removal, paper presented at the Workshop on Biological Nitrification/Denitrification of Industrial Wastes, Wastewater Technology Centre, CCIW, Burlington, Ontario.

Thomas, J.L. and Sanborn, D.A. (1973) Activated sludge–bio-disc treatment of distillery wastes, *Proc. 4th Nat. Symp. on Food Processing Wastes*, EPA-6601/2-73-031, Corvallis, Oregon, 352–375.

Williams, C.R., Burchett, M.E., Witley, R.D. and Walker, L.F. (1973) Results of pilot studies on biological treatment of combined food processing/domestic wastewater at Tracy, California, paper presented at the WWEMA Industrial Water and Pollution Conference and Exposition, Chicago, Illinois.

5 The treatment of wastes from the dairy industry

K R Marshall and W J Harper, *New Zealand Dairy Research Institute*

5.1 Introduction

This chapter reviews the sources and treatment or disposal of wastewaters from the processing of milk and milk products. Waste products from dairy farming are not considered.

The dairy industry is a major enterprise in many countries, occupying a significant place in the food supply and, in some countries, playing an important role as an earner of export income. For example, the dairy industry in the UK provides some 24% of the protein content of the national diet at a cost of approximately 18% of total food costs (Hemmings, 1980); the New Zealand dairy industry earned NZ 1.3×10^9 in overseas exchange in 1981–82 (New Zealand Dairy Board, 1982), representing 20% of the total overseas exchange of that country.

In many countries the dairy industry has been recognized as a significant contributor to the pollution of natural waterways. In the early 1960s a US Senate Committee noted that the dairy industry was the second most important single source of pollution in streams. However, much progress has been made by the industry in the intervening period to reduce the impact of the waste discharges and this review outlines the methods that are being employed.

Wastes from the dairy industry are mainly dilutions of milk or milk products. Hygiene requirements in the industry result in significant quantities of cleaning compounds and sanitizers being present in the wastes, although these generally have little effect on treatment or disposal. The wastes are essentially organic, and methods used for domestic sewage treatment have been adapted for handling dairy plant wastes. However, the dairy wastes tend to have higher organic solids concentrations when compared with domestic wastes. Treatment of dairy wastes is complicated by marked variations in hourly, daily and seasonal flow rates.

Large quantities of water are used in the dairy industry for cooling or in barometric condensers in evaporators. Normally this water is not contaminated and separate disposal is desirable. However, inadequate cooling can lead to contamination from high temperatures. In the past, liquids such as skim milk, buttermilk and whey were considered wastes from the dairy industry. They are now considered to be valuable materials and are processed to numerous products. Nevertheless, in many plants whey is still an embarrassment, despite the considerable advances in appropriate technology. The high cost of the processes and the lack of suitable markets for the whey products are limiting the rate of expansion of whey processing.

The nature of dairy wastes, their treatment and disposal have been the subject of a number of reviews over the last decade and these have formed the basis for this chapter. Further details should be sought from those reviews. They include Ministry of Agriculture, Fisheries and Food (1969), Arbuckle (1970), EPA (1971), An Foras Taluntais (1974), Jones (1974), and IDF (1974, 1978, 1980, 1981).

5.2 Milk

5.2.1 Composition of milk

Cow's milk is the liquid food secreted by the mammary gland for the nourishment of the newly born calf. Whole milk contains water, fat, protein, lactose, minerals (measured as ash, also known as salts) and vitamins. There is no one composition of whole milk, as the composition varies with the breed of cow, the stage of lactation, the plane of nutrition, the health of the cow and the conditions to which the milk is subjected after milking. A typical composition of milk in the USA is shown in Table 5.1. The concentration of milkfat in New Zealand milk is significantly higher than that in USA milk (Table 5.1) and the composition varies throughout the dairying season, reflecting the total pasture feeding of New Zealand cows.

Table 5.1 Composition (%) of normal bovine whole milk produced in the USA (Johnson, 1974) and New Zealand

	USA (average)	New Zealand	
		(average)	(range)
Milkfat	3.6	4.7	3.0 −7.0
Protein	3.5	3.6	3.0 −5.0
Lactose	4.9	5.0	4.7 −5.3
Minerals	0.7	0.75	0.75−0.80
Total solids	13.0	14.3	

Milkfat, a complex mixture of triglycerides (98−99%), phospholipids (1−2%) and traces of sterols, carotenoids and the fat-soluble vitamins A, D, E and K, exists as an oil-in-water emulsion in small globules (2−5 μm) in the milk serum. The milk proteins comprise two major classes, the casein and the whey proteins. The caseins (80% of the total protein) are of four major types, α_s-casein, β-casein, k-casein and γ-casein. The caseins exist mainly as suspended agglomerates known as micelles. The micelles are a colloidal calcium phosphate complex containing all the types of casein and about 7% inorganic material. The casein micelles can be destabilized and precipitated by the action of rennin or by isoelectric precipitation by lowering the pH to approximately 4.6. The whey proteins (α-lactalbumin, β-lactoglobulin, bovine serum albumin and immunoglobulins) are soluble over the complete pH range provided that they have not been exposed to temperatures greater than 65°C. Together with the lactose and minerals they form the milk serum in which the milkfat globules and casein micelles are suspended. Lactose, a disaccharide molecule, is a carbohydrate comprising one molecule of glucose and one of galactose. The major salts in the mineral component are those of calcium, sodium, potassium and magnesium, which occur as phosphates, chlorides, nitrates and caseinates. Sulphur, zinc, rubidium, silicon, bromine, aluminium, iron, etc., are present in trace quantities.

5.2.2 History of commercial processing

It is probable that the milking of animals to provide food for man predates the beginning of organized and permanent recording of human activities. Milk has a relatively short life and the making of products such as butter, dried milk, cheese and cultured products developed earlier than 2000 BC as a means of preserving the food value of milk. However, until the mid-19th century the factory system of milk processing and milk product manufacture was unknown. Prior to the 1850s each household maintained its family cows or secured milk from a neighbour. Butter, cheese and other dairy products were manufactured on the farm. As the population of the cities increased, fewer households could keep a cow and larger farms developed on the outskirts of cities. The concentration of population in cities, together with the development of the centrifugal cream separator (invented and placed on the market in 1878), the advent of pasteurization and the development of analytical tests for milkfat concentration (Babcock in the USA and Gerber in Europe) eventually led to the establishment of factory processing. In the USA the first commercial cheesemaking plant was started in Rome, New York, in 1851 and the first buttermaking plant was opened in Iowa in 1871. Similarly the first cooperative factory to manufacture cheese was established in New Zealand in 1871.

From the 1870s until about 1910 whole milk was hauled from the farm to the factory. If butter was manufactured, the milk was separated and the skim milk returned to the farm, generally for feeding to pigs. Whey from cheese was also returned to the farmer. Farm separation grew rapidly in the period 1905–20, and by 1930 cream only was being hauled to those factories where butter was being manufactured. The high cost of milk transport restricted the size of factories for cheese, which required whole milk. By the 1950s there was a

Table 5.2 Trend in the number of butter and cheese plants and average annual production for New Zealand dairy plants

	Butter		Cheese	
Year	(no. of plants)	(tonnes per plant)	(no. of plants)	(tonnes per plant)
1954/55	119	1569	209	491
1959/60	107	1943	160	583
1964/65	84	2923	123	870
1969/70	67	3529	85	1188
1974/75	37	5336	48	1704
1979/80	33	6191	30	1991
1981/82	25	8892	22	4932

Table 5.3 Comparative annual output (tonnes) of dairy products per dairy plant in various countries (New Zealand Dairy Board, 1980)

	France 1976	West Germany 1976	UK 1976	USA 1978	New Zealand[a] 1976/77
Butter	445	847	1254	1508	6220
Cheese	567	1059	1898	2052	1628
Skim milk powder	5800	5200	not available	3287	6410

[a] The relative daily throughputs in New Zealand plants are higher than is implied by these figures because of the seasonal nature of the milk flow.

return to the hauling of whole milk facilitated by improvements in transport systems and the introduction of farm milk cooling, and necessitated by the need to improve milk quality and to increase economic returns from the sale of products manufactured from all the components of milk.

Over the past 20 years, in most developed countries there has been a marked trend to fewer and larger processing plants. Data in Table 5.2, for New Zealand, are typical of trends in other countries (EPA, 1971), although the average outputs of New Zealand dairy plants, particularly of butter, are high compared with those of other dairying countries (Table 5.3).

Other trends over recent years which have significance to the nature of dairy wastes and waste treatment include:

- changes in the relative production of the various types of dairy foods each with a different level of waste load;
- automation of plant processes to an increased degree with increasing plant size and consolidation;
- a shift in location of new plant facilities.

5.2.3 Milk production

Annual world milk production in 1974–78 averaged 438 million tonnes, major producers being the European Economic Community (EEC), the USSR and

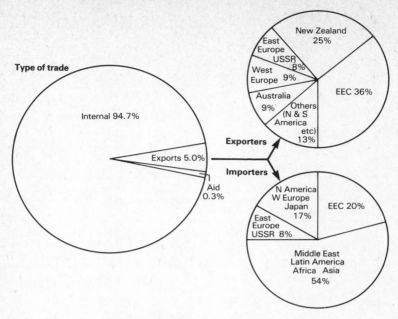

Fig. 5.1 World dairying trade (milk equivalent) 1974–78 (Ojala, 1982)

the USA (Table 5.4). Most of this milk (94.7%) was used in internal trade with only 5% exported, the major exporters being the EEC and New Zealand (Fig. 5.1).

World milk production grew at a rate of 1.7% per year between 1962 and 1977 and was projected to increase by 1.6% per year up to 1985 (FAO, 1979). The production of major dairy products in some countries in 1979 is shown in Table 5.5.

5.3 Milk products

The basic function of the dairy processing industry is the manufacture of foods based on milk or milk products. Individual plants, worldwide, are typically complex, manufacturing a multiplicity of products using a variety of unit operations. Dairy products are so diverse and many are so specific to an individual country that it is difficult to classify and describe the processes involved. However, the determination of the significant sources of wastes from dairy product manufacture requires an understanding of some of the processes and a brief outline is given of the major production methods. For detailed information see, for example, Harper and Hall (1976).

Table 5.4 Average annual milk production (million tonnes) in dairying countries in 1974/78 (Ojala, 1982)

European Economic Community	102.3
USSR	92.2
USA	54.0
India	25.5
Poland	17.1
Brazil	10.7
East Germany	8.0
Canada	7.6
New Zealand	6.2
Australia	6.1
China	5.5
Japan	5.3
Argentina	5.3
Romania	4.7
Mexico	4.4
Other West Europe	27.4
Other Latin America	10.5
Other East Europe	9.6
Remainder	35.5
Total	438.0

Table 5.5 Annual production of milk and major dairy products in various countries in 1979 (New Zealand Dairy Board, 1981)

	Australia	Canada	EEC	New Zealand (1979/80)	USA
Milk production (million tonnes)	5.65	7.55	101.93	6.6	56.17
Percentage used as fluid milk, cream etc.	28	37	35	8	44
Butter (thousand tonnes)	101	98	1918	254	447
Cheese (thousand tonnes)	141	167	3328	106	1686
Whole milk powder (thousand tonnes)	101		569	77	39
Skim milk powder (thousand tonnes)	75	114	1995	192	412

5.3.1 Dairy product manufacture

Figure 5.2 shows the relationships between some of the major products manufactured from whole milk and some typical yields from New Zealand milk.

Common procedures used in the manufacture of all dairy products include:

- *Receiving of milk* from the farm or receiving station. The milk is delivered in cans or tank trucks, although cans are much less common today.
- *Storage of milk:* the milk is transferred to storage tanks which are usually insulated and refrigerated.

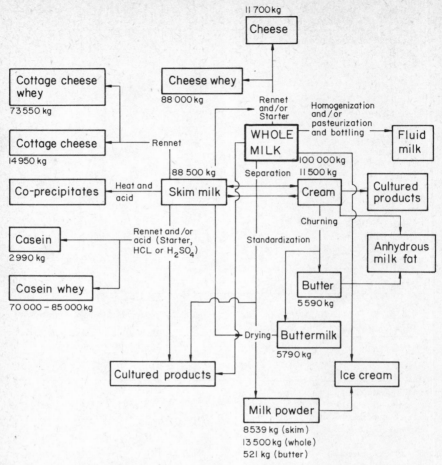

Fig. 5.2 Some of the major products manufactured from whole milk. Quantities given are typical yields from 100 000 kg of New Zealand whole milk

- *Thermization*, a technique used by some large dairies, in which milk is given a mild heat treatment (63°–65°C for 15 s) immediately after delivery so that it can be stored in silo tanks for hours or even days without deterioration of quality.
- *Pasteurization*, a heat treatment designed to kill non-spore-forming or vegetative pathogenic micro-organisms and to reduce the number of adventitious micro-organisms to a level which will minimize the adverse effects on milk quality. Older plants use batch pasteurization, heating the milk to 63°C in a vat and holding for 0.5 h. Modern plants use high-temperature short-time (HTST; 72°–75°C for 15 s) or ultra-high-temperature (UHT; 135°–150°C for 5–3 s) continuous techniques.

- *Clarification* (by centrifugal force of about 6000 g) to rid the milk of foreign particles and sediment, and *centrifugal separation* to skim the cream from the skim milk.
- *Standardization* of the milkfat concentration, by separating a part of the milk, removing the cream and returning the skim milk to the storage tank.
- *Homogenization* to reduce the size of the milkfat globules so that they will remain dispersed instead of rising to form a cream line on the surface.
- *Deaeration* to expel gases and malodorous volatile substances.

Market milk

Market milk (also known as liquid milk or drinking milk) products are liquid products made from whole milk, skim milk and cream and are normally used directly by consumers (Fig. 5.3). Generally, chilling, clarification and pasteurization are compulsory stages in the manufacture. Homogenization, standardization and deaeration are commonly practised. Market milk products can be packed under aseptic (or sterile) conditions using UHT methods to give products with a long shelf-life (up to 1 year).

Cream is produced from whole milk by centrifugal separation. The skim milk is packaged for consumer use, used for standardizing, used for cultured products or used for the manufacture of skim milk products (powder, casein,

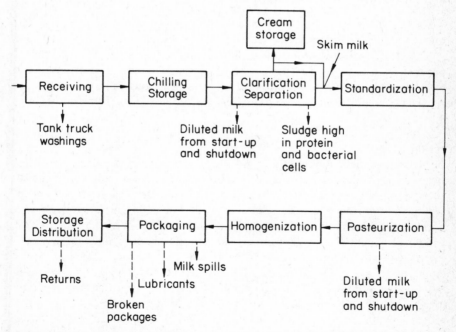

Fig. 5.3 Outline of the general processes used for market milk production, showing specific wastes produced at each stage. Other wastes produced at all stages include dilute milk, detergents and sanitizers

etc.). The liquid milk products are packaged in glass or plastic bottles and plastic or plastic-paper laminate containers. Polyethylene and polystyrene are commonly used materials (Tuszynski, 1978). Damaged packages and returned products can be significant sources of waste loads (EPA, 1971).

Cultured products

Cultured products are frequently manufactured by the market milk industry. Major products include yoghurt, ymer, acidophilus milk, kefir, cultured buttermilk, koumiss, sour milks and sour cream. Generally batch operations are used. The milk or cream is normally pasteurized and may be subjected to a preheat treatment. The milk may be partially concentrated by reverse osmosis or ultrafiltration, or the total solids may be supplemented by the addition of concentrated or dried milk.

The milk or cream is inoculated with a starter culture consisting of one or more strains of bacteria or yeast. The inoculated milk is held at the optimum temperature for the micro-organisms to grow, converting lactose to lactic acid and substances such as carbon dioxide, acetic acid, diacetyl, acetaldehyde or ethyl alcohol which impart the characteristic taste and aroma to the product. After the required time, the product is cooled and packaged. Some cultured products (e.g. set yoghurt) are packed immediately after inoculation with starter, and are incubated in the packages before cooling and distribution to the consumer. The products also frequently are flavoured or have fruit, berries, nuts or herbs and spices added (Kosikowski, 1977).

The viscosity of cultured products is very high and a considerable quantity of product remains on the surfaces of the processing equipment, with subsequent high waste loads in the rinsing and cleaning solutions. The wastes are of relatively low pH, because the pH of the cultured products is normally about 4.5–4.7.

Butter

Butter is essentially the fat of the milk and is available in two main categories, sweet cream butter and acidulated (soured or lactic) cream butter. Butter may also be unsalted or salted. Whey butter is manufactured from the cream separated from the whey after cheesemaking. The principal constituents of a normal salted butter are milkfat (80–82%), water (15.5–17.5%), salt (1–2%) and small quantities of milk proteins, calcium, phosphorus and vitamins.

Butter is manufactured in churns (batch process) or butter making machines (continuous, Fig. 5.4) from cream separated from whole milk. The cream is heat treated and deaerated (or deodorized) (vacreated) and ripened, generally for 12–15 h. During the ripening process the added starter produces acid and aroma components and the milkfat crystallizes. For sweet cream butter no starter is added and only crystallization occurs. The ripened cream is pumped to the churn or continuous buttermaking machine and churned by violent agitation to break down the milkfat globules, which coagulate into butter grains.

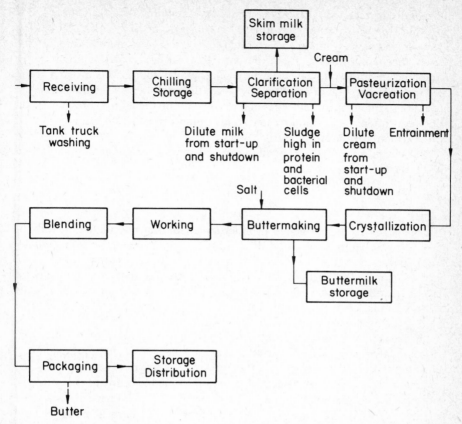

Fig. 5.4 Outline of the general process for the manufacture of sweet cream, salted butter using a continuous churn, showing specific wastes produced at each stage. Other wastes produced at all stages include dilute milk or cream, detergents and sanitizers

The grains are separated from the buttermilk. If the butter is to be salted, salt is spread over the grains in the churn or a salt slurry is added to the buttermaking machine. The butter is worked and discharged to the packaging unit. The butter is extruded into the required shape, cut to size and wrapped or cartoned. The wrapped butter is boxed and stored cold (McDowall, 1953).

Buttermilk is a possible major source of organic waste, but most dairy plants now process it to cultured buttermilk or buttermilk powder. Washing of butter granules, where practised, also produces a high-strength waste.

Anhydrous milkfat
Anhydrous milkfat (AMF or butter oil) contains greater than 99% milkfat and is produced from cream or butter. It is used for recombining or reconstitution

of milk in the ice cream and chocolate industries, and in warm countries, such as India (where it is known as ghee), for cooking. In the production of AMF from cream, the cream is concentrated and the milkfat globules are broken down mechanically, liberating the fat. The dispersed water droplets are separated from the fat phase, leaving a pure fat compound. When butter is the raw material, the butter is melted (by steam heating), heated, held to allow the protein to agglomerate, and separated to concentrate the fat to more than 99%. The fat is often washed and treated in a vacuum dryer. The AMF is packaged for distribution (Fjaervoll, 1970).

The serum from AMF production can be a significant source of organic load in the waste stream.

Cheese

Cheese is the fresh or ripened product obtained after coagulation of milk or cream and separation of whey (Fig. 5.5). Cheese contains protein, milkfat, water and minerals in varying amounts. There is an enormous variety of cheeses (some 2000 names (Scott, 1981)) with different characteristics of flavour, texture and shelf-life. Most types of cheese are produced by coagulation of milk with rennet or other proteolytic enzymes and the action of micro-organisms (Kosikowski, 1977; Scott, 1981).

Cheesemaking involves a number of steps common to most types. The milk is heat-treated and the milkfat concentration standardized before the addition of a starter culture. Rennet or other proteolytic enzyme is mixed with the pretreated milk. The enzyme activity causes the milk to coagulate to a solid gel, known as the coagulum. The coagulum is cut by special cutting tools into small particles of the desired size, to form curd grains. The curd is usually stirred mechanically and heated to produce syneresis (separation and expulsion of liquid whey from the curd). The curds are separated from the whey by draining and then, frequently, subjected to a knitting or reforming process (cheddaring). Salt is added to some kinds of cheese at this stage. The curds are placed in cheese moulds (characteristic of the type of cheese) and pressed to knit the curds together and expel more whey. The pressed cheese may be placed in a brine bath or dry salted. The cheese is usually stored at controlled temperature and humidity for curing or ripening. It is finally coated, wrapped or packaged.

Processed cheese is made from a blend of fresh curd and matured cheese. The cheese is cut and ground, melted, heated and blended with water, stabilizers, flavourings and preservatives. The cheese blend is cooled and packaged (Thomas, 1977).

Whey, inadvertently or deliberately wasted, is the most significant contributor to the effluent from cheesemaking. The pressings from the cheese, comprising whey with a high concentration of sodium chloride, are difficult to handle in biochemical effluent treatment processes. Cheese curd particles may result in a high concentration of settleable solids in the wastes.

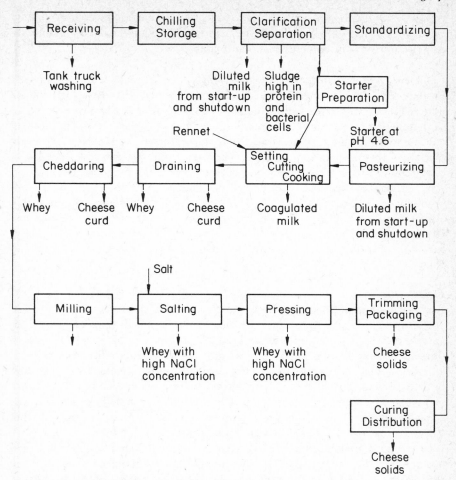

Fig. 5.5 Outline of a typical process for Cheddar cheese manufacture, showing specific wastes produced at each stage. Other wastes produced at all stages include dilute milk, detergents and sanitizers. Note that the whey from the draining, cheddaring and milling stages may be further processed

Casein

Casein is the major protein of milk. It is produced by destabilization of the casein micelles in skim milk either by the action of rennet at normal milk pH (pH 6.7) or by lowering the pH of the milk to the isoelectric point of the casein (pH 4.5–4.7). The pH is lowered by the addition of dilute mineral acid or by conversion of some of the lactose in the milk to lactic acid by suitable strains of bacteria. The coagulum formed by destabilization is heated and the curd which then separates from the whey is washed, dried, ground and bagged (Fox, 1970; Southward and Walker, 1980).

Whey, casein wash-waters and casein curd are major contributors to the waste load. Whey from acid casein manufacture is particularly difficult to process and treat in effluent treatment plants.

Condensed and evaporated (unsweetened) milk
Pasteurized milk is preheated and then concentrated to 20–40% total solids. The milk is concentrated by heating under vacuum. The vacuum is maintained so that boiling can occur at a temperature which is low enough to avoid thermal damage to the product. While some batch evaporators are still in use, most large modern plants use continuous units. The use of multiple effects and thermocompressors in the evaporaters improves the thermal efficiency. The evaporated product is homogenized, cooled and packaged. Sweetened condensed products are produced by blending sugar syrup with the pasteurized milk at the evaporation stage, and cooling and crystallizing before packaging (Hunziker, 1949).

Losses of concentrated milk are the main sources of organic load in the effluent.

Dried milk
A major method of preserving milk is to remove the water to produce a powder containing 2.5–5.0% moisture. The most common means is to remove water from clarified standardized milk by evaporation followed by drying. Drying is most commonly carried out in a spray drier in which the milk concentrate is atomized into a flow of hot air. The hot air removes the remaining water. Roller or drum drying is also used for some products: the concentrated milk is spread on a heated, rotating drum and the thin film formed is scraped off by stationary steel blades. The dried powder is ground, blended, sifted and packed into consumer packages, 20–25 kg bags or large (1 tonne) tote bins.

Whole milk powder is produced from standardized whole milk, skim milk powder (or non-fat dried milk) from the skim milk remaining after cream separation from whole milk and buttermilk powder from the buttermilk which is produced during buttermaking from cream (Hall and Hedrick, 1966; Masters, 1979).

Dried powder, if not swept up, can be a major source of waste.

Whey processing
In the manufacture of cheese or casein from milk, curds are formed by the action of rennet-type enzymes and/or acid. Whey is the liquid remaining after the recovery of those curds. The whey comprises 80–90% of the total volume of milk entering the process and contains more than half the solids present in the original whole milk, including 20% of the protein and most of the lactose, minerals and water-soluble vitamins. In general, the manufacture of 1 tonne of cheese or casein results in the production of 8 or 25 tonnes of liquid whey, respectively. Whey has been the most visible source of waste in the dairy

industry in recent times because the economics of whey processing have precluded manufacture of whey products. However, whey processing methods are evolving and whey is increasingly being recognized as a source of valuable constituents which can be used by the food and animal feed industries.

Whey is used, unconcentrated and concentrated, as an animal feed, particularly for pigs and calves. Whey cheese is produced in a number of countries by concentrating and heating whey. Whey powders are produced by drying whey, often after demineralizing by ion exchange or electrodialysis. Proteins are recovered from whey by a variety of techniques, including ultrafiltration, heat precipitation and adsorption onto suitable resins. Lactose is recovered from concentrated whey by crystallization and drying. The lactose in whey can be converted to other products by hydrolysis or fermentation (see Coton, 1980; Fox, 1970; Gillies, 1974; Short, 1978).

Because whey historically has been considered as a waste product, it is frequently mishandled during processing. The result is that significant quantities of whey may be discharged to the liquid waste with quite marked effects on the effluent treatment.

Ice-cream and frozen desserts

The ingredients of ice-cream are fat, milk solids-not-fat, sugar, water and various additives such as emulsifiers, stabilizers, flavourings, fruit, nuts and colourings. Most manufacturers make their mix from a combination of liquid milk and condensed or dry products. Quantities of the various ingredients are blended to a homogeneous mixture to form the ice-cream mix. The mix is pasteurized, homogenized and pumped to the freezers. In the freezer the mix is frozen on the surface of a refrigerated tube and scraped off by sharp blades rotating at high speed. The blades also whip air into the mix to give the ice-cream its characteristic texture. Fruit, nuts and other solid ingredients are injected into the frozen ice-cream stream. The ice-cream is fed to filling or portioning machines of various types for moulding or packaging. Finished products in the form of bars, cups, cones, bricks and large packs are then placed in cold storage, generally after a hardening process in continuous freezing tunnels (Arbuckle, 1972).

Ingredients such as sugar, nuts, etc., are non-dairy components in the waste from ice-cream plants. Spillage can be a major source of effluent load, because the organic strength of ice-cream is very high.

5.3.2 Process equipment

The process equipment used in dairy product manufacture is very diverse. Equipment common to most processes includes storage tanks (silos), processing vessels (vats), buffer or balance tanks, heat exchangers for heating and cooling (frequently plate-type), pumps (including centrifugal, liquid ring and positive displacement pumps), piping and associated valves, control equipment,

Table 5.6 Unit operations in dairy processing

Transportation of fluids	Size reduction
Mixing	Mechanical separation
Heat transfer	Evaporation
Drying	Gas absorption/desorption
Distillation	Leaching
Extraction	Crystallization
Humidification	Filtration
Ultrafiltration	Reverse osmosis
Ion exchange	Electrodialysis
Fluidization	Centrifugation
Settling	Fermentation
Enzyme reaction	

instrumentation, and equipment required for packaging, water supply and treatment, steam raising and compressed air. Particular operations require equipment associated with the wide variety of unit operations given in Table 5.6. Specialized equipment is used in the manufacture of many products, e.g. butter, cheese, ice-cream, and for such operations as pasteurizing, separating and homogenizing.

Because of the nature of milk and milk products, which are susceptible to microbial spoilage, equipment is characterized by designs which facilitate hygienic operation, easy cleaning and sterilizing. While many older plants use open equipment and batch processing, modern dairy food plants use closed systems operated continuously for periods up to 24 h. Shutdown for cleaning is generally required at least once per day. Most cleaning is carried out by cleaning-in-place (CIP) techniques (Heldman and Seiberling, 1976).

5.4 Dairy plant wastes

Dairy food plant wastes are generally dilutions of milk or milk products, together with detergents, sanitizers, lubricants, chemicals from boiler and water treatment, washings from tank trucks and domestic wastes. The waste may include water from condensing and heat transfer operations and storm-water, though good practice would normally divert these waters for reuse or separate disposal. The wastes are characterized by a relatively (cf. domestic wastes) high organic concentration and high initial total oxygen demand. They are also highly variable in quantity and composition.

5.4.1 Sources of dairy plant wastes
The processes and other sources of wastes that have a significant effect on the liquid effluent from all dairy plant operations include:

- Rinsing and washing of bulk tanks or cans in receiving operations.

- Rinsing of residual product remaining in or on the surfaces of all pipelines, pumps, tanks, vats, processing equipment, filling machines, etc.
- Washing of all processing equipment (note that rinsing and washing operations are performed routinely every processing cycle, and at least once per day).
- Water–milk solids mixtures discharged to the drain during start-up, product changeover and shutdown of pasteurizers, heat exchangers, separators, clarifiers, vacreators and evaporators.
- Carryover (entrainment) of droplets of milk into the tailwaters of vacreators, pasteurizers and evaporators.
- Sludges discharged from clarifiers.
- Fines from cheese and casein operations.
- Spills and leaks due to improper equipment operation and maintenance, overflows, freezing-on and incorrect handling.
- Wilful waste of unwanted byproducts (e.g. whey, buttermilk) or spoiled materials.
- Loss in packaging operations through equipment breakdowns and broken packages.
- Product returns.
- Lubricants from equipment, casers, stackers and conveyors.
- Washings from the outsides of tank trucks including dirt, stones and farm debris.
- Dust from coal and wood fuel and spills of fuel oil.
- Powder deposited from discharges from driers.
- Ash from boilers.
- Water and boiler treatment chemicals.
- Chemicals from the regeneration of ion-exchanger resins (significant in some plants).

In the past, and even in some older plants today, byproducts such as skim milk, buttermilk and whey were significant contributors to the waste load. However, the continuing desire to increase farmers' incomes by processing all of the milk constituents, the pressure of anti-pollution regulations, the increasing costs of effluent treatment and disposal and the consolidation of dairy processing into large units have resulted in effective utilization of these byproducts, particularly skim milk and buttermilk. But whey, particularly acid whey, is still a problem in many plants.

5.4.2 Composition of dairy fluids and products

If domestic wastes are excluded, the principal contaminants of dairy wastes are milk, milk fractions and milk products. It is generally conceded that 90% of the organic loading (BOD) in dairy processing wastes comes from these materials. Estimates of the average daily equivalent quantity of milk in the wastes from

Table 5.7 Typical composition of some USA dairy products (per 100 g product)

	Milkfat (g)	Protein[a] (g)	Lactose (g)	Lactic acid (g)	Total organic solids (g)	Ca (mg)	P (mg)	Cl (mg)	S (mg)	Total minerals (g)	Viscosity (cP at 20°C)
Skim milk	0.08	3.5	5.0	—	8.6	121	95	100	17	0.7	1.4
2% milk	2.0	4.2	6.0	—	12.2	143	112	115	20	0.8	2.4
Whole milk	3.5	3.5	4.9	—	11.9	118	93	102	19	0.7	2.2
Coffee cream	18.0	3.0	4.3	—	25.3	102	80	73	12	0.6	15.0
Cream	40.0	2.2	3.1	—	45.3	75	59	38	9	0.4	25.2
Chocolate milk[b]	3.5	3.4	5.0	—	18.5	111	93	100	19	0.7	15.0
Churned buttermilk	0.3	3.0	4.6	0.1	8.0	121	95	103	15	0.8	1.5
Cultured buttermilk	0.1	3.6	4.3	0.8	8.8	121	95	105	17	0.7	500.0
Sour cream	18.0	3.0	3.6	0.75	25.3	102	80	73	12	0.6	10 000.0
Yoghurt[c]	3.0	3.5	4.0	1.1	11.6	143	112	105	19	0.7	3 000.0
Evaporated milk	8.0	7.0	10.0	—	25.0	757	205	210	39	1.6	30.0
Ice-cream[d]	10.0	4.5	6.8	—	36.3	146	115	104	20	0.9	35.0
Sweet whey	0.3	0.9	4.9	0.2	6.3	51	53	95	8	0.6	1.4
Cottage cheese whey	0.08	0.9	4.4	0.7	6.1	96	16	95	8	0.8	1.3

[a] Total Kjeldahl nitrogen × 6.38.
[b] Sucrose added, 6%; chocolate solids added, 1%.
[c] Fruits added.
[d] Sugar added, 15%.

Table 5.8 Typical compositions of some New Zealand dairy products (per 100 g product)

	Protein (g)	Fat (g)	Carbohydrate (g)	Water (g)	Minerals (g)
MILKFAT PRODUCTS					
Creamery butter (salted)	1.0	81.8		15.7	1.5 (salt)
Lactic butter	0.6	82.7	0.8	15.8	0.1
Frozen cream	1.7	44.5	2.8	50.0	1.0
Anhydrous milkfat	trace	99.9		0.1	
Ghee	trace	99.9		0.1	
Whey butter	0.6	82.2		15.6	1.6 (salt)
CHEESE					
Cheddar	24.0	36.5		35.0	4.5
Cheshire	25.0	32.0		38.5	4.5
Colby	23.5	33.0		39.0	4.5
Egmont	26.5	31.0		38.0	4.5
Fetta	15.0	27.0		52.0	6.0
Gouda	25.5	30.0		40.0	4.5
Granular	24.5	35.5		35.5	4.5
Parmesan	40.0	22.5		32.0	5.5
Processed cheese and cheese spreads	21.5	32.0		41.2	5.3
MILK POWDERS					
Whole milk powder (spray and roller)	28.0	26.5	36.7	2.8	7.0
Skim milk powder (spray and roller)	38.0	1.0	49.5	3.5	8.0
Buttermilk powder (spray and roller)	34.5	9.0	45.5	3.5	7.5
Rennet casein whey powder	16.0	1.5	70.0	3.0	9.5
MILK PROTEIN PRODUCTS					
Acid casein	86.0	1.2	trace	11.0	1.8
Sodium caseinate (spray dried)	92.0	1.2	trace	3.5	3.5
Rennet casein	80.5	0.5	trace	11.0	8.0
Soluble lactalbumin	50–80	2–5	10–30	5.0	2–6
Insoluble lactalbumin	85.0	5.0	4.0	4.6	1.4

dairy processing plants (whey excluded) vary from 0.5 to 6% of the milk received at the plant. Thus a knowledge of the compositions of milk, dairy fluids and dairy products is a helpful guide to the typical composition of dairy wastes. The compositions of some typical dairy fluids are given in Table 5.7 and of dairy products in Table 5.8.

Milk and its products are organic materials and the organic waste parameters normally measured include the biochemical oxygen demand (BOD), chemical oxygen demand (COD) and total organic carbon (TOC). Published values of the five-day BOD (BOD_5) have been summarized (EPA, 1971) (Table 5.9). Table 5.9 also gives some COD values.

Table 5.9 Biochemical oxygen demand (EPA, 1971) and chemical oxygen demand (IDF, 1980) of some dairy fluids

	BOD_5 (mg l^{-1})	COD (mg l^{-1})
Skim milk	67 000	100 000
2% milk	100 000[a]	
Whole milk	104 000	210 000
Coffee cream	206 000	
Cream	399 000	860 000
Chocolate milk	145 000[a]	
Churned buttermilk	68 000	110 000
Cultured buttermilk	64 000[a]	
Sour cream	218 000[a]	
Yoghurt	91 000[a]	
Evaporator milk	208 000	
Ice-cream	292 000	
Sweet whey	34 000	75 000
Cottage cheese whey	31 500	

[a] Value calculated from BOD_5 values of the components.

Whole milk has a high BOD (about 100 000 mg l^{-1}), the actual value depending on the composition (see Table 5.1). Thus, even relatively dilute milk solutions have a marked polluting effect. The major constituents which contribute to the BOD of dairy wastes are lactose, milkfat, protein and lactic acid, and the reported average values are 0.65, 0.89, 1.03 and 0.63 kg BOD_5 per kg component, respectively. The individual reported values show a wide range reflecting differences in measurement techniques, particularly the concentration in the test (EPA, 1971). Nevertheless, the average values were used in EPA (1971) to calculate BOD_5 values for various dairy fluids which, except for whey, were in good agreement with the reported measured values.

BOD_5 is the standard expression for effluent pollution potential and is the normal parameter for the design of effluent treatment plants. However, it has a number of disadvantages, including the time required to carry out an analysis (normally five days, although some countries use seven days) and poor reproducibility ($\pm 20\%$). The COD estimation is rapid (2.5 h), relatively simple and cheap and the reproducibility is generally good (±5%). This test is being used by many dairy plants as a routine monitor of the strength of dairy effluents. The estimation of TOC is very rapid (minutes) and accurate, although competent operators are required and the high cost of the instrument is a disadvantage. There is a large number of publications devoted to discussing the results of studies to determine the ratio of COD and TOC to BOD. However, the values of the ratios vary considerably with the chemical composition of the milk and wastes and no one value applies in all plants. Ratios established in one dairy plant cannot be used in another plant.

The BOD_5/COD ratios for casein, lactose and milkfat are 0.46, 0.53 and 0.79 respectively. The BOD_5 for untreated dairy wastes is most frequently 50 to 70% of the COD value, but percentages as high as 80 and as low as 20 have been reported (IDF, 1981). Values which are outside the range of 50 to 70% probably indicate wastes of an unusual nature such as those contaminated by ammonia or glycol from refrigerant leaks or by the presence of some compounds toxic to the BOD_5 test. Few data have been reported on TOC values. Ratios of TOC to BOD_5 reported by IDF (1981) range from 0.30 to 0.90. Values from a limited number of determinations at the New Zealand Dairy Research Institute are shown in Table 5.10.

Table 5.10 Some values of BOD_5/COD and TOC/COD ratios measured by the New Zealand Dairy Research Institute (unpublished data)

	BOD_5/COD	TOC/COD
Whole milk	0.69	
Skim milk	0.63	0.34
Buttermilk	0.66	
Whey	0.52	
Casein	0.46[a]	0.38
Lactose	0.53[a]	0.40
Whey protein	0.23	
Milkfat	0.79[a]	
Casein whey		0.76
DAIRY PLANT WASTES		
Butter and buttermilk powder	0.52–1.13	0.24–0.35
Lactic casein	0.53–1.13	
Cheddar cheese	0.33–0.78	
Whole milk powder		0.21–0.37
Multi-product		0.25–0.45

[a] Value taken from EPA (1971) — note that the value for milkfat as published in EPA (1971) was a misprint.

COD or TOC measurements can be used directly for the routine monitoring of the amount of waste produced by a dairy plant. However, the values must be used with some caution in making comparisons between plants or between different processes within plants. BOD_5 measurements are required by legislative authorities and are needed as a basis for design of treatment systems.

EPA (1971) discusses the compositions of detergents, sanitizers and lubricants used in dairy processing. Wetting agents and surfactants have BOD_5 values in the range $0.05-1.2\,kg\,BOD_5$ per kg product and the acids used in dairy plant detergents have values in the range $0.25-0.85\,kg\,BOD_5$ per kg product. Detergents and sanitizers are, potentially, significant contributors to BOD_5, refractory COD and phosphate concentrations in dairy plant wastes. EPA (1971) concludes that it is likely that the concentration of components from detergents and sanitizers from normal operations will be too low to cause toxic effects in treatment plants. Subsequent work (Harper and Chambers, 1978) suggests that surface-active compounds may have an adverse effect on the microflora of activated sludge plants. Further study is required.

5.4.3 Composition and characteristics of dairy plant wastes

The nature of the wastes generated by the various processes operated by the dairy industry is generally quite similar, reflecting the overwhelming effect of the wasted milk and dairy products. However, different processes affect the detailed composition. Waste from a milk-receiving or bottling plant contains mostly whole milk; that from a butter factory, mixtures of whole milk, cream, buttermilk and sodium chloride; that from a cheese or casein plant, whole milk and whey, etc. Thus the strength and volume of wastes from any plant depends on the processes carried out, on the volume of milk handled, on the condition and type of equipment, the waste reduction practices, the attitudes of management and staff and the amount of water used in cooling and washing. For any

Table 5.11 Characteristics of dairy plant wastes in the USA (EPA, 1971 and Kearney, 1973)

Characteristic	Concentration ($mg\,l^{-1}$)	
	Range	Mean
Biochemical oxygen demand	40–48 000	2300
Chemical oxygen demand	80–95 000	4500
Suspended solids	24–4500	820
Total solids	135–8500	2500
Nitrogen	1–180	64
Carbohydrate	250–930	520
Fat	35–500	209
Calcium	55–115	37
Sodium	60–810	320
Potassium	10–160	70
Phosphorus (as PO_4)	9–210	48
Chloride	48–1930	480
BOD_5 coefficient (kg BOD_5 m^{-3} milk)	0.2–7.1	5.8
Wastewater volume coefficient (m^3 water m^{-3} milk)	0.1–7.1	2.4
pH	4.4–9.4	7.2
Temperature (°C)	18–55	35

Table 5.12 Characteristics of dairy plant wastes in New Zealand (New Zealand Dairy Research Institute data)

Characteristic	Concentration (mg l^{-1})
Biochemical oxygen demand	90–12 400
Chemical oxygen demand	180–23 000
Suspended solids	7–7200
Nitrogen	1–70
Fat	0–2100
Phosphorus (as PO$_4$)	4–150
BOD$_5$ coefficient (kg BOD$_5$ m^{-3} milk)	0.2–28
Wastewater volume coefficient (m^3 water m^{-3} milk)	0.7–4.4
pH	3.0–13.2
Temperature (°C)	11–72

one plant, published data can give only an approximate guide to the expected waste discharge. Good information can be obtained only by actual measurement.

The characteristics of the wastes from the USA dairy industry are summarized in Table 5.11, adapted from EPA (1971) and from Kearney (1973). Data obtained from New Zealand plants during 1972–80 are given in Table 5.12. Table 5.13 is from a UK publication (Ministry of Agriculture, Fisheries and Food, 1969) and summarizes information collected from practical experience over a number of years. In this table the data for BOD$_5$ coefficients (mass of BOD$_5$ per volume of milk processed) do not include significant contribution from spillages, leaks, etc. or from direct discharge of surplus low-value materials such as whey or skim milk. The figures give a guide to the assessment of possible minimal waste loads from individual dairy operations.

Volume of waste

Wastewater volume coefficients (volume of wastewater per volume of milk) show a wide range (Tables 5.11 and 5.12). Data published by IDF (1981) for a number of countries show a range from 0.5 to 37 m^3 water per m^3 milk. High values reflect the presence of cooling and evaporator tailwaters and, perhaps, storm-water. Typically, coefficients of 0.5 to 2.0 m^3 water per m^3 milk can be achieved.

Biochemical and chemical oxygen demand

The ranges of BOD$_5$ and COD concentrations are very large; typical values for BOD$_5$ are about 2500 mg l^{-1}, the majority of plants having values in the range 500–4000 mg l^{-1}. Low values are associated with milk-receiving operations and high values reflect the presence of whey or waters from the washing of cottage cheese or casein curd. Similarly, typical values for COD are about 4000 mg l^{-1}.

The BOD$_5$ of whole milk is approximately 100 000 mg l^{-1}, thus a BOD$_5$

Table 5.13 Wastes arising from various dairy plant operations (IDF, 1978; originally adapted from Ministry of Agriculture, Fisheries and Food, 1969)

The suggested target values should be achievable by reasonable effort. The measurement values do not include any significant contribution from spills, leaks or waste of whey or skim milk.

Operation	kg BOD_5 m^{-3} of milk processed		
	Average	Range	'Target'
1 Milk reception, can washing, cleaning up	0.26	0.11–0.66	0.11
2 Cooling raw milk, storage, washing tanks and pipelines	0.19	0.07–0.31	0.07
3 Washing tankers	0.25	0.10–0.40	0.10
4 Separating, storage of skim milk and cream	0.14	0.09–0.24	0.10
5 (4), plus cream pasteurizing	0.66	0.46–1.20	0.30
6 Churning and washing butter	0.46	0.25–0.30	0.30
7 Evaporating skim milk to low total solids	0.23	0.16–0.30	0.16
8 Evaporating skim milk to high total solids and spray drying	0.74	0.14–1.50	0.30
9 Roller drying	0.53	0.25–1.30	0.30
10 Pasteurizing milk, and storage	0.29	0.10–0.54	0.10
11 Bottling pasteurized milk	0.11	—	0.10
12 Bottle washing	0.23	0.05–0.37	0.15
13 (10) to (12) inclusive	0.85	0.49–1.70	0.35
14 Clotted cream	1.20	—	0.60
15 Cream pasteurizing and packing	0.79	—	0.40
16 Cheesemaking (hard-pressed)	0.89	0.23–2.00	0.50
17 Cottage cheese (washed curd)	15.00	—	2.00
18 Condensing fresh whey to low total solids	0.25	—	0.25
19 Condensing sweetened separated condensed milk	1.40	1.20–1.70	0.60
20 Full cream evaporated milk with canning	0.75	0.50–1.00	0.50

coefficient of 1 kg BOD_5 m^{-3} milk is equivalent to 1% of the milk received. An average BOD_5 coefficient of 5.8 kg BOD_5 m^{-3} milk (Table 5.11) implies that 5.8% of the milk solids received are wasted. This estimate of the average percentage of milk solids wasted is unduly high because of the presence of whey in many of the effluents. Nevertheless milk solids losses of up to 4% of the weight received are not uncommon.

Rate of deoxygenation

Dairy wastes exert a high rate of deoxygenation; the initial oxygen demand is high, about 50% of the total oxygen demand being exerted in 24 h. The rate of deoxygenation is more than twice that of domestic sewage (Fig. 5.6). Table 5.14 presents some data, calculated from laboratory BOD measurements on some dairy fluids, for the deoxygenation coefficient, k, in the equation:

$$BOD_t = BOD_{ult}(1 - e^{-kt})$$

where BOD_t, BOD_{ult} are the BOD at time t and the ultimate BOD, respectively.

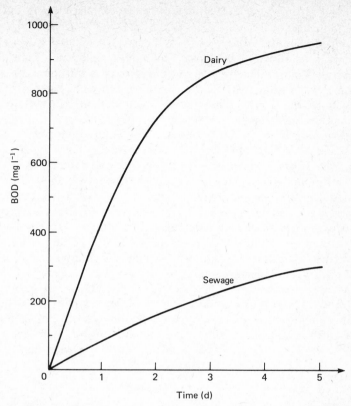

Fig. 5.6 BOD curves for domestic sewage and a dairy plant waste (Strom, 1974)

Table 5.14 Deoxygenation coefficients k (d^{-1} (base e)) calculated from laboratory BOD measurements (Barnett *et al.*, 1982)

Whole milk	0.42
Skim milk	0.39
Whey powder	0.51
Multi-product factory wastewater	0.44–0.62

Chemical composition

Few systematic studies of the chemical composition of dairy plant wastes have been reported, and the values in Tables 5.11 and 5.12 are based on limited data. In general, nitrogen-to-carbohydrate (lactose) ratios and fat-to-carbohydrate ratios are significantly different from those for whole milk. Milkfat or lipid losses are, in general, quite well controlled in most dairy plants. Nevertheless losses can be quite high, reported concentrations in wastes ranging up to $2100\,\mathrm{mg\,l^{-1}}$. It should be noted that milkfat is highly biodegradable, as illustrated by its high BOD_5 value, high BOD_5/COD ratio and high BOD_5 rate constant (0.31 (Brown and Pico, 1980)). The floating oil and grease (FOG) values, as measured by non-aqueous solvent extraction from wastewaters, make no distinction between petroleum-based oils and edible oils and fats. The former are much more difficult to treat biochemically than edible oils and fats such as milkfat. Thus surcharges (see later) which include both BOD_5 and FOG concentrations may be unwarranted for dairy wastes.

Phosphate values, which vary independently of BOD values, are in excess of $10\,\mathrm{mg\,l^{-1}}$ and offer a potential source of phosphate for enhancing algae growth in natural waterways. While a major source of phosphate is the cleaners used in the dairy plant, non-phosphate cleaners are not as effective and add to the cost of cleaning because they require higher concentrations, longer cleaning cycles and sometimes the addition of an acid cleaning cycle. Phosphate concentrations vary widely, probably reflecting differences in the amount of phosphate in the cleaning compounds used. High concentrations of acid whey (from cottage cheese or casein operations) also result in high concentrations of phosphate.

Variations in the concentrations of magnesium and calcium are generally similar to the variations in BOD_5, but those of sodium and potassium are unrelated to BOD_5 changes, reflecting the differences in the sodium and potassium composition of cleaning compounds and water treatment chemicals. Generally, in USA practice, 90% of the sodium, 50% of the magnesium and 25% of the calcium in dairy wastes appear to come from non-milk sources.

Solids

The reported suspended solids concentrations (material removed by filtration) of raw dairy wastes vary widely ($7-7200\,\mathrm{mg\,l^{-1}}$) although most plants have values in the range $400-2000\,\mathrm{mg\,l^{-1}}$. The suspended solids in dairy wastes are mainly organic in nature; volatile suspended solids, as a percentage of the total suspended solids, range from 68% to 98% with an average of 85% (Kearney, 1973). The proportion of non-organic solids in milk is about 13%, which suggests that most of the suspended solids in dairy food plant wastewaters are of a dairy food origin. An increase of non-organic matter in the total solids reflects a contribution of non-organic material from detergents, sanitizers and lubricants.

Dairy wastes generally have low concentrations of settleable solids (material removed by gravity settling) although wastes from cottage cheese and casein operations, with high concentrations of fines, wastes from ice-cream plants where fruits and nuts are used and wastes from drying plants may be exceptions. Washings from the exterior of tanker trucks and from yards may contribute also to settleable solids concentrations.

pH and temperature

The daily average pH of dairy food plant wastewaters varies from about 4.4 to 9.4 with a median value around pH 7.2. Hourly pH variation over a day's operation can range from 2 to 11. Although it would be expected that cheese plants might have a lower pH wastewater than other types of plant, because of the presence of whey, this has not been found. The major factor affecting the pH of dairy plant wastes is the cleaning compound, either acid or alkali.

Only limited studies have been made of the temperature of dairy wastewaters as they leave the dairy plant. Reported values ranged from 11° to 72°C, although values above about 35°C indicate excessive losses of energy.

Variations in waste characteristics of a single plant
In practically all dairy plants there is a marked fluctuation in the rate of flow, strength, temperature and other characteristics of dairy wastes. Wide variations occur within minutes throughout the day, depending on the processing and cleaning operations taking place (Figs 5.7 and 5.8). Furthermore there are

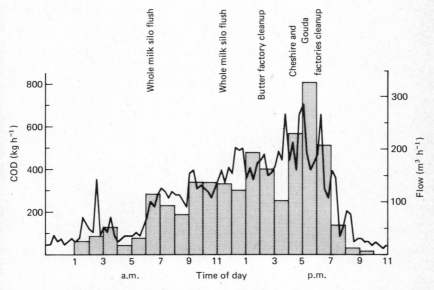

Fig. 5.7 Twenty-four-hour fluctuations in flow rate of wastes (—) and quantity of COD (bar graph) discharged from a dairy plant manufacturing milk powders, butter and cheese

usually substantial daily and seasonal fluctuations depending on the types of product manufactured, production schedules, maintenance operations, etc. (Fig. 5.9).

Some dairy plants operate only five days a week, causing major flow and strength variations. In some countries (e.g. New Zealand, Australia, Republic of Ireland) dairy processing is seasonal, with complete shutdown of the plants for two to three months during the winter, followed by a rapid rise in milk flow in the spring (Fig. 5.10).

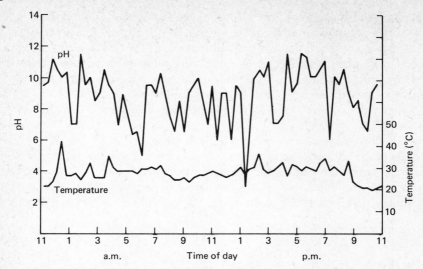

Fig. 5.8 Twenty-four-hour fluctuations in pH and temperature of wastes from a dairy plant manufacturing milk powders, butter and cheese

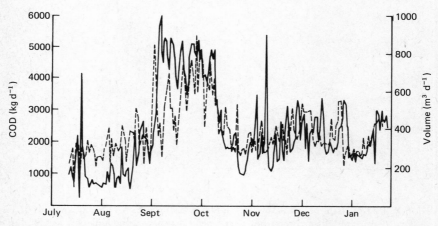

Fig. 5.9 Seasonal fluctuations in volume (- - -) and strength (———) of dairy wastes from a New Zealand dairy plant

Summary of dairy waste characteristics
The data for all parameters of dairy plant waste show a very wide variation and summaries such as those given in Tables 5.11–13 should be used as a guide only to the characteristics of dairy plant effluent. EPA (1971) have shown that the quantity and strength of wastes are not related to plant size or degree of automation. Plants with fully automated product processing and CIP systems frequently have above-average waste coefficients (see also Lytken, 1974). Plants

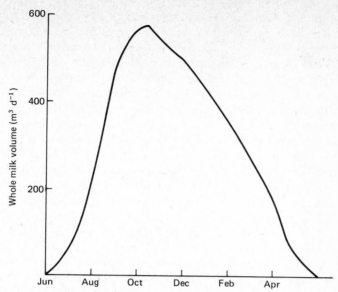

Fig. 5.10 Typical seasonal trend in the whole milk intake of a New Zealand dairy plant

processing milk to cottage cheese or casein may be expected to have significantly higher BOD_5 coefficients than any other type of operation because of the presence of whey. Ice-cream plants also tend to have high BOD_5 coefficients.

From all the data available it appears that the main factor controlling wastewater volume and BOD_5 coefficients is the attention that management pays to the control of waste production. Plants with very low coefficients (0.5 m^3 water m^{-3} milk; 0.5 kg BOD_5 m^{-3} milk) generally have excellent management control, whereas plants with high coefficients (both >3.0) have been shown, generally, to have poor management supervision of waste-generating practices.

In comparison with domestic sewage, dairy wastes have a high BOD_5 concentration. They exert a rapid oxygen demand in BOD tests and hence may be expected to have a similar effect in natural waterways or in treatment plants. Dairy wastes also show much greater variations in strength and quantity than domestic wastes. Dairy wastes do not contain significant quantities of toxic compounds, although accidental spillage of cleaning and sanitizing chemicals may constitute a hazard. Generally the pH and temperature of dairy wastes are higher than those of domestic wastes and show marked variations throughout the day.

5.5 Waste reduction

Milk is an expensive raw material and effluent from a dairy plant represents not only a waste of that raw material but also a cost for treatment. An integral part of any programme for the treatment and disposal of dairy waste is the control of waste within the manufacturing process. This control involves monitoring waste levels and taking steps to reduce the wastage to an unavoidable minimum. Despite the considerable emphasis given to this control, both in the literature and by authorities in waste treatment, many dairy plants have excessive milk losses, frequently without knowing their full extent or realizing the impact of such losses on the economics of milk processing (Chambers, 1980). Authorities agree that a waste survey of most dairy plants would reveal substantial unsuspected losses of milk and milk products and excessive use of water.

Harper and Chambers (1978) report that in the USA the average dairy plant that has not been concerned previously with waste control could reduce wastewater volume by as much as 50% and the quantities of BOD_5 and suspended solids by 33%. For example, in one multi-product plant receiving about 500 m^3 milk per day, the introduction of improved management waste control measures reduced daily BOD_5 from 5500 kg to 2800 kg with little expenditure. Such reductions have a two-fold benefit: savings are derived from the increased yield and from the more efficient handling of water, raw materials, products and byproducts; there is also a reduction in the costs of providing, operating and maintaining the waste treatment and disposal facilities necessary to meet statutory requirements.

Details of methods of monitoring wastes, determining milk and product losses and reducing losses to an unavoidable minimum are contained in Ministry of Agriculture, Fisheries and Food (1969), EPA (1971), An Foras Taluntais (1974), Jones (1974), IDF (1974, 1978, 1980), Carawan *et al.* (1979), Chambers (1980) and Galpin and Parkin (1981).

5.5.1 Use of effluent data to compute losses and yield

Before effective action can be taken to reduce waste, company management must know the extent of the losses and the economic consequences. Unfortunately, the traditional method by which dairy plant management determines plant productivity and yield information — i.e., comparing quantities of product packed with quantities credited to the plant as raw materials — is inherently too inaccurate to provide the necessary data on a daily basis. While these traditional methods will continue to be used for accounting purposes, daily direct measurement of losses gives management a more accurate estimate and identifies the sources of the losses so that corrective action can be taken.

Galpin and Parkin (1981) illustrate the advantages of loss measurement with an example from the determination of the yield of butter. The quantity of milkfat delivered to a plant can be estimated with a single determination of the volume of whole milk and the concentration of milkfat in a representative sample, to an accuracy of ±3.3%. The quantity of product can be estimated with an accuracy of ±1.0%. Thus if 1000 units of raw material are measured as received on one day and 980 units are measured in the product, the estimated yield (Equation 5.1) is $98 \pm 4\%$ and the estimated loss is 20 ± 43 units. The inaccuracies in the measurement of the losses are greater than the actual estimated loss.

$$\text{Yield} = \frac{P_o \times 100}{P_i} \quad (5.1)$$

where P_i is the potential quantity of butter in the whole milk received and P_o is the quantity of butter packed.

On a daily basis, a comparison of this type will not reliably reveal quite significant losses.

If all the losses are measured (in this case, these include loss to the effluent, failure to attain moisture specification, excess milkfat in the buttermilk and excess butter in the final container — Galpin and Parkin (1981) showed an average total value of 1.72%) to an accuracy of, say, $\pm 20\%$, the estimated loss is 20 ± 4 units. Direct measurement of the losses, even with a quite low degree of accuracy, leads to an order of magnitude increase in the degree of accuracy of the estimation of the losses. A mass balance equation leads to

$$P_o = P_i - L \quad (5.2)$$

where L is the losses. Rearrangement of Equation (5.1) gives

$$\text{Yield} = \left(1 - \frac{L}{P_i}\right) \times 100 \quad (5.3)$$

or

$$\text{Yield} = \left(1 - \frac{L}{P_o + L}\right) \times 100 \quad (5.4)$$

Using Equation (5.3), the estimated yield is $98 \pm 0.47\%$; using Equation (5.4) it is $98 \pm 0.43\%$.

In detailed studies over ten years the New Zealand Dairy Research Institute has applied this technique to most dairy products manufactured in New Zealand. The technique gives a reliable estimate of losses and provides managers with information very quickly, so that appropriate action can be taken. The studies also show that the greatest avoidable loss is to the liquid effluent (in New Zealand about 65% of the total measured losses). A similar approach was used by Goldsmith and Watson (1980), who assigned BOD_5 values to food-processing raw materials and products by means of a formula relating composition to equivalent BOD_5 and used this equation to relate waste loads to production activities for waste management.

5.5.2 Waste monitoring

A waste monitoring system, as a measure of efficient operation, is an integral part of the manufacturing process. Once incorporated into the production system, it is an invaluable check on the overall efficiency of plant operations as well as an aid in meeting legal requirements. The monitoring programme also provides basic data that will be valuable in the design of a wastewater treatment system to meet regulatory requirements.

Initial survey

To initiate any waste survey, the first step is a complete inventory of the factory plant and processes. Drainage plans, if available, are thoroughly reviewed and each drain is physically examined to determine its exact course and destination. The precise source at which each waste stream enters the main effluent drain is determined. Where plans of drainage and storm-water systems do not exist or are archaic, dye traces such as Rhodamine B or fluorescein are used to detect the existence of any cross-linkages in the drainage system. All storm-water and uncontaminated discharges within the factory are isolated from the main effluent drain.

In the establishment of a waste monitoring system, the logical place to start is at each main outfall, and samples collected from these places will characterize the total plant effluent. In an old plant with multiple outfalls from the factory, a study is made of the feasibility of developing a sampling station that would combine all outfalls to provide a single point for measuring the total waste from the plant.

The sources of all clean water entering the plant are also isolated and a detailed water line plan is made. This is necessary for later isolation of individual departments for determining water use.

Flow measurement

The nature of the waste production from dairy processing requires the use of continuous flow measurement techniques capable of accurate measurement of variable flow rates over 24 h. Spot measurements over a short time period will give inaccurate and misleading data. The *Handbook for Monitoring Industrial Waste Water* (EPA, 1973) gives details of suitable devices (see also IDF, 1980; Carawan *et al*., 1979). Equipment that has been used in dairy plants includes various types of open-channel and closed-channel flumes and weirs and in-line flow meters (turbine, electromagnetic, etc.).

The selection of a flow measurement device must take into account the maximum flow rates that will be experienced in 24 h. The device should be sited so as to minimize restrictions to the normal flow and to ensure strict adherence to the hydraulic conditions required for accurate operation. The effects of suspended solids or milkfat on the operation of the device need to be assessed.

Sampling

Of equal importance is the problem of obtaining a truly representative sample of the effluent stream. Samples may be required for determination not only of the 24 h effluent load but also of the peak load concentrations, the duration of peak loads and the occurrence of variation throughout the day. Dairy wastes are so variable in composition that grab samples are useless as an estimate of waste strength. A number of sampler types are available (EPA, 1973; IDF, 1980; Carawan et al., 1979) and they operate in one of the following ways:

- A time-proportional method, which involves taking a sample at a set time interval, e.g., every 1 min, every 30 min. The greater the frequency of sampling, the more representative will be the sample.
- A volume-proportional method, in which a small sample is taken from the drain after a known volume, e.g., 1000 litres, has passed through the flow-measuring device.
- A flow-proportional method, in which a sample is taken from the drain at a particular time but proportional to the flow passing the device at the time the sample is taken.

The sampler is installed so as to obtain a truly representative sample. Care must be taken to ensure that suspended solids and fat concentrations are not biased by the method of sampling. Samples are taken into refrigerated containers and correctly preserved until analysis (APHA, 1980).

Analysis

Analysis of dairy wastes is carried out using standard methods (APHA, 1980; IDF, 1980). The main analysis required to assess the quantity of product loss in the liquid effluent are COD, milkfat and total nitrogen concentrations. The most informative test is the COD, because of its relative accuracy, simplicity and speed. Since the COD test is of short analytical duration, it is a useful tool to monitor plant effluent on a daily basis. However, as it does not measure the actual oxygen depletion characteristics of wastewater in streams (which can only be done biochemically) it is not a direct substitute for the BOD_5 test. The COD test can be used to establish trends, to pinpoint spills promptly and to identify other non-routine problems. The test can be used to monitor BOD_5 test results. Unusual BOD_5/COD ratios may indicate problems such as inadequate BOD seed, poor analytical techniques or equipment malfunctions.

The concentration of total solids is not a reliable indicator of losses. In a comparison of five methods of measuring the quantity of skim milk in the effluent from a drying process, the estimate based on total solids was 50% higher than the values estimated from the other parameters (Table 5.15). The high results for total solids were due, in part, to the large quantities of cleaning compounds in the effluent (Parkin et al., 1977).

COD analysis is appropriate when the nature of the material in the effluent is similar to that of the final product, i.e. for milk powder production (Table 5.16). Total nitrogen and milkfat analyses are required to determine the losses from the manufacture of butter. The COD test cannot be used because of the markedly different COD values of the protein, milkfat and lactose, the relative concentrations of which can vary in the effluent. The quantity of milkfat in the

Table 5.15 Comparison of five methods used to estimate the quantity of skim milk in the effluent from a drying process (Parkin et al., 1977)

Parameter	Estimated quantity of skim milk (kg d^{-1})
COD	470
BOD$_5$	470
Lactose	480
Protein	~500
Total solids	730

effluent determines the loss from butter manufacture, while knowledge of the protein concentration allows the buttermilk loss to be calculated.

Casein losses in the whey, wash-water and floor effluents are readily determined by measuring the concentration of suspended solids (casein fines). Some account may need to be taken of bacterial cells in the suspended solids centrifuged from the effluent (Thomas, 1982). The COD test is not applicable because the variable amounts of whey present in the wash-water and floor effluents swamp the COD value due to the presence of casein.

The COD (or milkfat and protein) concentration in the waste is converted to

Table 5.16 Appropriate analytical methods to estimate losses from various manufacturing processes (Parkin, M.F., private communication)

Raw product	Type of process	Nature of effluent constituents	Analytical method to assess loss
Whole milk or skim milk	Tankering, separating, evaporating and drying	Whole or skim milk	COD
Cream	Buttermaking	Butter (fat) Buttermilk	Milkfat Protein
Skim milk	Casein making	Casein protein Whey	Casein fines, dry mass at pH 4.6
Whole milk	Cheesemaking	Cheese curd Whole milk Whey	Milkfat Curd fines
Ice-cream	Mix preparation Freezing	Ice-cream mix Milkfat Nuts, fruits	COD Milkfat
Cultured products	Culturing of whole milk, skim milk or cream	Whole milk Skim milk Cream	COD
Byproducts from:			
Cream processing	Buttermilk powder drying	Buttermilk	COD
Skim milk processing	Whey powder drying	Whey	COD

an equivalent product concentration from a knowledge of the COD value of the product. The mass of product in the waste is calculated from the measured concentration in a representative sample and the total volume of waste.

Where the plant treats its own wastes, analyses of BOD_5, settleable solids and suspended solids concentration will be necessary also.

Continued monitoring
Once an initial survey is completed and steps have been taken to reduce preventable losses, a permanent monitoring system is established to provide data on yields on a routine basis. These data can be used to promote further waste prevention and to keep management informed on performance and productivity.

5.5.3 Control programme
A majority of the avoidable wastes in dairy plants can be traced to staff who have been insufficiently trained, supervised and motivated. Also, a large proportion of losses in a number of plants can be traced to poor maintenance of equipment or maintenance–operator interrelated problems that again have their base in ineffective communications.

The control programme should focus on *product loss, water loss, packaged material loss* and *energy loss*: these constitute the greater part of the economic losses and they are interrelated in that about 75% of the preventable loss can be associated with product that has been processed and up to 25% of the product loss may occur during or after packaging.

The operation of a successful control programme entails the following:

- The commitment of management to the programme.
- The establishment of a loss control (yield improvement) team including members from maintenance and plant operators from each department and shift.
- The installation of proper equipment to monitor product losses.
- The establishment of a cost accounting system to evaluate the monetary costs of the resource losses, the capital that can be expended and the pay-back from the allocated capital.
- The use of the findings to assess major loss areas and the action or engineering modifications that could be employed to reduce the losses.
- The establishment of a continuing education programme for managers, supervisors and operators to develop an understanding of the need for waste control, the economic benefits to be derived and the factors involved in water and waste control.
- The setting of goals to be achieved in the control programme.
- The establishment of a scheme for continual monitoring of losses and use of resources, with feedback of the data to the operators.

- Daily attention to both preventive maintenance and good processing practices.

Implementation of a control programme is facilitated by the appointment of a Waste (or Resources) Control Supervisor with responsibility and authority assigned by top management. Job descriptions for all dairy plant staff should include a description of responsibilities for resource control.

5.5.4 In-plant control

Knowledge of the amount of waste, particularly when the waste is expressed in financial terms, provides an incentive to institute adequate in-plant controls. Detailed methods of reducing waste in dairy processing are discussed by Royal (1978), Carawan *et al.* (1979), Chambers (1980) and IDF (1980).

Production scheduling
Production scheduling is important in the control of wastes. Careful planning, particularly in coordination with the marketing department, can help to minimize 'crisis' production or manufacture of excess product with subsequent spoilage. Production scheduling can:

- eliminate over-production and resultant excessive product return;
- minimize the number of start-ups and shutdowns required on waste-generating operations such as separation, pasteurization and evaporation;
- minimize the number of shutdowns of operations due to insufficient supply of product or similar stops due to improper production planning. Almost all equipment stoppages, especially in filling operations, cause an increase in waste discharges;
- optimize the sequence of processing to avoid unnecessary clean-up between products;
- optimize the utilization of equipment such as vats, lines and processing units;
- minimize peaks in wastewater volume and concentration flows. Proper staggering of process operations in clean-up may be able to reduce hydraulic and organic loading on the waste treatment facilities.

Quality control
The control of product quality is an indirect asset in waste management. It helps to:

- reduce the quantity of returned products that have to be dumped or disposed of in some manner;
- reduce fouling of some specialized equipment (e.g. in membrane processes), thus reducing wastes associated with excessive cleaning of the equipment;

- reduce the frequency of manufacture of low volumes of product, thus providing for more efficient processing and less waste discharge per unit quantity of product processed;
- optimize production scheduling.

Alternative use of wasted products
Potential waste products include returned products, products resulting from overflows, leaks and accidental spills, whey, buttermilk, and residual products left in tanks, lines, etc.

Returned products, recoverable rinses, cheese curd, nuts and fruits from ice-cream operations, spilt frozen novelties and frozen desserts, material from the drip shields of filling operations and drips collected from major leaks may all be collected and disposed of as animal feed in many instances. Provided that hygienic collection methods are used, material from the start-up and shutdown of pasteurizers, heat exchangers, evaporators, separators, etc., or initial rinsings from some plant processes may be recycled or used in the manufacture of ice-cream, etc.

Process and product related waste control practices
In general, waste reduction involves the elimination of deliberate wastage, drips, leaks, overflows, spills. The following specific recommendations are adapted from IDF (1980):

(a) *Leakage.* Due to the very high BOD_5 of milk and milk products, even small leaks cause a remarkable pollution load. Therefore leaks should be immediately repaired. If this is not possible, the wastes should be collected in suitable containers and either reprocessed or disposed of for animal feeding. Any such containers should be checked frequently to ensure that overflow of the collected milk or product does not occur. All fittings, valves, cocks and seals should be regularly checked for leaks. Manually operated valves should be readjusted if necessary and defective seals replaced. Valves, seals, fittings, pump parts and parts of fillers have to be treated carefully when being dismantled, cleaned and reassembled so that the finished surface is not damaged. Damaged surfaces should be repaired.

When pipelines are being altered or repaired, their contents should be collected in a bucket. With long pipelines, shut-off devices should be mounted at intervals to reduce the product quantity when emptying. Furthermore, the pipelines should not have any expansions, narrow parts, syphons or dead ends which might cause increased product wastage. Instead of being fitted together with threaded fittings, pipe connections should be welded, as far as possible, in order to prevent leaks.

(b) *Spillage.* In order to avoid overflows, tanks, containers and cans should be constantly checked while being filled. Therefore, it is useful to provide level alarms and suitable high-level control devices (e.g. filling level regulator with automatic pump interruption) where overfilling may occur. Tanks and vats containing milk and milk products should never be filled to such a level that the product overflows during agitation. Tanks should have well-rounded corners, be well sealed and designed for ease of emptying and rinsing. It is useful to provide standby pumps to prevent uncontrolled loss of milk and milk products when the normal pump fails. The provision of a standby generating plant should not be overlooked.

All floor channels, drains and outlets should be maintained in good order, and be of sufficient capacity and fall to prevent wastewater from backing up into the drain lines or on the floor during treatment or processing. They should be covered with suitable grids which should be inspected and cleaned regularly to avoid blockage. Each opening to a subsurface drain should be fitted with a basket in conjunction with a trap which should be accessible and properly constructed to prevent solid material from entering the sewage system. Each such trap should be cleaned regularly (not less than once daily) and be maintained in good condition. It is particularly important not to wash away any spilt dry products. They should be swept up and removed in a dry state.

(c) *Foaming.* Excessive foaming of milk, skim milk, whey, etc., is often responsible for overflow and thus the actual cause of loss of milk solids. Foaming is often caused by open-type separators and incorporation of air during the filling of tanks. Foaming can also be caused by excessive agitation of liquid products while being mixed. To prevent foaming, the connections of the pipelines on the suction side of the pumps must be tight. Foaming during the filling process can also be reduced by a suitable arrangement of the filler pipelines. Where skim milk or whey is used for animal feed, foaming associated with filling either storage tanks or tankers may be overcome by the addition of a suitable anti-foaming agent prior to or during filling.

(d) *Processing and cleaning losses.* Processing losses are generated during sludge discharge from separators, start-up, shutdown and product changeover in heat exchangers, evaporator entrainment, discharges from washers, spattering or breakage of containers in packaging equipment and product changeover in filling machines. Various steps can be taken to reduce these losses (IDF, 1980).

Water conservation is an important aspect of waste treatment and is discussed in detail by Hills (1978) and Holmstrom (1978).

Engineering approaches to waste reduction
Many equipment, process and systems improvements can be made within dairy plants for better control of wastage and for yield improvement. In many cases these improvements become obvious only after a waste control programme has

been put into operation and specific data have been obtained from the monitoring programme. An accounting evaluation can then be used to determine the feasibility of such changes and the projected payback. EPA (1971) and Carawan *et al.* (1979) have detailed numerous suggestions for waste reduction innovations, many of which are or have been utilized in the industry. The most common engineering improvements include:

   Automated cleaning systems
   HTST product recovery system lines
   Automated process systems

Many of the suggestions are capital intensive and utilize some degree of automation. Such installations require superior maintenance and well educated plant operators to realize fully the system's potential. Improper operation of automated recovery and control systems can actually markedly increase wastes and decrease yields.

Maintenance programme
The level of control of water and wastes in a dairy food plant can be correlated directly with the maintenance programme in operation in the plant. A good waste management control programme cannot be achieved without a good maintenance programme. It should be stressed that a maintenance programme involves more than the engineering department. Effort by operators of equipment is also essential to the maintenance programme. Communication between operators and plant engineers is essential. The larger the plant, or the more shifts in operation, the more difficult is the achievement of good waste control and a continuing good preventive maintenance programme. Recognition must be given to the need for adequate supervision of second and third shifts in order to obtain the greatest efficiency from the engineering crew.

Ideally a maintenance programme is separated into four parts: (1) emergency repair, (2) 'chronic' repair or minor repair, (3) operator-associated maintenance, and (4) preventive maintenance. Of these, the operator-associated maintenance and the preventive maintenance programmes are the most important to waste control but are usually the ones given the least attention.

Education
Carawan and Jones (1977) outline a four-phase educational programme on the management of water and wastes in dairy processing. Where it is not feasible to have fully trained staff in the plant, external extension personnel can be extremely useful in providing expertise and educational programmes for individual plants. This is of particular value to small plants. Carawan *et al.* (1979) have provided a number of special training programme booklets for use by extension personnel in assisting individual plants in loss (yield) control programmes.

5.5.5 Achievable levels of losses

A waste control programme often has more chance of success if realistic target values for losses are set. These provide objectives for management and operators. EPA (1971) suggests that excellent management can achieve waste coefficients of 0.5 m^3 water m^{-3} milk and 0.5 kg BOD m^{-3} milk (i.e. 0.5% loss of raw material to the liquid waste). The Ministry of Agriculture, Fisheries and Food (1969) lists target values which experience in the UK showed could be reasonably achieved by minimizing waste, replacing or redesigning inefficient plant and collecting stronger wastes and rinsings for reprocessing or separate disposal (Table 5.13). In general, the target values are similar to the lowest values achieved by operating dairy plants.

In the Netherlands, the Gelders–Overijsele Zuivelbond in Zutphen measures the wastewaters from 85 dairy plants (Baltjes, 1978). Such measurements allow comparisons between separate plants. These comparisons are made by computing the waste load as a percentage of a standard (a population equivalent (p.e.) based on the volume of wastewater and the COD and nitrogen concentrations). The standard, which is reviewed every few years, is set at a value that is bettered by 30% of the plants; thus the other 70% of the plants, which exceed the standard, are set a goal. The standards, in milk loss equivalents, operating in 1980 were:

Milk reception	0.20%
Cheesemaking	0.20%
Market milk	1.9%
Condensed milk	0.48%
Whole milk drying	0.64%
Buttermaking and skim milk drying	0.60%

The effect of setting these standards has been quite marked. In 1971 the standard for butter manufacture was 140 p.e. per 1000 kg butter; in 1974 the value was 80 p.e. per 1000 kg, in 1980, 40 p.e. per 1000 kg and in 1983, 20 p.e. per 1000 kg (Baltjes, J., private communication). Similarly the standard for bottled milk was halved between 1971 and 1980. The dairy industry in the Netherlands has reduced the pollution load of its effluents from 2.5 million p.e. in 1969 to less than 0.5 million p.e. in 1981, despite a 60% increase in the quantity of milk processed (Baltjes, J., private communication).

5.6 Treatment of unavoidable wastes

Despite the best efforts at in-plant control, some wastes from milk processing are inevitable. These wastes are treated and disposed of in the most economical manner consistent with local environmental legislation. Disposal methods used for untreated, partially treated or treated wastes include discharge to natural waterways, use of municipal sewage systems and disposal on land by spray irrigation, ridge and furrow methods or application from trucks. Some byproducts and strong wastes are returned to the farmer for animal feed. Treatment methods include modified activated sludge, extended aeration, aerated lagoons, stabilization ponds, trickling filters and rotating biological contactors. Chemical precipitation and anaerobic methods are used infrequently. Costs, capital and operating, the availability of suitable staff and the needs of the legislation determine the choice of system.

Wherever possible, discharge to natural waterways is the method of choice and in many cases is a major factor in the siting of the dairy plant. However, larger factories and increasingly restrictive legislation are forcing dairy plants to seek alternative methods. Treatment with municipal effluents is widely practised in some countries, but increasing surcharge costs are making this less attractive without some pretreatment. Some treatment and disposal methods are more popular in some countries than in others, reflecting economic factors and the personal preferences of waste treatment advisers to the particular industries. Thus spray irrigation disposal is widely used in New Zealand, alternating double filtration and high rate filters in the UK, and extended aeration in the Netherlands, the Federal Republic of Germany, France and Poland.

Effluents from dairy food plants are of four types:

(1) Clean water from indirect heating and cooling of condensing and heat treatment plants.
(2) Slightly polluted water such as that from the final rinsing of processing plant, condensate containing some entrainment, or storm-water.
(3) Polluted water containing waste milk or milk products or plant and product rinsings.
(4) Domestic wastes.

In general, treatment of effluents from dairy plants is facilitated by keeping each type of waste separate from the others. The clean water may be reused or may require cooling before disposal to a natural waterway. The slightly polluted water can often be disposed of directly to a natural waterway or to a municipal treatment plant with minimum pretreatment and cost. Domestic waste is frequently treated in a municipal plant or anaerobically in septic tanks.

Effluents of type 3 are of greatest concern to dairy company management. These wastes contain materials of mineral and organic origin, but generally are free of toxic materials which would adversely affect the operation of conventional methods of biochemical treatment. Under normal control the concentrations of cleaning reagents, detergents and bactericides will not disrupt the operation of biochemical systems.

Characteristics of dairy plant wastes that require special consideration in the design of treatment plants and disposal methods include:

- The wide variations in volume, flow rate, organic strength and composition during the day, from day to day and throughout the dairying season.

- Temperature.
- pH.
- Concentrations of milkfat, curd, organic matter and settleable solids.
- Nutrient (nitrogen and phosphorus) balance.
- Initial high rate of oxygen demand.

5.6.1 Pretreatment

Pretreatment of fluid wastes is frequently useful to reduce the load on the dairy plant's treatment facility or to reduce treatment costs where wastes are discharged to municipal treatment plants. In either case, useful techniques include balancing flows, fat removal and screening or settling.

In the USA and Europe where sewer charges are high or where municipalities impose limits on BOD_5, suspended or settleable solids and fat concentrations, preliminary biochemical treatment may be used to bring the BOD_5 down to levels of $400-600\,mg\,l^{-1}$. The type of biochemical treatment to be used must be established for each individual plant. For a plant with a BOD_5 of $2000\,mg\,l^{-1}$ the reduction required would be about 70%.

Balancing

To reduce the impact of variations in flow rate and strength of effluents, and to provide for pH adjustment, use of balance tanks frequently precedes disposal to natural waterways or municipal and biochemical treatment. Some treatment systems, such as lagoons, have balancing built in to their basic design. The actual fluctuations in flow rate and strength must be known to design the size of the balance tank (Adams and Eckenfelder, 1974). A typical design criterion is a capacity corresponding to 4 to 6 h of the average hourly flow rate. Many authors suggest that the maximum volume should allow less than 6 h retention time, whereas others recommend up to 24 h retention time. The latter would be more expensive because of the greater size and would require aeration to prevent septic conditions. If the balancing tank is too small, fluctuating conditions will occur in the treatment plant. If the retention time in the balancing tank is too long, anaerobic conditions will cause offensive odours, and solubilization of suspended material will result in an increased organic load on the treatment plant. In practice, retention times range from 0.5 h to 72 h. Effluent is discharged from the balance tanks by overflow or the use of a floating-arm take-off, either by gravity or pumping.

Both floating and submerged aerators are used to aerate balance tanks. Quoted power inputs range from 0.4 to $8\,W\,m^{-3}$ of tank volume with typical values of $3-4\,W\,m^{-3}$. Various materials of construction are used, including concrete, concrete with a corrosion-inhibiting liner, mild steel, galvanized steel, stainless steel and plastic.

Some balance tanks incorporate facilities for the automatic control of pH (generally in the range pH 6.5–9.0) by dosing with sodium hydroxide or acid.

While sulphuric acid is used for pH adjustment in many plants because of its low cost, some authorities recommend that other acids should be used because of the danger of corrosion of concrete by sulphates (at concentrations >1000 mg l^{-1}) or the production of foul-smelling sulphides. In some plants, the exhaust gases from the boilers are used to reduce the effluent pH (Anon., 1978).

Provision is made to remove floating and settled material from balance tanks. Good practice suggests that this desludging should be carried out daily, although in fact many tanks are desludged weekly and some annually. The latter practices are not recommended.

Balancing is often associated with fat removal and solids screening facilities.

Fat removal
While every precaution should be taken not to waste milkfat during dairy processing, dairy plant effluent inevitably contains free fat. Removal of this fat is a necessary precursor to most treatment and disposal methods. There is little reliable data on design criteria for gravity fat traps for dairy effluent. Trebler and Harding (1955) recommend a depth of at least 0.75 m, a length-to-width ratio of at least 2 to 1 and a tank surface area of 0.4 m^2 m^{-3} maximum hourly effluent flow. A commonly used rule-of-thumb suggests using a baffled tank, less than 1 m deep, with a total volume such that the residence time is about 0.3 h at the maximum expected flow rate.

A common problem in dairy effluent treatment plants is precipitation of fat as a result of the lower pH and temperature in the treatment plant. The operation of fat traps is enhanced by ensuring that the effluent is as cool as possible. Milkfat is a liquid at temperatures greater than 35°C and is difficult to remove in a fat trap. More fat is removed in fat traps at acid pH values. At pH values above 8.5 the milkfat tends to be emulsified or saponified and therefore is not removable in the fat trap.

Dissolved air flotation for fat and suspended solids removal is incorporated in modern dairy waste treatment to improve efficiency and to lower capital costs.

Screening and settling
Suspended material is removed from dairy plant effluents as soon as possible to prevent solubilization of BOD from the solid material. However, most solid constituents of dairy effluent are in solution so sedimentation gives little reduction in BOD$_5$. Mechanically brushed and inclined screens with 1–5 mm openings typically are used to remove curd particles and large solid objects. Where washings from yards, paved areas or outsides of road tankers contribute to the effluent, a detritus chamber or solids settling tank is incorporated to remove grit.

The use of desludging centrifuges within the dairy plant for treatment of whey or wash-waters from cottage cheese or casein manufacture materially reduces the BOD_5 concentration in such effluents by removing fat and curd fines. These fines are often recycled or recovered as animal feed.

Temperature
High-temperature effluent can adversely affect natural waterways and the operation of fat traps, and may be detrimental to biochemical treatment systems. A high temperature also represents a waste of energy. Where possible, high-temperature liquid streams are segregated for reuse (where the water is relatively clean) or cooled by heat recovery.

Nutrient addition
Some dairy plant effluents are deficient in nitrogen and/or phosphorus, impairing the efficiency of operation of some biochemical treatment processes, e.g. high-rate filtration. A few dairy plants incorporate dosing of ammonium sulphate or sodium phosphate to give a ratio of BOD_5 to available nitrogen and phosphorus of 100:5:1.

5.6.2 Discharge to natural waterways

Where direct discharge of wastes to natural waterways (streams, rivers or the sea) is allowed by legislation, dairy plant wastes have little effect on the waters provided that sufficient dilution is available. The principal effect of dairy wastes on natural waterways is a reduction in the dissolved oxygen concentration. This lowering of the oxygen concentration results from the activity of micro-organisms growing on the nutrients in the milk wastes and, as a result of their metabolic activity, consuming oxygen at a rate greater than it can be absorbed from the atmosphere by natural reaeration. The maximum effect of the waste on the oxygen concentration may not be evident until some considerable distance downstream of the discharge. It is generally considered that, provided that the oxygen concentration in the natural waterway does not fall below 5 $mg\,l^{-1}$, small amounts of dairy waste are actually beneficial to fish life because the nutrients stimulate the growth of organisms upon which the fish feed.

Horton and Trebler (1953) quote from Pien to show that wherever sufficient dilution is available to reduce the initial BOD_5 to 20–30 $mg\,l^{-1}$, natural stream reaeration is able to maintain the dissolved oxygen concentration at an adequate level. However, present-day legislation in most countries is unlikely to allow such high BOD_5 concentrations in natural waters, other than the sea. For example, in the Republic of Ireland the BOD_5 of the effluent is required to be such that admixture with receiving waters will not increase the BOD_5 concentration of the waters by more than 1 $mg\,l^{-1}$ and the maximum allowed BOD_5 is not to be increased above 5 $mg\,l^{-1}$. A Water Right granted to a dairy company in New Zealand (where discharges are based on receiving water standards) allows the discharge of wastes to a river so that the maximum mean daily and night-time BOD_5 concentration does not exceed 5 $mg\,l^{-1}$ at a site downstream of the 5 km reach into which there are three major discharges (dairy plant, meat works and city) (Currie and Rutherford, 1980). For the dairy plant, the allowable discharge of BOD_5 is based on the equation

$$B = 180.6\,Q + 950.3$$

where B is the quantity of BOD_5 which can be discharged ($kg\,d^{-1}$) and Q is the flow in the river ($m^3\,s^{-1}$, with a minimum of $13\,m^3\,s^{-1}$). Spray irrigation land disposal is used when the river is unable to accept all the effluent from the plant.

Excessive quantities of dairy waste are detrimental to normal aquatic life because of oxygen depletion. The presence in the water of chironomid larvae and turbificid worms, which are tolerant of adverse conditions, is normally an indication of heavy pollution. If all of the dissolved oxygen is removed, the lactose in the waste will be converted to organic acids and casein precipitated, forming black benthal deposits.

Any suspended solids in the wastes will settle and smother the bed of a slow moving stream. Nutrients (phosphorus and nitrogen) may also have an effect, causing the growth of certain organisms such as algae and, particularly, that known as 'sewage fungus'. These growths are aesthetically unpleasant and, when they break off, their decomposition further deoxygenates the stream. The discharge of hot wastes, even if otherwise uncontaminated, can cause changes in the biological balance of the river. A particular problem caused by the discharge of milk wastes is that even quite large dilutions of milk are turbid so that relatively small quantities can cause cloudiness in natural waterways.

The effects of large quantities of milk wastes on rivers are demonstrated by data obtained following the intermittent discharge of a considerable volume of whole milk to a number of rivers over a period of five days. One river was flowing at $200\,m^3\,s^{-1}$. The discharge of milk caused noticeable discoloration and scums in the river for approximately 20 km. The measured average increases in COD and BOD_5 concentrations in the river were $12-18.5\,mg\,l^{-1}$ and $8-12\,mg\,l^{-1}$ respectively. A maximum oxygen depletion of $3\,mg\,l^{-1}$ was observed, but at no stage did the dissolved oxygen concentration fall below $5.5\,mg\,l^{-1}$. No permanent damage to aquatic life was observed. Another river, flowing at $25\,m^3\,s^{-1}$, received similar quantities of milk. The measured peak COD and BOD_5 concentrations in the river were $45\,mg\,l^{-1}$ and $35\,mg\,l^{-1}$ respectively. Two slugs of water, one 20 km long and the other 5 km long, contained zero dissolved oxygen. The slugs took five days to travel to the confluence with a much larger river. These conditions resulted in large numbers of fish and eels being killed. Nevertheless, two weeks after the milk discharges ceased, aquatic life was reverting to normal.

Discharge to natural waterways, where allowed, is the cheapest solution to waste disposal. Frequently the only pretreatment required is removal of floating fat and gross suspended solids. In other cases, aerobic, chemical or other treatments may be used to reduce the BOD_5 and suspended solids concentrations to levels which can be assimilated by the river.

5.6.3 Municipal treatment

Milk wastes are frequently treated in municipal treatment plants. About 95% of USA dairy plants producing ice-cream and fluid milk products discharge their wastewater to municipal treatment plants (Chambers, 1980). Similar values apply in many countries, e.g. various authors in IDF Doc. 104 (1978) report treatment by municipal plants of 85% of dairy wastes in Spain, 75% in Germany, 74% in Denmark, 72% in Canada, 60% in France and 50% in Northern Ireland.

In the USA the 1977 Amendments (PL95-217) significantly altered the thrust of the Clean Waters Act of 1972 (PL92-500) and, among other things, required the US Environmental Protection Agency to focus on priority pollutants that are discharged to 'Publicly Owned Treatment Works', and to differentiate between conventional and non-conventional pollutants. BOD_5, suspended solids, extremes of pH, fats, oils and greases, phosphorus and COD (parameters characteristic of dairy wastes) are classified as conventional pollutants, and are effectively removed in well operated 'Publicly Owned Treatment Works' employing conventional secondary treatment or best practicable control technology. Brown and Pico (1980) demonstrate and reaffirm that dairy plant wastes are compatible with 'Publicly Owned Treatment Works', can be treated satisfactorily in typical biochemical treatment systems and generally do not require pretreatment, unless the municipal plant is of inadequate size or the surcharge costs are high. It is desirable to remove fat and to provide balancing facilities at the dairy plant. If the municipal plant is discharging to eutrophic waters it may be necessary to reduce the concentration of phosphorus in the dairy waste. Only small quantities of whey can be tolerated by most municipal facilities.

When the BOD_5 contributed by milk solids exceeds about 50% of the total BOD_5, or the BOD_5 contributed by whey exceeds 10% of the total BOD_5, the treatment in the municipal plant takes on the characteristics of a dairy wastes treatment system. The addition of dairy wastes to municipal treatment facilities has caused operating problems as a result of inadequate size of the treatment plant, insufficient balancing, or a design which does not allow for the high and rapidly expressed oxygen demand of the dairy waste. In some cases, the most economical disposal system involves pretreatment at the dairy plant to reduce the BOD_5 concentration to a value in the range 300–600 mg l^{-1} before discharge to a municipal plant. High-rate aerobic treatment (residence times of about 4 h) or anaerobic processes are suitable pretreatments.

In the USA, in particular, treatment in a municipal plant used to be considered the most economical option available to a milk processor. High surcharges are causing some reconsideration of this. Nevertheless, skilled supervision is more likely to be provided at a municipal plant than at a relatively small dairy waste treatment facility and it is likely that municipal treatment will continue to be the method of choice for many dairies.

5.6.4 Land treatment

Dairy processing wastes are particularly suitable for disposal onto land. The balanced mineral profile, the presence of nitrogen, the lack of toxic substances and the irrigation benefits have promoted the widespread use of land disposal in some countries. Methods used include spray irrigation, ridge and furrow, overland flow, flooding and truck distribution.

Irrigation systems provide a very high treatment efficiency, have low capital and operating costs, can provide some return by way of irrigation and fertilizer value and have no sludge disposal problems. They require large areas of land and are often some distance from the dairy plant. They are not suitable for areas subject to freezing temperatures. When improperly designed and operated they cause odours, ponding, surface run-off and burning of crops or pasture, and attract flies and insects. Some systems have a high labour requirement.

Spray irrigation

Spray irrigation is widely used in New Zealand where effluents from 45 of 81 dairy plants, representing 52% of dairy product manufacture, are disposed of by spray irrigation (Galpin, 1981). The practice of spray irrigation disposal of dairy factory effluent is reviewed by EPA (1971) and Parkin and Marshall (1976). In common with many other authors, they concluded that spray irrigation is a successful low cost method of disposal of dairy factory effluent which provides positive benefits to farming.

A typical spray irrigation disposal system comprises:

- An effluent collection system within the dairy processing plant
- Pretreatment facilities
- Flow-balancing
- Pump and pipeline from the dairy plant to the irrigation site
- The irrigation site
- An irrigation pipe network
- Spray nozzles

For land disposal it is desirable to separate all clean and slightly polluted waters for alternative disposal. Domestic wastes must be excluded. The pretreatment facilities comprise standard fat and solids removal. Generally, sufficient flow balancing is provided to balance excessive variations in pH and organic strength. The balance tank has a volume equivalent to 4 to 6 h of the maximum flow rate. The offtake of the balance tank is normally through a floating-arm pipe and facilities are provided to allow the removal of settled solids. It is highly desirable that the balance tank and associated pipework are emptied every day and flushed with clean water. This prevents the onset of anaerobic acidic conditions which cause burning of grass or crops and unpleasant odours.

In New Zealand, the typical irrigation site is pasture used for grazing dairy cows, beef cattle or sheep. Land used for trees and grains has also been utilized for spray irrigation disposal (Watson et al., 1977). Individual dairy plants in New Zealand have land areas ranging from 2 ha to 225 ha available for irrigation.

The irrigation network is generally of unplasticized polyvinylchloride (uPVC) or aluminium pipe. In large systems all the pipework is underground, spray nozzles being connected to permanent hydrants in the underground grid. Smaller or less sophisticated systems usually have movable lateral pipes connected to a central pipeline. Travelling irrigators, with connection to a small number of hydrants, are also used. In operation, an area of the disposal site is irrigated for the desired time, then the sprinklers are moved to a new area. Typically, the size of the irrigation site is such that each area is irrigated approximately every 14–21 days. This return cycle or resting period allows the soil bacteria to decompose the organic components of the effluent and the pasture to grow, utilizing the nitrogenous and mineral nutrients applied with the effluent. The pasture is grazed prior to irrigation.

In New Zealand the amount of effluent applied in a single irrigation operation varies from $130\,m^3\,ha^{-1}$ to $1500\,m^3\,ha^{-1}$, with an average of $450\,m^3\,ha^{-1}$. Typical total applications in a year are $3000\,m^3\,ha^{-1}$ to $15\,000\,m^3\,ha^{-1}$. The rate of spraying is equivalent to a rainfall of between $2.5\,mm\,h^{-1}$ and $30\,mm\,h^{-1}$ with an average of $9\,mm\,h^{-1}$. Modern practice favours the use of small flows from nozzles <8 mm diameter, so that the rate of application is low ($<8\,mm\,h^{-1}$). This ensures that the infiltration rate of the soil is not exceeded so that the effluent is absorbed by the soil as it is applied. If effluent is applied at a rate which exceeds the infiltration rate of the soil, the resulting ponding causes burning of the pasture and unpleasant odours, may attract flies and can cause run-off with consequent pollution of natural waterways. For some soil types, where the infiltration rate is low, underground tile drains have been installed to give a satisfactory disposal system.

Even relatively strong wastes, such as whey (McDowall and Thomas, 1961; Parkin and Marshall, 1976; Watson et al., 1977), have been applied to land at rates up to $100\,m^3$ per ha dose (i.e., an average of $7\,m^3\,ha^{-1}\,d^{-1}$ when the return cycle was 14 days). Strong wastes are mixed with other more dilute plant wastes before application (e.g. by the provision of adequate balancing facilities). High concentrations of minerals cause burning of pasture because of high osmotic pressures. High concentration of sodium (from CIP fluids) can cause the breakdown of the soil structure, particularly in clay soils, as a result of the exchange of the natural calcium with sodium in the effluent. This can be prevented by the application of lime. The recommendations of Wilcox (1948) should be followed, i.e., the total concentrations of the cations calcium, magnesium, potassium and sodium in the effluent should not exceed 25 milli-equivalents per litre and the ratio of the concentration of sodium to that of the total cations (in milli-equivalents/l) should be less than 0.8.

A properly designed and managed spray disposal system improves the biological and chemical environment in the soil. In particular, microbial and earthworm activities are stimulated, and plant-available nutrient levels are

Table 5.17 Typical effect on soil pH of spray irrigation with effluent from casein processing (Factory A) and spray-dried milk powder processing (Factory B) (Parkin and Marshall, 1976)

Period of operation (years)	Soil pH Factory A	Factory B
0	5.5	5.7
1	5.7	5.8
2	6.0	5.8
3	6.0	5.9
4	5.9	6.0
5	5.9	5.9

Table 5.18 Typical effect on soil constituents of spray irrigation with effluent from casein manufacture (Parkin and Marshall, 1976)

Period of operation (years)	Calcium (me %)[a]	Magnesium (me %)	Potassium (me %)	Phosphate (me %)
0	4.0	6	4	15
1	5.0	10	22	70
2	6.0	14		120
3	7.5			130
4	8.5	18		140
5	10.0	26	34	260

[a] Milli-equivalents per hundred parts of soil.

Table 5.19 Typical effect on soil constituents of spray irrigation with effluent from milk powder manufacture (Parkin and Marshall, 1976)

Period of operation (years)	Calcium (me %)[a]	Potassium (me %)	Phosphate (me %)
0	11	9	15
3	14	13	15
5	12	10	17

[a] Milli-equivalents per hundred parts of soil.

increased (McAuliffe et al., 1979; Parkin and Marshall, 1976). The effects on soil fertility, as measured by tests for pH and concentrations of calcium, magnesium, potassium and phosphorus, can be separated into two classes depending on whether whey is included in the effluent or not. Soil pH tends to rise slightly, despite wide variations in the pH of the effluent (Table 5.17). Soils receiving regular applications of whey show marked increases in cation concentrations (Table 5.18), whereas soils receiving only equipment and floor wash-waters show little change in concentration of cations (Table 5.19).

A two-year study at a spray irrigation site receiving whey and general wash-waters (26 000 kg BOD_5 ha^{-1} $year^{-1}$) concluded there was no measurable contamination of surface or underground water. Application volumes of up to

Table 5.20 Chemical concentration, pH and coliform count of groundwater from a spray irrigation site during the fourth year of irrigation (Parkin and Marshall, 1981)

	At beginning of dairying season	Mean value during dairying season (July–March)	At end of dairying season
Nitrate (g m^{-3})	5.2	4.7	3.8
Ammonia (g m^{-3})	0.01	0.01	0.01
Total Kjeldahl nitrogen (g m^{-3})	0.13	0.41	0.44
Dissolved reactive phosphate (g m^{-3})	0.002	0.004	0.008
Total phosphate (g m^{-3})	0.062	0.106	0.073
pH	5.9	5.6	5.6
Coliforms (no. per ml)	4.6	<1	<1

Table 5.21 Increase in milkfat production and number of cows per ha due to spray irrigation with dairy factory effluent (Parkin and Marshall, 1976)

Period of operation (years)	Area of farm[a] (ha)	Number of cows	Milkfat production per annum (kg)
0	80	150	20 000
1	80	150	19 500
2	80	200	29 100
3	93	225	34 100
4[b]	93	225	23 200
5	93	225	34 900

[a] The area available for irrigation was 57 ha.
[b] Severe drought.

65 mm per dose were being applied. The farm had been irrigated with up to 700 mm of wastes per year for three years prior to the study. The groundwater receded to a minimum of 2.2 m below the surface during the late summer after being at 1.7 m at the beginning of the spring. There was sufficient soil to prevent serious leaching of nutrients to the groundwater. Concentrations of ammonia, nitrate, phosphate and nitrogen were measured in the groundwater (Table 5.20). Concentrations in the groundwater were low and changes through the season were insignificant. It was notable that the high availability of water and nutrients at the disposal site increased grass production, particularly over the dry summer period. This, in turn, resulted in increased milk production from the pasture-fed cows.

Table 5.21 (Parkin and Marshall, 1976) gives a further example of the increase in production resulting from increases in pasture growth. The carrying capacity of the farm increased from an average of 1.87 cows per ha to 2.42 cows per ha and the milkfat production increased by 75% over a period of five years.

Problems which occur with spray irrigation (run-off, ponding, offensive odours, flies, marked loss of soil infiltration rate and loss of cover crops) are

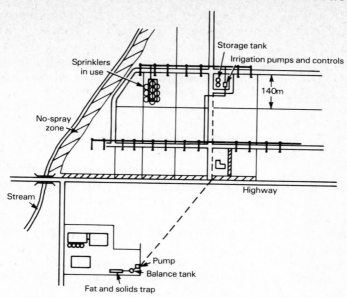

Fig. 5.11 Spray irrigation layout (not to scale)

caused by poor initial design or incorrect operation of the scheme. The total area available for irrigation is often too small for the volume of effluent. This results in water saturation of the soil or burning of the pasture by the strong effluent. The rate of spraying must be less than the infiltration rate of the soil. Incomplete removal of fat and suspended solids causes plugging of the soil (McAuliffe *et al.*, 1982) and blockages of the spray nozzles. A common cause of offensive odours, breeding of flies and damage to the pasture is the failure to drain and clean regularly (preferably daily) the balance tanks and irrigation lines. Pasture grass is preferably grazed relatively short before the application of effluent. Conscientious labour and knowledge of the principles of spray irrigation disposal considerably enhance the prospects of success.

Spray irrigation disposal is not readily applicable to areas which show significant freezing in winter months. Storage in lagoons for disposal when the land thaws is possible.

There is no evidence of health hazards to humans or animals as a result of irrigation with dairy plant wastes (Magnusson, 1974). Nevertheless, good practice requires spraying in a manner which minimizes the risk of spray drift to neighbouring houses, and a resting period of at least three to four days before the pasture is grazed.

The following is a description of a recently designed spray irrigation disposal system for a dairy plant processing $550 \, m^3 \, d^{-1}$ of whole milk to produce butter, skim milk powder and acid casein (Fig. 5.11). The effluent includes some of the casein whey. The system is designed to accept $670 \, m^3 \, d^{-1}$ of effluent containing $4500 \, kg \, BOD_5$. The effluent from the dairy plant is passed through a 1.8 mm gap wedge-wire screen and a fat trap to a balance tank ($380 \, m^3$, normally operated at 50% capacity) at the dairy plant. It is then pumped to a tank ($150 \, m^3$) on the disposal farm, 1.5 km from the processing plant. The farm has a total area of 95 ha and is used for dairying. Topographical features, including a stream, excessively sloping ground and the need for a buffer zone around buildings, reduce the sprayable area to 65 ha. From the balance tank at the disposal site, effluent is pumped to the spray nozzles. Normally 16 nozzles are in use simultaneously. Each nozzle is 7 mm diameter and, at a pressure of 400 kPa, distributes effluent at the rate of $1.2 \, l \, s^{-1}$ over a diameter of 30–35 m. This gives a dosage rate of $4-6 \, mm \, h^{-1}$ or $40-60 \, m^3 \, ha^{-1} \, h^{-1}$. Spraying is continued for 5 h at one site, then for a further 5 h at another site, again using a set of 16 nozzles. Thus each daily dose is $200-300 \, m^3 \, ha^{-1}$. After each period of spraying, pipelines and nozzles are rinsed with clean water. In good weather conditions, the nozzles are shifted to a new site after three days; if rainfall is high, the nozzles may be moved daily. Each area is grazed two or three days prior to spraying. All pipes (uPVC) on the farm are underground and terminate in a bayonet-type fitting into which is fitted a sprinkler comprising an upstand and a nozzle. Automated controls operate the irrigation pumps according to the level in the balance tank. The rinsing cycle is also automatically controlled. Sensors are installed to monitor pressure and temperature in the irrigation line and are connected to switches to stop the irrigation pumps automatically if the temperature or pressure exceeds safety limits for the plastic pipe. The irrigation pump is also automatically stopped if low pressure is sensed, signifying a break in the irrigation line or nozzles.

Ridge and furrow
Ridge and furrow irrigation is another system of land disposal (Schraufnagel, 1957). Ridge and furrow irrigation involves gravity-feeding or pumping wastes to the disposal site and allowing them to flow into a channel. By the use of suitable dams the wastes are caused to flow into the furrows, which are higher than the main ditch. The furrows are filled with liquid to within 50–70 mm of their tops before the flow is diverted to another area. Typically, the furrows are 2 to 5 m apart, 0.3 to 0.6 m deep and 0.3 to 1 m wide, practically level or with a slight slope down towards the feed channel to prevent ponding at the far end.

Usually the wastes are absorbed by the soil in less than 24 h. The quantity of effluent that can be applied is dependent on soil type, drainage, crops and climate. Schraufnagel (1957) reports rates between 40 and $280 \, m^3 \, ha^{-1} \, d^{-1}$. Harding (1968) reports that ridge and furrow seepage trench systems initially may handle daily flows as high as 120 to 200 litres per m^2 wetted area ($1200-2000 \, m^3$ per ha of wetted area) but often, due to microbial growth, can handle only 40 to $60 \, l \, m^{-2} \, d^{-1}$ (400 to $600 \, m^3 \, ha^{-1} \, d^{-1}$) after one to three years. Sludge can accumulate in the furrows ($0.12 \, m^3$ wet sludge per m^2 wetted area).

Tile drains permit higher loadings, particularly in naturally poorly drained soils. Drains at a depth 750 mm below the furrows resulted in BOD_5 values for the liquid in the tile drains in the range $20-50\,mg\,l^{-1}$ in winter and early spring, and lower in summer (Schraufnagel, 1962). Reed canary grass growing on the ridges is the cover crop favoured for ridge and furrow systems because it is permanent, reproduces itself, has a substantial root structure and endures water and ice cover well (Schraufnagel, 1957).

In an example quoted by Schraufnagel (1957) a small creamery butter plant (daily milk processed 22 700 l) used about 1 ha divided into three sections by means of check dams. Infiltration rate in the clay loam was approximately $45\,mm\,h^{-1}$ with an effluent volume of $190\,m^3\,d^{-1}$, estimated BOD_5 of $210\,mg\,l^{-1}$. BOD_5 reduction as measured in the liquid in a tile drain was about 90%.

Schraufnagel (1957) compared spray irrigation with ridge and furrow irrigation. The spray system had a lower capital cost, posed less danger of causing a nuisance, required less land preparation, was easier to expand, easier to crop and more suitable for woodland and hill slopes than ridge and furrow. The ridge and furrow system's comparative advantages were that it needed less land, was cheaper to operate and maintain, less difficult to use in winter and might not require pumping energy. A disadvantage of the ridge and furrow system is the need to remove sludge from time to time. This can be achieved by reploughing and/or reforming the furrows.

Flooding

Flooding of land has been used as a means of disposal of dairy effluents. The principles are similar to those for ridge and furrow irrigation, although, in general, waste land is used. The system is also similar to lagooning. High cropping rates can be expected from land which has been flooded in previous years.

Truck distribution

A simple, flexible method of waste disposal, truck distribution is used by many dairy companies to dispose of waste during peak seasonal flows or to relieve overloaded treatment systems. The principles and loading rates are similar to those used for spray irrigation. The use of trucks specially designed to apply low force per unit area facilitates spraying onto soft wet soils (Cunningham *et al.*, 1980).

5.6.5 Aerobic biochemical treatment

Dairy wastes consist almost entirely of solutions of organic nutrients and thus they are easily treated by aerobic processes. In aerobic systems the provision of oxygen and a bacterial culture accelerates the process of biochemical oxidation which would take place naturally in an oxygenated stream or river. Generally, aerobic dairy waste treatment follows accepted sewage treatment practice, except that conventional activated sludge has not been popular. Extended aeration activated sludge systems, aerated lagoons and trickling filters are the prevailing systems.

General considerations of aerobic treatment

In the aerobic biochemical treatment of dairy wastes three distinct phases have been identified (Hoover, 1953; Porges *et al.*, 1960):

- *Rapid initial incorporation* of the milk constituents, both dissolved and suspended, into the biomass floc. This is considered to be an adsorption phenomenon or, possibly, the formation of a complex between the bacterial cells in the floc and the protein and lipids. The extent of this rapid incorporation or adsorption depends on the sludge loading ratio (mass of BOD_5 in the waste per mass of suspended cells) and on the ecological condition of the sludge.
- *Oxidation of the organic material*, both adsorbed and in solution, by bacterial assimilation. The organic material in the dairy waste is partially synthesized to new cell material and partially oxidized to supply the energy needed for the growth of new cell material. The rate of assimilation (oxidation) is only about 10% of the rate of adsorption in the first stage. Adamse (1968) reports that the protein in milk wastes is not utilized until all the lactose has been degraded in this phase.
- *Endogenous respiration* in which the substance of the bacterial cells is oxidized, with the production of simple compounds such as water, ammonia and carbon dioxide. The rate of oxygen consumption in this phase is 10% of that during the assimilation phase. Destruction of cells by endogenous respiration occurs at the rate of 20% per day at 32°C and 10% per day at 20°C.

Generally it has been found from laboratory studies that each kg of milk solids requires about 1.25 kg of oxygen for complete oxidation. Of this, 37.5% is required during the assimilative phase and 62.5% in the endogenous phase.

Studies have shown that the microflora in dairy food plant waste treatment systems differ from the microflora of municipal systems (see Adamse, 1966, 1968, 1974; EPA, 1971; Goronszy and Barnes, 1980). The presence of Gram-negative organisms in aerobic treatment systems increases the rate of metabolism of the milk wastes and produces good sludge characteristics, i.e. rapid rate of settling and low sludge volume index (volume/unit mass) (Harper and Chambers, 1978). For Gram-negative organisms to be able to dominate in an activated sludge treatment system, low loading, good oxygen transport and a uniform 'food supply' appear to be essential. Generally these organisms do not store carbohydrates as polysaccharides and cannot compete with the filamentous Gram-positive organisms when the food supply becomes limiting, such as may occur when the dairy plant is not operating for two or more consecutive days per week.

A high concentration of filamentous Gram-positive organisms leads to sludge 'bulking' with consequent poor settling and loss of sludge in the discharge from the treatment system. Chambers (1975) made a study in one commercial plant, operating on a five-day week, where he seeded an acclimatized pseudomonad culture several times a week. This resulted in an increase in the removal of BOD_5 and a marked decrease in sludge volume. Similar seeding techniques have been used to speed up the commissioning of a combined municipal–dairy waste plant (Parkin, M.F., private communication).

There are many reports of whey, even quite small quantities, upsetting the operation of lagoons, trickling filters and activated sludge aerobic systems. This is often attributed to an undesirable BOD_5-to-nitrogen ratio in the whey. However the ratio of BOD_5 to nitrogen in whey is close to 20:1, a value normally considered satisfactory. Adamse (1968) has suggested that the nitrogen in whey is not readily available to the micro-organisms, whose growth is promoted by a high BOD_5 loading in the form of lactose. Either these filamentous bacteria cannot use the amino acids present as a nitrogen source or the extracellular enzymes needed to break down the whey protein are not present in a sufficiently high concentration. It is also possible that the rate of transport of lactose across the membranes of the bacterial cells is limiting. Harper and Chambers (1978) report that increased prehydrolysis of the lactose to glucose and galactose resulted in increased rates of BOD_5 reduction in a laboratory activated sludge plant with a microflora of mixed *Pseudomonas*. The BOD_5 reduction rate was further increased by the addition of proteolytic enzymes.

Other mechanisms suggested as reasons for the difficulty of whey breakdown include feedback inhibition by lactate, complex formation between whey proteins and the cell wall interfering with permease activity, and the removal of essential growth factors from the milk during the processes which produce whey.

There are few data on the effect of cleaning compounds, sanitizers and surfactants on the operation of aerobic treatment systems. It is generally conceded that the concentrations of these compounds are too low to have any toxic effect, although it has been noted that dumping of batches of cleaning materials does have a marked adverse effect. In a laboratory study Harper and Chambers (1978) report that concentrations of surfactants as low as $13\ mg\ l^{-1}$ altered the microflora, suppressing the growth of the Gram-negative organisms. Further study of the effects of these materials is desirable.

Activated sludge
Activated sludge systems used commercially for dairy wastes include fill-and-draw, high rate, deep shaft, extended aeration (oxidation ditch) and contact stabilization. Activated sludge treatment involves an aeration process followed by a clarification step and a return of varying amounts of active biomass (sludge) to the aeration tank. The types of activated sludge plants differ principally in the

method adopted for aeration of the mixed liquor, the design of the reactor (tank, basin, ditch, etc.), the period of contact of the waste with the activated sludge and the concentration of micro-organisms maintained in the reactor. Two handbooks (MAFF, 1969; Porges *et al.*, 1960) provide basic data for the design of activated sludge systems. The presence in the waste of free and emulsified fat, lubricants, cleaning compounds or whey often alters the operation of the plant from that predicted in these handbooks, but the books are a useful first guide.

The activated sludge process provides aerobic biochemical treatment employing suspended growths of bacterial flocs in a fully mixed reactor. The bacterial flocs are maintained at a high concentration by return of some of the sludge separated from the treated effluent by sedimentation. Normally, long hydraulic retention times (10 h to 5 days) are used to treat dairy wastes, although shorter design times ($\geqslant 4$ h) have been reported. While some authorities report that the daily applied load is normally in the range $0.3-1.2$ kg BOD_5 m^{-3} of reactor volume, practice in Europe favours a lower loading of less than 0.2 kg BOD_5 m^{-3} of reactor volume to avoid sludge settling problems (Scheltinga, 1972). High-rate systems use the higher loadings.

Typically the concentration of the bacterial floc (mixed liquor suspended solids (MLSS) or volatile suspended solids (VSS)) is maintained in the range 1000 to 8000 mg l^{-1}. Aeration is provided by one of the numerous types of diffused air or mechanical aeration systems. Coarse bubble diffusion devices are commonly recommended for dairy wastes as they clog less frequently and require less maintenance than the fine bubble systems. Jet aerators, shear devices, surface turbines, draft-tube aerators, rotors and brushes are popular mechanical devices. Oxygen transfer rates are dependent upon the aerator selected, the geometrical configuration of the reactor and the wastewater characteristics. The aeration is sufficient to keep the solids and bacterial floc in fully mixed suspension.

The long retention times and the fully mixed regime used tend to reduce the need for balancing tanks for flow equalization and only very severe pH fluctuations will cause difficulties (Doedens, 1974*a*). The effluent from the reactor is passed through a settling tank and the sludge is removed. Some of the sludge is recycled to the reactor. The clarified liquid is treated further, if necessary, and discharged.

EPA (1971) provides tabulated information from a wide range of activated sludge plants, including laboratory and industrial units. Some of the latter were operating poorly. The data show BOD_5 reductions ranging from 30 to $>99.5\%$ (mean value 86%). The average loading was 0.46 kg BOD_5 m^{-3} d^{-1}, with an average MLSS of 5260 mg l^{-1}. Detention times ranged from 1.4–50 h. In general, superior performance was reported at the longer detention times, although conscientious operation of the plant was the factor most frequently correlated with successful performance. The EPA (1971) data are difficult to interpret for design purposes. However, experience suggests that with good operation BOD_5 removal efficiencies are in excess of 90%. This may not be sufficient to reach effluent standards acceptable to some legislative authorities — a 90% reduction for a raw dairy waste of 2000 mg BOD_5 l^{-1} gives an effluent of 200 mg l^{-1}, considerably in excess of the 20 to 30 mg l^{-1} values set for many waterways. Two-stage systems are frequently needed to meet such standards. The actual design loadings for an activated sludge plant depend on the required effluent standard and the proposed other stages of treatment.

Activated sludge systems show good BOD reduction, good operating flexibility, reasonable resistance to shock loads and have low land requirements. They require a substantial capital investment, have high operating costs, require continuous good supervision and frequently present problems associated with sludge disposal. The system tends to recover slowly from an excessive shock load or following a 2 or 3 day shutdown.

The most common difficulty with activated sludge systems is the tendency of the sludge to 'bulk' or to decrease in bulk density, i.e. the sludge volume index rises to the region of $100-300$ ml g^{-1} of sludge. When this occurs the sludge does not settle efficiently but is lost with the final effluent. This results in a discharge of excessive suspended solids and a reduced concentration in the sludge recycled to the aeration tank. This in turn leads to a progressive decrease in the concentration of activated sludge in the mixed liquor. The conditions under which sludge bulking occurs are difficult to define; it is often associated with the treatment of effluents of low pH and high carbon-to-nitrogen ratios. Inadequate initial aeration is a common cause. It is less likely to occur at either relatively low sludge loadings (<0.1 kg BOD_5 kg^{-1} MLSS d^{-1}) or relatively high sludge loadings (>0.6 kg BOD_5 kg^{-1} MLSS d^{-1}). Sludge bulking is caused by the preferential development of filamentous microorganisms which have the ability to accumulate storage compounds for continued growth after assimilable nitrogen compounds become available (Mechsner and Wuhrmann, 1974).

Pinpoint floc and floating sludges or foams are also a common problem. Foams are often associated with whey discharges to activated sludge systems. EPA (1971) reports that foams studied contained predominantly β-lactoglobulin, a surface-active whey protein, which was not readily degraded by the microflora normally present in activated sludge systems.

Deep shaft activated sludge units (Hines *et al.*, 1975) have been used to treat dairy wastes (Gallo and Sandford, 1979) although few data have been reported. They are generally high-rate systems with relatively low BOD_5 removal efficiencies. They may find application for dairy plants with limited land available which are required, or find it more economical, to pretreat wastes prior to discharge to a municipal plant.

Extended aeration systems use relatively long residence times (2–10 days). They are employed often without primary removal of suspended solids. Low loading rates (0.1–0.3 kg BOD_5 m^{-3} d^{-1}) and high MLSS concentrations (4000–8000 mg l^{-1}) (<0.05 kg BOD_5 kg^{-1} MLSS d^{-1}) are employed to

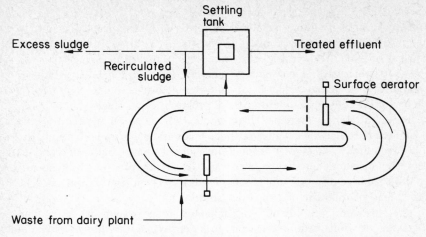

Fig. 5.12 Typical oxidation ditch for activated sludge treatment

minimize sludge production from the system. Most of the settled sludge is recycled to the aeration tank. Because of the high degree of oxidation of the solids, the effluent is often difficult to clarify, resulting in a relatively high ($\leqslant 50\,\mathrm{mg\,l^{-1}}$) solids carryover with the final effluent.

A common form of extended aeration is the oxidation (or Pasveer) ditch, used particularly in Europe (Scheltinga, 1972). The reactor takes the form of a ditch or channel, typically an elongated oval with a central island several metres wide and $1.0-1.5\,\mathrm{m}$ deep (Fig. 5.12). One or more electrically driven brush rotors, agitating the surface of the liquid, aerate and circulate ($\geqslant 0.3\,\mathrm{m\,s^{-1}}$) the contents of the ditch and keep the activated sludge in suspension. In the bends at each end of the ditch, baffles assist the maintenance of streamline flow. The height of the liquid in the ditch is adjusted to alter the rate of aeration, maintaining the dissolved oxygen concentration in the range $0.5-2.5\,\mathrm{mg\,l^{-1}}$ and providing $1.5-2.5\,\mathrm{kg\,O_2\,kg^{-1}\,BOD_5}$. Power loadings are about $8\,\mathrm{W\,m^{-3}}$ tank contents. Typically, sludge production is 20–30% of the BOD_5 load. The sedimentation tank or clarifier has a minimum retention time of 1 h at a loading of about $0.8\,\mathrm{m^3\,m^{-2}\,h^{-1}}$. In some small systems operating in Poland and in Australia, the agitation is stopped once or twice per day to allow sludge to settle, and a portion of the clarified liquor is discharged. Adjustable-height rotors are needed in this application.

A simple form of extended aeration (Fig. 5.13) uses a tank, normally square, of depth about 20% of the length of a side. Typical depths are 5 m; floating turbine aerators ($15\,\mathrm{W\,m^{-3}}$ reactor volume) provide aeration and agitation.

A 'Carrousel' oxidation system (Fig. 5.14) combines the principles of ditch and tank aeration. A small rectangular tank is placed at one end of the oxidation ditch ($2.5-4.0\,\mathrm{m}$ deep) and a turbine aerator in the tank provides aeration,

THE DAIRY INDUSTRY   353

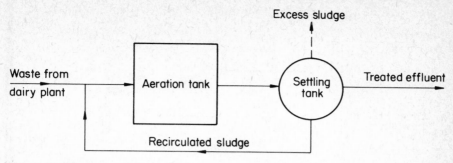

Fig. 5.13 Simple extended aeration system for activated sludge treatment

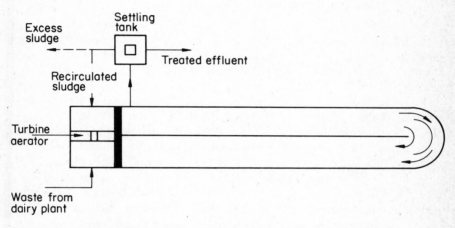

Fig. 5.14 Carrousel extended aeration activated sludge system

agitation and streamline flow. Power input is less than that required for other extended aeration systems ($3-5\,\text{W}\,\text{m}^{-3}$ reactor volume).

The contact stabilization activated sludge process takes advantage of the rapid initial assimilation of organic material from dairy wastes (Fig. 5.15). Waste is in contact with activated sludge for a short period ($0.5-2.0\,\text{h}$). The mixed liquor is passed to a settling tank where the sludge is separated and the clarified final effluent is discharged or passed to a further treatment stage. The separated sludge is passed to a reaeration or stabilization tank where it is aerated for a relatively long period ($3-5\,\text{h}$) during which oxidation of the adsorbed material takes place. A large proportion of the activated sludge is then recycled to the initial contact stage and the surplus sludge is passed to another stage for further aeration before disposal (about $0.5\,\text{kg}$ sludge per kg BOD_5 removed).

Hydraulic loadings of $5\,\text{m}^3\,\text{m}^{-3}\,\text{d}^{-1}$ and BOD_5 loadings of $7.5\,\text{kg}\,\text{m}^{-3}\,\text{d}^{-1}$ are recommended (Ministry of Agriculture, Fisheries and Food, 1969). The BOD_5

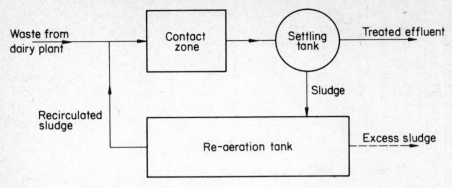

Fig. 5.15 Contact stabilization activated sludge system

(1500 mg l^{-1} feed) removal is relatively low (60–70%) and the process tends to produce 'bulking' sludge. The method is used in a high-rate first-stage treatment where sludge loadings are maintained at a high level (0.6 kg BOD$_5$ kg^{-1} MLSS d^{-1}).

The simplest activated sludge plants are batch-operated fill-and-draw systems. Such a system, operating in Finland, is described by Pankakoski (1978). Waste from the dairy plant is collected in a balancing tank. Once a day the effluent is drawn into an aeration tank which has a minimum volume of twice the expected daily volume of wastes. Over a period of 20 h the effluent is aerated through a coarse-bubble porous sparger to supply oxygen at 1.5 times the BOD$_5$ load. The aeration is switched off for 2.0–2.5 h to allow sludge settling and the clarified liquor is discharged. More effluent is introduced from the balancing tank and the process is repeated. Because of strict controls on phosphorus discharge, 60–100 g m^{-3} of ferrosulphate is added each day. Data for three plants treating 13–110 m^3 d^{-1} are given. BOD$_5$ removal is generally over 95%, phosphorus 87%, and total nitrogen 75%. Sludge is removed two or three times a year.

Goronszy and Barnes (1980) describe a continuously fed intermittently decanted single vessel activated sludge plant treating 60 m^3 d^{-1} of dairy waste containing 1100 mg l^{-1} BOD$_5$. Flow of waste to the treatment plant is continuous over 10–14 h. The plant is aerated for 10.5 h, settling occurs over 1 h and treated effluent is discharged over 0.5 h. Aeration is by means of two 4.6 kW blowers through 20 jet nozzles. Nitrogen and phosphorus are added as supplementary nutrients. At 35°C the BOD$_5$ concentration in the effluent from the treatment plant is consistently less than 20 mg l^{-1}.

Johnston (1978) describes a two-stage activated sludge plant constructed in mild steel, the first stage being high-rate and the second stage the normal low-rate system (Fig. 5.16). The plant was designed to treat 364 m^3 d^{-1} of effluent from a butter plant. The average BOD$_5$ concentration is 1190 mg l^{-1}. The effluent passes from the butter plant to a fat trap, then to an aerated

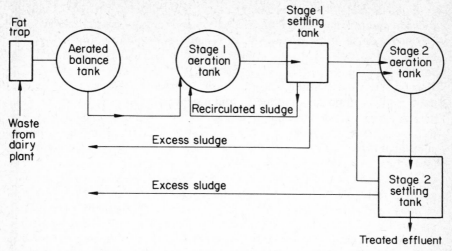

Fig. 5.16 Two-stage activated sludge plant using a high-rate system for the first stage and a low-rate system for the second stage (adapted from Johnston, 1978)

balance tank (total capacity 260 m^3, working capacity 90 m^3). The first stage of the activated sludge has a capacity of 170 m^3 and is aerated and mixed by a fixed bridge type two-speed aerator set off-centre. An adjustable outlet weir controls the liquid level in the tank and thus the dissolved oxygen concentration of the liquor. The effluent overflows the weir into a settling tank (hopper bottom type, 4.1 m square, 3.7 m deep). Sludge is recirculated to the aeration tank to maintain the MLSS concentration between 2000 and 2500 mg l^{-1}. Overflow from the first-stage settling tank enters the second-stage aeration tank (117 m^3). Sludge from the second-stage settling tank (4.6 m square, 4.6 m deep) is recycled to maintain the MLSS concentration between 3500 and 5000 mg l^{-1}. Excess sludge, 80–90 kg d^{-1} (70% from the first stage) at a daily flow of 105 m^3, is removed by tanker for land disposal. Total BOD$_5$ reduction is 97–99% and discharge BOD$_5$ and suspended solids concentrations are <20 mg l^{-1} and <30 mg l^{-1} respectively.

Guo *et al.* (1979) evaluated the performance of an oxidation ditch treating 140–220 m^3 d^{-1} of dairy effluent containing 380–1900 mg BOD$_5$ l^{-1} (average 950 mg l^{-1}). The oxidation ditch is 36 m long × 15 m wide with a divider 17.7 m long × 2.7 m wide. The total volume is 495 m^3. A brush-type aerator (3.7 m long, 1.1 m dia.) driven by a 18.7 kW motor maintains the oxygen concentration between 0.5 and 6.0 mg l^{-1}. Average temperatures are 8.6°C in winter and 21.9°C in summer. The MLSS concentration varies between 3000 and 6000 mg l^{-1} with an average value of 4000 mg l^{-1}. Average daily loading conditions are 0.08 kg BOD$_5$ kg^{-1} MLSS and 0.32 kg BOD$_5$ m^{-3}, indicating slight overloading. The plant gives 99% removal of BOD$_5$ and 98% removal of suspended solids. Problems experienced include unsteady flow of effluent to the oxidation ditch because of inadequate balancing, and floating sludge in the clarifier because of the lack of sludge scraping and scum removal.

Lagoons

Lagoons are used in a number of countries to treat high-strength dairy wastes and in others to polish an effluent from a conventional treatment plant. They are also used as buffer tanks, particularly over periods when the normal treatment or disposal is inadequate, e.g. to store effluent during the winter for subsequent spray irrigation. They are appropriate in areas where land costs are low, where the system can be installed some distance from neighbours and climatic conditions are suitable. Lagoons are of four main types: aerobic, aerated, facultative (combined aerobic–anaerobic) and anaerobic. Anaerobic lagoons tend to produce strong offensive odours if improperly designed or operated.

Lagoons facilitate the natural biochemical breakdown of the organic material in the waste. Bacteria break down organic matter, using oxygen and producing carbon dioxide. The available carbon dioxide, and some organic matter, is then utilized by algae which liberate oxygen, allowing aerobic bacterial activity to continue. In some lagoons low or zero dissolved oxygen concentrations near the bottom of the lagoon result in anaerobic decomposition of the solids produced by the bacterial activity. Thus the biological process in lagoons may include aerobic oxidation and growth of aerobic bacteria, growth and photosynthesis of algae, anaerobic oxidation and growth of methane-producing bacteria. Controlling factors for these processes are the oxygen concentration and sunlight intensity. In general, increasing the depth and/or increasing the loading will shift a lagoon from aerobic to facultative and finally to anaerobic operation. Table 5.22 summarizes data on the design characteristics of lagoons given in EPA (1971).

Table 5.22 Summary of data on operating characteristics for lagoons (EPA, 1971)

	Aerobic	Aerated	Facultative	Anaerobic
Depth (m)	0.18–0.30	1.8–4.6	0.60–1.5	2.4–3.0
Residence time (days)	2–6	2–10	7–30	30–50
BOD_5 loading (kg ha^{-1} d^{-1})	20–40		4–9	55–90
BOD_5 removal (%)	80–95	55–90	75–85	50–70

Lagoons, or stabilization ponds, are used for full treatment of wastes from dairy plants in Czechoslovakia (Svoboda *et al.*, 1966; Svoboda, 1974). The lagoons (1.0–1.5 m deep, 17.8–26 ha) receive effluent at a rate of 20 kg BOD_5 ha^{-1} d^{-1} or 1.5 g BOD_5 m^{-3} d^{-1} and, with detention times of 105–160 days, achieve a 95% reduction in BOD. Water leaving the lagoons has a BOD_5 concentration of 9–175 mg l^{-1}. Carp are fished from the lagoons with increases in fish weight of up to 700 kg ha^{-1}. The same reports contain descriptions of a three-stage stabilization basin (2 ha, 2–3 m deep) reducing 150–300 kg d^{-1} BOD$_5$ to 4 kg d^{-1}. Average daily surface loadings in the three basins are 217–449 kg BOD$_5$ ha^{-1}, 62–93 kg BOD$_5$ ha^{-1} and 24–35 kg BOD$_5$ ha^{-1} respectively. Similarly the daily volumetric loadings are 20–40 kg BOD$_5$ m^{-3}, 7–10 g BOD$_5$ m^{-3} and 3–4 g BOD$_5$ m^{-3}. Other lagoons have been described (An Foras Taluntais, 1974) with 10–40 days retention time, depth <1 m and loadings ≤70 kg BOD$_5$ ha^{-1} d^{-1}.

Aerated lagoons are a form of extended aeration treatment but sludge is not recycled. Air for aerated lagoons is supplied mechanically either by surface or submerged aerators. The lagoons are 2–4 m deep and normally have 2–10 day detention times, although up to 100 days has been reported. Frequently, the lagoons consist of two or more compartments. They operate at lower suspended solids concentrations than activated sludge systems.

The Kent Cheese Company (EPA, 1974) operates a system consisting of two aerated lagoons in series, one aerobic and one facultative. Each lagoon (48 m × 37 m × 3.7 m) has a working volume of 3600 m^3 and treats an average flow of 64 m^3 d^{-1} of effluent with an average BOD$_5$ of 1900 mg l^{-1}. The retention time in each lagoon is 45 to 80 days, depending on the wastewater flow. Air is supplied to the first lagoon by two rotary blowers via 13 Helixor aerators and to the second lagoon by 3 Helixor aerators. Air supply for each aerator is 0.4 standard m^3 per minute at a power input in the primary lagoon of 1.7 W m^{-3}. Overall BOD$_5$ removal efficiency averages 97.3%. Performance is affected by temperature, with 95% removal efficiency in winter. In spring, the turnover of solids results in zero dissolved oxygen in the lagoon and a higher rate of aeration has been recommended (EPA, 1974).

Tanaka (1974) describes an extended aeration multi-stage lagoon. The lagoon is 44 m × 88 m with a depth of 4 m, and is divided into equal basins, 6500 m^3 each. The total detention time for the 2000 m^3 d^{-1} of effluent is about 7 days. The BOD$_5$ loading is 0.06 kg BOD$_5$ m^{-3} d^{-1}. Agitation and aeration are supplied by two floating aerators, each with a 22 kW motor. Treated effluent is discharged once daily. The aerator in the second lagoon is stopped to allow the activated sludge to settle, and the clarified liquor is discharged. A high dissolved oxygen content of 7–9 mg l^{-1} is maintained in the lagoon. The MLSS vary between 400 mg l^{-1} (summer) and 800 mg l^{-1} (winter). BOD$_5$ removals exceed 95% and suspended solids in the discharge from the lagoon are less than 20 mg l^{-1}. The lagoons are claimed to be efficient even at atmospheric temperatures as low as −20°C. Similar lagoons operating in the UK have much higher MLSS concentrations (2000–7000 mg l^{-1}).

The main operational problems for lagoons appear to be odour control, algae control, temperature–performance control and a less serious problem regarding control of weeds, erosion and insects. Algae growth and species control does not yet appear to be on a scientific basis; depth control and skimming of algae mats are possibilities. Operation of parallel ponds followed by one or more additional ponds permits the control of loading and residence time.

Lagoons in series permit the higher efficiencies of staged operations for BOD_5 and algae control. The provision for longer residence times by putting longer parallel ponds in series, or by increasing depth to avoid discharge, offers a means to compensate for upsets or reduced performance in the winter. Solids build-up is still an undefined problem for dairy wastes; a high rate of build-up is expected in systems for cheese fines, fruits and nuts or for whey.

Aerated lagoons provide good BOD_5 reduction, show good resistance to shock loads, have a relatively low capital cost, require less supervision and have fewer sludge problems than activated sludge or trickling filters. However, they require large areas of land and performance is adversely affected by low temperatures. Stabilization ponds are particularly suitable as a pretreatment system to prevent shock loads proceeding to subsequent treatment systems. They have low capital and operating costs with few sludge problems. They provide good BOD_5 reductions in warm climates, but BOD_5 reduction efficiency is considerably reduced in colder temperatures. They have large land requirements, and frequently present problems of insects and odours; thus their possible locations are restricted.

Trickling filters and rotating biological contactors
Trickling filters take advantage of the rapid adsorption of organic material onto microbial flocs. They employ a fixed medium to support the growth of a film of active micro-organisms: a balanced flora of bacteria, fungi, protozoa, larvae, flies and worms builds up. Older plants employ rock, slag or other low-cost materials (such as brushwood, Johnston (1978)) to provide a support with a large specific surface area plus a high void volume. Typically, filter depths vary from 1.0 m to 2.5 m. More modern plants use various types of low mass per unit volume, high specific surface area ($80-230\,m^2\,m^{-3}$ media) plastic media either in sheet form or randomly packed. Plastic packing depths up to 12 m have been used (Hemmings, 1980; Seyfried, 1974).

These tower-type filters are intended mainly as a high-rate first treatment stage, accepting organic loads of the order of three times that of mineral media. The dairy waste flows over the medium, on which a microbial slime develops. The organic matter from the waste is absorbed or adsorbed into the microbial film and is subsequently oxidized. The effluent is collected in an underdrain system. The air for oxidation passes through the void spaces between the slime-covered packing, either by natural convection or by a forced airflow. The filter is operated in a manner to prevent the void spaces being clogged by solids in the dairy waste or by excessive slime growth. Excess slime sloughs off the medium and is removed in clarifiers following the filter. The settled solids are discharged. Time of contact between the waste and the microbial slime is normally short, although it can be increased by recirculation of partially treated waste through the filter or by having a number of filters in series.

Three basic types of operation are used: (i) low-rate filtration and (ii)

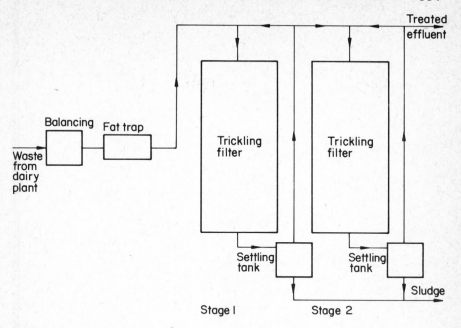

Fig. 5.17 Trickling filter system

high-rate recirculating filters, both types either singly or in pairs in series (Fig. 5.17); (iii) alternating double filtration (Fig. 5.18) using two filters in series, settling after each filter and with periodic reversal of the order of use after 7–15 days.

Trickling filtration is frequently used as a roughing treatment prior to a further form of treatment such as activated sludge or lagoons. Conventional filtration is occasionally used as a second-stage treatment following aeration, activated sludge or chemical flocculation.

For dairy wastes, single-stage filtration plants typically have maximum BOD loadings of 0.12 kg BOD $m^{-3} d^{-1}$. High-rate conventional filters operate at BOD loadings of $0.6-0.9$ kg $BOD_5 m^{-3}$ filter d^{-1}, but large-grade media (75–100 mm diameter) and slowly rotating distributors (4 rph) giving high instantaneous doses of effluent are used to control film growth. Alternating double filtration systems typically operate with maximum BOD_5 loadings of 0.3 kg $BOD_5 m^{-3} d^{-1}$ and volumetric loadings of $1 m^3 m^{-2} d^{-1}$ when the incoming dairy waste is diluted to $200-300$ mg l^{-1} BOD_5 by means of recirculation. High-rate filters, using plastic media, operate at BOD_5 loadings up to 3.0 kg $m^{-3} d^{-1}$. Rates as high as 10 kg $BOD_5 m^{-3} d^{-1}$ have been reported, although at high rates the efficiency is reduced (Hemmings, 1980).

The efficiency of trickling filters increases with increasing depth of the filter

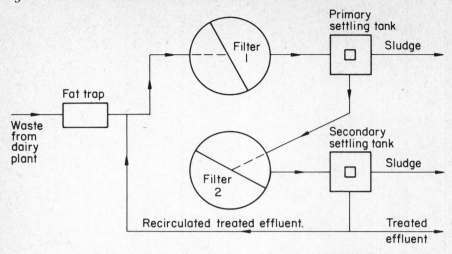

Fig. 5.18 Alternating double filtration system: after 7–15 days the order of the filters is reversed, with Filter 2 receiving the raw waste diluted with recirculated treated effluent, and Filter 1 receiving the overflow from the primary settling tank

bed and decreases with increasing hydraulic flow rate per unit cross-sectional area. The depth of the filter is limited by physical loading considerations. Particularly for plastic filter media, the hydraulic flow rate has a lower limit controlled by the need to maintain a flow sufficient to keep the medium wetted (typically $1.4\,m^3\,m^{-2}\,d^{-1}$). However, greater recirculation than the minimum required for wetting is not particularly helpful (Wheatland, 1974). Ventilation of trickling filters is important in maintaining the aerobic conditions necessary to ensure effective treatment. This is especially true for dairy wastes with their initial high oxygen demand. Poor air supply causes heavy biological growths, clogged drains and channelling of the wastes. These result in unpleasant odours and poor effluent quality. High concentration of salt or lactose can be detrimental to the operation of filters.

Shock loading can affect filter operation but there are many reports of the performance recovering readily after a shock load. However, fluctuations in hydraulic and organic loads can cause problems, especially from high organic loads, extremely low hydraulic loads (allowing the slime layer to dry out), very high hydraulic loads (causing washout of the slime) and extremes of pH (<4.5, >9.5). Balancing of the flow to trickling filters is beneficial.

Often effluent is recirculated from the overflow of the clarifier to the inlet of the filter to dilute the organic concentration of the incoming wastes, to dampen out variations in the organic load, to increase the hydraulic flow over the packing so that the medium is adequately scoured and wetted, to improve distribution over the filter surface and to increase the contact time, improving efficiency of removal. Alternating double filtration systems use recirculation of treated effluent to reduce the incoming BOD_5 concentration to 200–300 mg l^{-1}.

Trickling filters show good BOD_5 reduction and their operating costs are lower than those of activated sludge or lagoon systems. However, they require a substantial capital investment and have significant land requirements. Sludge production is high and disposal is often difficult. Good supervision is required for adequate performance. When poorly designed or operated they are subject to ponding, fly and odour problems.

Table 5.23 summarizes operating data for an alternating double filtration plant and another plant consisting of three stages in series — stages 1 and 2 using high-rate filters and stage 3 two conventional filters used as alternating double filtration; there are settling tanks between each stage.

Rotating disc biological contactors, in principle, are similar to trickling filters. The biological film grows on plastic discs which are rotated in the waste liquid to be treated. There are a few reports of the use of rotating discs to treat dairy wastes (Antonie and Hynek, 1973). A series of plastic discs (12.5 mm thick and ≤4 m diameter), closely spaced (12–25 mm gaps) on a horizontal shaft, is mounted above a tank with the shaft supported just above the surface of the liquid. The lower portion of each disc extends into the waste so that about 40% of the disc area is submerged while the upper portion rotates in the air. In rotation the discs carry a film of wastewater into the air, where it trickles down the surface of the discs and absorbs oxygen. At the same time organisms in the biomass remove dissolved oxygen and organic material from the wastes. The faces of the discs support the biological growth. The discs are rotated so that the peripheral velocity is about 0.3 m s^{-1}, sufficient to slough off excess filter growth.

Daily volumetric loading rates of $0.02-0.04$ m^3 m^{-2} of disc area are used for dairy wastes. The disc contactors are used in a series of stages, 1 h to 2 h retention time in each stage. The number of stages is determined by the final effluent quality required. Flow balancing of the raw waste is normally used and pH adjustment (pH 6.5–8.5) and nutrient addition may be desirable. One commercial dairy installation treats an average waste flow of 950 m^3 d^{-1} with BOD_5 concentrations varying from 230–2700 mg l^{-1}. Pretreatment includes a 2 h detention time flow-equalization tank, pH adjustment and an aerated 12 h detention time balance tank. Four 7400 m^2 rotating biological disc contactors in series, rotating at 1.6–2.0 rpm, reduce the BOD_5 concentration to 3–35 mg l^{-1}. The installed power is 60 kW for the contactors.

Rotating disc biological contactors are also used to treat dairy wastes to 200–300 mg BOD_5 l^{-1} before discharge to a municipal treatment plant.

While capital costs are relatively high, rotating disc systems require less area and have lower operating costs (particularly power) than activated sludge units. Because of the large quantity of biomass they also withstand shock loads reasonably well.

Table 5.23 Designs of two trickling plants

Plant A consists of two trickling filters used in alternating double filtration (ADF).
Plant B comprises three stages, two high-rate filters followed by two filters used in ADF, with settling between each stage.

	Plant A	Plant B		
	ADF	Stage 1 High rate	Stage 2 High rate	Stage 3 ADF
Volume of waste (m^3 d^{-1})	590	3500	3500	3500
BOD$_5$ of waste (mg l^{-1})	600	1500	675	236
Volume of filters (m^3)	Two of 835 each	1460	1050	Two of 1470 each
Dimensions of filters (m)	23 (dia) × 2	12 × 18.7 × 6.5	12 × 12.6 × 6.6	30.6 (dia) × 2
Organic loading (kg m^{-3} d^{-1})	0.21	3.6	2.4	0.28
Irrigation rate (m^3 m^{-2} h^{-1})	0.14	1.5	1.5	0.2
BOD$_5$ of final effluent (mg l^{-1})	15	675	236	<20
Settling tank dimensions (m)		9.2 (dia) × 2.2	9.2 (dia) × 2.2	Two of 13.6 (dia) × 2.0

5.6.6 Anaerobic treatment

Many small dairy plants have used septic tanks to treat dairy wastes anaerobically. However, the effluent quality from anaerobic processes has generally been below that required for surface water discharge. The process is more sensitive to shock loading and low temperature than the more commonly utilized aerobic processes. In anaerobic treatment a mixed microbial population, in a series of sequential reactions, reduces complex organic molecules such as fats and proteins to simpler structures such as amino acids, organic acids, aldehydes and alcohols. Lactose is converted to lactic acid. Subsequently these intermediates are fermented to hydrogen, carbon dioxide and acetic acid. Acetic acid is the primary precursor of the final product, methane. Because of the complex ecosystem, control of pH is necessary to keep the complex reactions in balance. Accumulation of intermediate acids, with a drop in pH, disrupts methane production.

Anaerobic systems must be enclosed, with suitable arrangements for the collection or burning-off of methane and the prevention of venting of malodorous gases, such as hydrogen sulphide.

The anaerobic treatment of excess sludge or concentrated milk wastes is practical in conventional anaerobic digesters at municipal treatment plants and has been reported to give good gas production.

In the past few years there has been renewed interest in direct use of anaerobic systems for treating dairy wastes and whey (Vandamme and Waes, 1980). The increasing cost of fuel, improved understanding of anaerobic mechanisms, advances in reactor systems and the lower sludge yields from anaerobic reactions than from aerobic reactions have been factors in this increased interest. In addition, anaerobic systems appear to be a useful pretreatment prior to discharge to a municipal sewer or prior to aerobic polishing.

The development of improved high-rate anaerobic systems has resulted in increased efficiency of COD removal and appreciable reduction in retention times required for municipal and some industrial wastes. The efficacy of these systems for dairy wastes remains to be fully developed. Recent advances include the development of sludge blanket and fluidized bed systems. In an upflow anaerobic system, a 90% reduction of COD was achieved with a simulated dairy waste (dilute skim milk; COD 1500 mg l^{-1}, space loading 7–8 g COD m^{-3} d^{-1}, temperature 30°C) (Le Hinga *et al.*, 1980). In an anaerobic fluidized bed reactor, an 80% reduction of COD of dilute whey was achieved (COD 8000–14 000 mg l^{-1}, loading 40 kg COD m^{-3} d^{-1}, temperature 35°C) (Hickey and Owens, 1981).

Anaerobic treatment appears to have some potential for use in pretreatment systems, reducing BOD$_5$ to, say, 300–600 mg l^{-1} prior to discharge to a municipal plant or other aerobic treatment.

5.6.7 Chemical precipitation

Chemical precipitation, coagulation or flocculation using calcium hydroxide, sodium aluminate, aluminium sulphate, ferric sulphate, ferric chloride and synthetic polymers has been practised for the treatment of dairy wastes. Reported BOD_5 removal efficiencies range from 5 to 80%, depending on the proportion of BOD_5 present in the feed effluent as colloidal or particulate matter. The soluble BOD_5 fraction is not affected. The technique is used very little because of the high cost, the low treatment efficiencies and the large volumes of sludge produced.

Chemical coagulation has been used to assist the precipitation of sludge from some biochemical treatments.

Calcium hydroxide, aluminium sulphate or ferric chloride is added to treated effluents to remove phosphorus in tertiary treatment. Typically, 5–7 mol Al^{3+} per mol of P at pH 7.0, or 3–5 mol of Fe^{3+} per mol of P at pH 5–6, will remove 90–100% of the phosphorus.

5.6.8 Tertiary treatment methods

While spray irrigation, biological filtration, activated sludge and aerated lagoons are the most common methods of effluent treatment, the need in some countries to achieve very high standards in effluents being discharged to natural waters and the cost of fresh water are focusing attention on tertiary or effluent-polishing techniques in the dairy industry. Cooper (1974) includes micro-strainers, sand filters, clarifiers, grass plots, lagoons and chemical treatment in his list of tertiary treatment methods. Other processes he mentions include activated carbon adsorption, electrodialysis, ion exchange and reverse osmosis. However, these are too expensive for use at present, although reverse osmosis has been studied as a means of upgrading dairy plant effluents (Eriksson et al., 1978).

5.6.9 Sludge disposal

Sludges are produced in a number of operations in the dairy plant. The main sludges are from the separation of milk and from the chemical or physical treatment of fresh water and raw or treated wastewaters.

Separator sludge

In the separation of cream and milk in centrifuges (separators) a sludge is obtained. In the older type of separator the sludge accumulates in the bowl and is removed manually after stopping the machine. In self-desludging separators the sludge is removed automatically at intervals (30–60 min) during continuous operation. The sludge volume varies from 0.01 to 0.1% of the milk processed. The sludges from each type of machine have a similar dry composition but the sludge from the self-desludging machine is more dilute (Table 5.24). Depend-

Table 5.24 Composition of sludges from milk separators

	Separator type	
	Batch	Self-desludging
Total solids (%)	25–40	1.9–8.5
Protein (%)	20–30	1.0–4.5
Fat (%)	1.4–3.5	0.14–1.3
Ash (%)	2.6–5.0	0.25–0.75
BOD_5 (mg l^{-1})		10 000–100 000

ing on the quality of the milk, the sludges contain various concentrations of microbial organisms.

Because of the risk of the presence of pathogenic or disease-causing organisms, in some countries the disposal of sludge from separators is regulated by legislation. For example, in Czechoslovakia, Denmark, Japan, the Netherlands, Norway and Sweden, sludge from the older style discontinuously operated separators must be burnt, buried or dumped at sea. In, for example, Denmark and the Federal Republic of Germany, sludge from self-desludging machines must be heat treated (80°–90°C for 20 min, possibly after the addition of 2% NaOH) or rendered before being fed to animals. In Sweden, sludge from self-desludging units can be disposed of with the rest of the dairy waste, provided that biochemical treatment is used. However, in other countries (e.g. Australia, Finland, New Zealand) sludge is fed, apparently untreated, to animals, particularly to pigs. Modern standards of hygiene on dairy farms has reduced the risk of disease from separator sludge to negligible levels, and special legislation on disposal probably is no longer necessary.

Raw water and waste sludges
The sludge from the treatment of raw water is normally buried or dumped as sanitary landfill. Settling of solids from raw dairy wastes is rarely practised at dairy plants. Chemical flocculation, particularly by reducing the pH to less than 4.5, produces relatively large volumes of sludge, mostly protein and fat. The sludge is generally buried or burnt. Chemical treatment of dairy wastes is not common because of the cost. Fat removed from fat traps is also usually buried or burnt, although in some places it is fed to pigs.

Sludge from biochemical treatment
Biochemical treatments of dairy waste produce sludge solids which require treatment and disposal. Cooper (1978), Doedens (1978), Doedens (unpublished draft for IDF, 1981), Salplachta (1978) and Scheltinga (1978) discuss the handling of sludge from dairy effluent treatment plants. The nature and quantity of sludges produced vary with the metabolic characteristics of the various micro-organisms present and depend on factors such as the BOD_5 and suspended solids concentrations of the raw waste, the type of treatment method, the degree of treatment, the method of operation, temperature, etc.

Table 5.25 Characteristics of sludge obtained at a number of dairy effluent treatment plants in the UK (Cooper, 1978)

Type of treatment	Total solids (%)	Organic matter (% dry weight of solids)	Sludge production (kg solids per kg BOD_5 applied)	BOD_5 removal (%)
High-rate filtration (plastic media)	2.2–4.1	81	0.51–0.83	60
High-rate filtration (stone media)	2.0–2.6	80	0.81–1.06	70–80
Single filtration	1.0–2.4	78	0.41–0.52	98
Alternating double filtration	1.1–2.6	77	0.48–0.67	98
Conventional aeration	0.5–1.4	77	0.46–0.59	98
Extended aeration	0.4–1.5	70	0.20–0.36	98
Extra-extended aeration	0.3–1.5	57–66	0.11–0.23	99

(Table 5.25). Doedens (1978) and Salplachta (1978) show that the quantity of sludge per mass of BOD_5 removed increases with the organic loading (mass BOD_5 per mass VSS per day) and decreases with the temperature. For example, from an extended aeration unit excess sludge production is 0.10–0.20 kg VSS kg^{-1} BOD_5 removed under summer conditions and 0.35–0.50 kg VSS kg^{-1} BOD_5 removed under winter conditions.

Excess sludge represents 1 to 5% of the volume of waste treated (or 20–70 litres sludge kg^{-1} BOD_5 removed), has a total solids concentration of between 0.6 and 2.0% and an organic matter content of 75 to 85% on a dry basis. The organic matter is usually in an unstable form with a high BOD_5 and this can give rise to odour problems. Sludges from low-rate trickling filters, from the second stage of an alternating double filter and from extended aeration processes are more stable (less organic matter) than those from high-rate trickling filters or highly loaded activated sludge plants. Extended aeration treatment produces less sludge than alternating double filtration, which in turn produces less than high-rate filtration. Sludge solids production ranges from 0.1 to 1.0 kg dry sludge solids per kg applied BOD_5, depending on the treatment method (Table 5.25). It should be noted that less sludge is produced, by most treatment methods, from dairy wastes than from domestic sewage (0.7–0.9 kg dry sludge solids per kg BOD_5 removed).

If the treatment plant is operated correctly, the sludge can be concentrated to 2–4% total solids by settling for 12–24 h and decanting. Typical decanters are about 4 m deep and daily sludge loading is 12–30 kg sludge m^{-3} tank. The decanted liquid is returned to the treatment plant.

Dissolved air flotation can give more reliable performance and higher total solids in the separated sludge than settling tanks (Pengilly, 1978). Conditioning chemicals markedly assist sludge removal. Ferric chloride (20–60 g per kg dry sludge solids) followed by lime (CaO or Ca(OH)$_2$) (75–200 g per kg dry sludge solids) is commonly used. Polyelectrolytes are also used (typically 1–6 g per kg dry sludge solids), the quantity depending on the sludge condition and the type of polymer.

Table 5.26 Throughputs and expected final solids concentrations for dairy waste sludge mechanical dewatering devices (Doedens, unpublished draft for IDF, 1981)

Higher throughputs and solids concentrations are reached only with the use of conditioning chemicals.

	Throughput	Final solids concentration (%)
Vacuum filter	10–30 kg sludge solids $m^{-2} h^{-1}$	20–33
Belt filter	100–200 kg sludge solids m^{-1} width of belt h^{-1}	20–40
Centrifuges	5–50 $m^3 h^{-1}$	20–40
Filter press	4–10 kg sludge solids $m^{-2} h^{-1}$	25–60

Mechanical dewatering of sludge is practised by some dairy plants, although the cost of such treatment is high (Cooper, 1978). Hemmings (1980) states that mechanical dewatering can represent 50% of the total operating cost for effluent disposal and up to 50% of the capital cost of the oxidation stage. Equipment used for mechanical dewatering includes vacuum filters, belt filters or band presses, centrifuges (decanters) and filter presses. Typical throughputs and final solids concentrations are given in Table 5.26. The vacuum filter requires a higher energy input (6 kWh m^{-3} sludge) than the other mechanical dewatering devices (1–2 kWh m^{-3} sludge) but is relatively easy to operate, has low maintenance costs and provides a filtrate with a low total suspended solids concentration. Mechanically dewatered sludges can be dried further on sludge beds (an annual loading of 0.5–5.0 m^3 sludge m^{-2}).

Sludges which are not dewatered are putrescible because of the high concentration of organic material, and may produce strong odours after a few hours. Thus sludges from dairy waste treatment should be disposed of quickly, or further stabilized to minimize problems from the offensive odours. Stabilization can be accomplished by separate anaerobic or aerobic processes to reduce the organic content by 30–60%.

Sludge as produced is disposed of on land by flooding, truck spreading or spray irrigation at annual rates of 60–120 $m^3 ha^{-1}$ (solids concentration, 4%; 14 day rest period between applications, with the lower rate applying for less well stabilized sludges). Alternatively, dewatered sludge is disposed of by land spreading (at solids rates equivalent to those used for liquid sludges), dumped as landfill (layers up to 2 m deep) or transported by barge for dumping at sea. Thermal drying and incineration are other possible, but expensive, methods of disposal.

Sludge from dairy waste treatment is an economic form of fertilizer with good concentrations of nitrogen (7.8% of dry matter), phosphorus (1.0%), potassium (0.4%) and humus (Cooper, 1978). There are negligible hygienic risks from pathogenic organisms in the use of dairy waste treatment sludge as a fertilizer (Cooper, 1978; Hayashi and Ishioka, 1978; Scheltinga, 1978) and heavy metal concentrations are low enough not to cause concern (Scheltinga, 1978). Scheltinga (1978) concludes that very small health risks and positive nutritional advantages are associated with the use of sludges from dairy waste treatment as a feed for pigs.

Comparison of treatment methods

Table 5.27 briefly compares some of the characteristics of processes to treat dairy wastes in the USA to an effluent standard of 30 mg l^{-1} BOD$_5$ consistently.

Comparisons between treatment methods will be different in other countries and for achieving different effluent standards. Combinations of methods may also give different comparisons and, frequently, a combination of two or more systems will give the lowest costs and the greatest reliability. For example, Scheltinga (1972) states that in the Netherlands, for BOD$_5$ loads greater than 1000 kg d^{-1}, a first stage consisting of high-rate filtration followed by extended aeration would give the optimum treatment. Many dairy plants find the best combination to be pretreatment to BOD$_5$ of 200–600 mg l^{-1} and discharge to a municipal sewer.

Provided that suitably trained staff are available and the treatment method will meet the legislative requirements, total cost, capital and operating, is the main criterion for choice of treatment method. Costs of treatment of wastes depend on so many factors that detailed cost estimates must be obtained for each individual case. The costs vary with the treatment method, the organic and volumetric load to be treated, the effluent standard to be achieved and the country in which the system is to be installed. In general the initial capital costs are made up of the costs of the site, the mechanical equipment, civil works and any buildings. Synnott *et al.* (1978) state that capital costs for effluent treatment plants in Ireland are very often of the order of 15 to 20% of the total cost of a milk processing installation. Doedens (1974b) estimates that the cost of effluent treatment accounts for 5 to 10% of the capital costs of new dairy plants and that waste disposal constitutes 1 to 5% of the dairy plant operating costs. Operating costs include interest on capital, energy, labour (including laboratory staff for analytical work), chemical additives, maintenance and sludge disposal.

An Foras Taluntais (1974), Scheltinga (1972) and Stanley Associates (1979) give some cost data for dairy waste treatment. Hemmings (1980) gives a relative capital cost ratio in the UK (excluding sludge disposal) of 1.0:1.62:0.87 for conventional activated sludge, alternating double filtration (mineral media) and high-rate biofiltration followed by either activated sludge or alternating double filtration; operating costs (including sludge disposal) were in the ratio 1.0:0.30:0.45 respectively. Riddle and Chandler (1974) estimate the relative costs (capital and operating) of activated sludge, trickling filter and aerated lagoons to be in the ratio 1.0:0.9:0.5. Buxton *et al.* (1977) compared the cost of treating wastes from various dairy plant operations by stabilization pond, aerated lagoon combined with irrigation, and ridge and furrow. The ridge and furrow capital and operating costs were about half those of the other two treatment methods.

Costs of treatment in municipal sewers is now significant in many places.

Table 5.27 Comparison of treatment methods for USA dairy wastes to achieve a consistent effluent quality of 30 mg l^{-1} (Boyle and Polkowski, 1973)

3 — Excellent, 2 — Good, 1 — Fair, 0 — Poor
H — High, A — Average, L — Low

Treatment	Effluent quality BOD and suspended solids	Phosphorus	Reliability	Capital cost	Operating cost	Land requirement	Response to shock	Economic life (years)
Activated sludge (extended aeration)	3	1	3	H	H	L	3	15
Oxidation ditch	3	1	3	H	H	A	3	15
Aerated lagoon	3	1	2	A	A	H	3	20
Stabilization pond	1	1	1	L	L	H	2	30
Trickling filter	2	0	2	H	H	A	2	15
Biological rotating disc	3	1	3	H	L	L	3	15
Anaerobic processes	1	0	1	L	L	L	1	20
Irrigation	3	3	2	A	A	H	3	20

Surcharges for BOD_5, suspended solids and phosphate concentrations in excess of those found in domestic sewage are now so high for some dairy plants that treatment at the dairy plant is justified.

Staff required for treatment plants depends on the type and size of the plant. An Foras Taluntais (1974) states that a full-time operator is required for the larger plants and smaller plants need 2–4 man-hours per day for cleaning and process control. Laboratory staff time for analysis would also be required.

As an indication of power requirements, data quoted by An Foras Taluntais (1974) show average power consumptions per kg BOD removed of 1.1 kWh for activated sludge, 0.7 kWh for plastic medium filtration followed by activated sludge and 0.3 kWh for plastic medium filtration followed by alternating double filtration. Other power consumptions are quoted earlier in this review.

5.7 Conclusion

Treatment of dairy wastes is expensive. Of even greater importance is the cost of the raw material and product lost in the effluent. Substantial savings can be made by conducting and maintaining a campaign of good housekeeping and minimizing water use, raw material and product loss. Direct daily measurement of losses in the effluent is the only reliable method of assessing the costs of such losses. Reducing losses to a minimum also lowers the capital and operating costs of the treatment and disposal system required to treat the unavoidable losses.

Dairy wastes are organic in nature, with no significant toxic components. They are thus amenable to biochemical treatment. Wherever possible the natural assimilative capacity of natural water should be used, but not to the detriment of the environment or other water users. Complete disposal on land makes good use of the fertilizer value of dairy wastes and is applicable where suitable areas of land are available. Treatment in municipal plants is widely practised, but increasing surcharges will cause the trend to pretreatment at the dairy plant to continue. Biochemical treatment at the dairy plant requires conscientious operation to minimize the problems. Designers need to take account of the nature of dairy wastes and modify designs used for treatment of domestic sewage.

Acknowledgement

We acknowledge, with gratitude, the assistance of our colleagues at the New Zealand Dairy Research Institute, particularly Messrs M.F. Parkin and D.B. Galpin.

References

Adams, C.E. Jr and Eckenfelder, W.W. Jr (1974) *Process Design Techniques for Industrial Waste Treatment*, Enviro Press, Nashville, Tennessee.

Adamse, A.D. (1966) Bacteriological studies on dairy waste activated sludge, *Meded. Lands Hogesch., Wageningen*, **66**(6), 1–80.

Adamse, A.D. (1968) Formation and final composition of the bacterial flora of a dairy waste activated sludge, *Water Res.*, **2**, 665–671.

Adamse, A.D. (1974) Some bacteriological aspects of dairy waste activated sludge, in *Dairy Effluent Treatment*, Doc. 77, IDF, Brussels, 60–74.

An Foras Taluntais (1974) *Methods of Treatment of Milk Processing Wastes*, An Foras Taluntais, Dublin.

Anon. (1978) Neutralization of carbonic acid in smoke gas with the help of alkaline waste water, *Nord. Mejeritidsskr.*, **44**(4), 104–106.

Antonie, R.L. and Hynek, R.J. (1973) Operating experience with Bio-Surf process treatment of food processing wastes, *Proc. 28th Purdue Ind. Waste Conf.*, Ann Arbor Science, Ann Arbor, Michigan, 849–860.

APHA (1980) *Standard Methods for the Examination of Water and Wastewater*, 15th edn, American Public Health Association, American Waterworks Association and Water Pollution Control Federation, Washington, DC.

Arbuckle, W.S. (1970) Disposal of dairy wastes, in *Byproducts from Milk*, Webb, B.H. and Whittier, E.D. (Eds), 2nd edn, AVI Publishing, Westport, Connecticut, 405–422.

Arbuckle, W.S. (1972) *Ice Cream*, 2nd edn, AVI Publishing, Westport, Connecticut.

Baltjes, J. (1978) Modern waste water control in the dairy industry, *Proc. IDF seminar on Dairy Effluents*, Doc. 104, IDF, Brussels, 28–39.

Barnett, J.W., Parkin, M.F. and Marshall, K.R. (1982) The characteristics and oxygen demand of New Zealand dairy food plant effluent discharges, *Proc. Aquatic Oxygen Seminar*, Water and Soil Misc. Publ. No. 29, National Water and Soil Conservation Organization, Wellington, New Zealand, 49–53.

Boyle, W.C. and Polkowski, L.B. (1973) Alternate methods of treating or pretreating dairy plant wastes, paper presented at USEPA Technology Transfer Program on the Treatment of Dairy Plant Wastes, Madison, Wisconsin.

Brown, H.B. and Pico, R.F. (1980) Characterization and treatment of dairy wastes in the municipal treatment system, *Proc. 34th Purdue Ind. Waste Conf.*, Ann Arbor Science, Ann Arbor, Michigan, 326–334.

Buxton, B.M., Ziegler, S.J. and Moore, J.A. (1977) *Implication of Water Quality Regulations for Minnesota Dairy Processing Plants*, Agricultural Experimental Station Bulletin No. 520, University of Minnesota, St Paul.

Carawan, R.E., Chambers, J.V. and Zall, R. (1979) *Dairy Processing—Water and Wastewater Management*, Extension Special Report No. AM-18B, North Carolina Agricultural Extension Service, Raleigh, North Carolina.

Carawan, R.E. and Jones, V.A. (1977) Water and waste management educational program for dairy processing, *J. Dairy Sci.*, **60**, 1192–1197.

Chambers, J.V. (1975) Engineering control of a dairy waste treatment system — a concept, *Italian Cheese J.*, **4**(2), 1; 3–6; 8.

Chambers, J.V. (1980) Cost of dairy wastewater, disposal and management, *Am. Dairy Rev.*, **42**(11), 27; 28; 32; 36; 40.

Cooper, J.S. (1974) Research and development within dairy effluent treatment, *Dairy Effluent Treatment*, Doc. 77, IDF, Brussels, 3–17.

Cooper, J.S. (1978) Sludge handling and utilization in the United Kingdom, *Proc. IDF Seminar on Dairy Effluents*, Doc. 104, IDF, Brussels, 235–242.

Coton, S.G. (1980) The utilization of permeates from the ultrafiltration of whey and skim milk, *J. Soc. Dairy Technol.*, **33**, 89–94.

Cunningham, J., Nemke, J.L. and Marske, D. (1980) Land treatment uses sludge at Madison, Wisconsin, *Water Sewage Works*, **127**(3), 28–29; 49.

Currie, K.J. and Rutherford, J.C. (1980) Management of BOD in the lower Manawatu River, *Aquatic Oxygen Seminar Proceedings*, Water and Soil Miscellaneous Publication No. 29, National Water and Soil Conservation Organization, Wellington, New Zealand, 110–114.

Doedens, H. (1974a) Aerobic biological treatment of dairy wastewater — excess sludge and effects of variations in pH, *Dairy Effluent Treatment*, Doc. 77, IDF, Brussels, 108–121.

Doedens, H. (1974b) Reduction of effluent pollution in dairies and the problem of charges for effluent disposal, *Dte Milchw.*, **25**, 545–548.

Doedens, H. (1978) Studies on excess sludge in activated sludge plants handling dairy effluents, *Proc. IDF Seminar on Dairy Effluents*, Doc. 104, IDF, Brussels, 243–247.

EPA (1971) *Dairy Food Plant Wastes and Waste Treatment Practices*, Harper, W.J., Blaisdell, J.L. and Grosskopf, J. (Eds), USEPA 12060 EGU 03/71, US Environmental Protection Agency, Washington, DC.

EPA (1973) *Handbook for Monitoring Industrial Waste Water*, US Environmental Protection Agency Technology Transfer, US Govt Printing Office.

EPA (1974) *Treatment of Cheese Processing Wastewaters in Aerated Lagoons*, Environmental Protection Technology Series, EPA-660/2-74-012, US Environmental Protection Agency, Washington, DC.

Eriksson, G., Levin, A.-K. and Samuelsson, G. (1978) Dairy rinse water — experimental evaluation and recovery by reverse osmosis, *Proc. IDF Seminar on Dairy Effluents*, Doc. 104, IDF, Brussels, 90–92.

Fjaervoll, A. (1970) Anhydrous milk fat, manufacturing techniques and future applications, *Dairy Inds*, **35**, 424–428.

Food and Agriculture Organization (1979) Milk and milk products — supply, demand and trade projections 1985, *Dairy Inds Int.*, **44**(12), 37–47.

Fox, K.K. (1970) Casein and whey protein, in *By-products from Milk*, Webb, B.H. and Whittier, E.D. (Eds), 2nd edn, AVI Publishing, Westport, Connecticut, 311–355.

Gallo, T. and Sandford, D.S. (1979) The application of deep shaft technology to the treatment of high-strength industrial wastewaters, *AIChE Symp. Ser.*, **76**, 197; 288–300.

Galpin, D.B. (1981) Effluent disposal from New Zealand dairy plants, *N.Z.J. Dairy Sci. Technol.*, **16**, 289–292.

Galpin, D.B. and Parkin, M.F. (1981) Estimating the yield of butter by the measurement of losses, *N.Z. J. Dairy Sci. Technol.*, **16**, 231–241.

Gillies, M.T. (1974) *Whey Processing and Utilization — Economic and Technical Aspects*, Noyes Data Corporation, Park Ridge, NJ.

Goldsmith, H.I. and Watson, K.S. (1980) Use of equivalent BOD as a new waste management tool, *J. Water Pollut. Control Fed.*, **52**, 372–380.

Goronszy, M.C. and Barnes, D. (1980) Continuous single vessel activated sludge treatment of dairy wastes, *Water — 1979, AIChE Symp. Ser.*, **76**, 271–277.

Guo, P.H.M., Fowlie, P.J.A., Cairns, V.W. and Jank, B.E. (1979) *Performance Evaluation of an Oxidation Ditch Treating Dairy Wastewater*, Report No. EPS 4-WP-79-7, Technology Development Report, Canada Water Pollution Control Directorate, Ottawa.

Hall, C.W. and Hedrick, T.I. (1966) *Drying Milk and Milk Products*, AVI Publishing, Westport, Connecticut.

Harper, W.J. and Chambers, J.V. (1978) Upgrading dairy waste treatment systems, *Proc. IDF Seminar on Dairy Effluents*, Doc. 104, IDF, Brussels, 173–175.

Harper, W.J. and Hall, C.W. (1976) *Dairy Technology and Engineering*, AVI Publishing, Westport, Connecticut.

Hayashi, H. and Ishioka, Y. (1978) Reuse of effluent in the dairy industry, *Proc. IDF Seminar on Dairy Effluents*, Doc. 104, IDF, Brussels, 81–89.

Heldman, D.R. and Seiberling, D.A. (1976) Environmental sanitation, in *Dairy Technology and Engineering*, Harper, W.J. and Hall, C.W. (Eds), AVI Publishing, Westport, Connecticut, 272–321.

Hemmings, M.L. (1980) The treatment of dairy wastes, *Dairy Inds Int.*, **45**(11), 23–28.

Hickey, R.F. and Owens, R.W. (1981) Methane generation from high strength industrial wastes with the anaerobic biological fluidized bed, *Biotech. Bioeng. Symp.*, **11**, 399–413.

Hills, J.S. (1978) Water conservation, *Proc. IDF Seminar on Dairy Effluents*, Doc. 104, IDF, Brussels, 74–80.

Hines, D.A., Bailey, M., Ousby, J.C. and Rousler, F.C. (1975) The ICI deep shaft aeration process for effluent treatment, in *The Application of Chemical Engineering to the Treatment of Sewage and Industrial Liquid Effluents*, IChemE Symposium Series, **41**, D1–D10.

Holmstrom, P. (1978) Recirculation of evaporator condensates, *Proc. IDF Seminar on Dairy Effluents*, Doc. 104, IDF, Brussels, 65–73.

Hoover, S.R. (1953) Biochemical oxidation of dairy wastes, V, A review, *Sew. Ind. Wastes*, **25**, 201–206.

Horton, J.P. and Trebler, H.A. (1953) Recent developments in the design of small milk waste disposal plants, *Can. Dairy Ice Cr. J.*, **33**(4), 38–41; 78.

Hunziker, O.F. (1949) *Condensed Milk and Milk Powder*, 7th edn, Author, La Grange, Illinois.

International Dairy Federation (1974) Dairy effluent treatment, *Proc. IDF Symposium held in Denmark, 1973*, Doc. 77, IDF, Brussels.

International Dairy Federation (1978) *Proc. IDF Seminar on Dairy Effluents, Warsaw, October, 1976*, Doc. 104, IDF, Brussels.

International Dairy Federation (1980) *Guide for Dairy Managers on Wastage Prevention in Dairy Plants*, Doc. 124, IDF, Brussels.

International Dairy Federation (1981) *Dairy Effluents: Legislation on Water Conservation — Water to Milk Ratios — BOD, COD and TOC Relationships*, Doc. 138, IDF, Brussels.

Johnson, A.H. (1974) The composition of milk, in *Fundamentals of Dairy Chemistry*, 2nd edn, Webb, B.H., Johnson, A.H. and Alford, J.A. (Eds), AVI Publishing, Westport, Connecticut, 1–57.

Johnston, M.A.H. (1978) Some experiences of treating dairy effluent in Northern Ireland, *Proc. IDF Seminar on Dairy Effluents*, Doc. 104, IDF, Brussels, 177–182.

Jones, H.R. (1974) *Pollution Control in the Dairy Industry*, Noyes Data Corporation, Park Ridge, New Jersey.

Kearney, A.T., Inc. (1973) *Development Document for Effluent Limitation Guidelines and Standards of Performance, Dairy Product Industry*, USEPA 68-11-1502, US Environmental Protection Agency, Washington, DC.

Kosikowski, F.V. (1977) *Cheese and Fermented Milk Foods*, 2nd edn, Edward Bros, Ann Arbor, Michigan.

Lettinga, G., van Velsen, A.F.M., Hobma, S.W., de Zeeuw, W. and Klapwijk, A. (1980) Use of the upflow sludge blanket (USB) reactor concept for biological wastewater treatment, especially for anaerobic treatment, *Biotech. Bioeng.*, **23**, 699–734.

Lytken, E. (1974) Relations between production size and effluent pollution, *Dairy Effluent Treatment*, Doc. 77, IDF, Brussels, 24–28.

McAuliffe, K.W., Earl, D.D. and Macgregor, A.N. (1979) Spray irrigation of dairy factory wastewater onto pasture — a case study, *Prog. Water Tech.*, **11**(6), 33–43, Pergamon, London.

McAuliffe, K.W., Scotter, D.R., Macgregor, A.N. and Earl, K.W. (1982) Casein whey wastewater effects on soil permeability, *J. Environ. Qual.*, **11**, 31–34.

McDowall, F.H. (1953) *The Buttermaker's Manual*, New Zealand University Press, Wellington, New Zealand.

McDowall, F.H. and Thomas, R.H. (1961) *Disposal of Dairy Wastes by Spray Irrigation on Pasture Land*, Pollution Advisory Council Pub. No. 8, Wellington, New Zealand.

MAFF (1969) *Dairy Effluents*, Ministry of Agriculture, Fisheries and Food, HMSO, London.

Magnusson, F. (1974) Spray irrigation of dairy effluent, *Dairy Effluent Treatment*, Doc. 77, IDF, Brussels, 122–130.

Masters, K. (1979) *Spray Drying Handbook*, 3rd edn, George Godwin, London.

Mechsner, K. and Wuhrmann, K. (1974) Ecological considerations and an exemplification on biological treatment for dairy wastes, *Dairy Effluent Treatment*, Doc. 77, IDF, Brussels, 75–84.

NZ Dairy Board (1980) *The New Zealand Dairy Industry — A Survey*, 4th edn, New Zealand Dairy Board, Wellington, New Zealand.

NZ Dairy Board (1981) *Annual Report for Year Ended 31 May 1981*, New Zealand Dairy Board, Wellington, New Zealand.

NZ Dairy Board (1982) *Annual Report for Year Ended 31 May 1982*, New Zealand Dairy Board, Wellington, New Zealand.

Ojala, E.M. (1982) Dairy marketing prospects — long range market determinations, *Agricultural Economist*, **31**(1), 7–9.

Pankakoski, M. (1978) Experiences of a batch type treatment plant for dairy effluents in Finland, *Proc. IDF Seminar on Dairy Effluents*, Doc. 104, IDF, Brussels, 187–194.

Parkin, M.F., Clark, J.N. and Burton, C.E. (1977) Estimation of the quantity of milk solids in liquid effluent, *New Zealand Dairy Research Institute Annual Report*, 113–114.

Parkin, M.F. and Marshall, K.R. (1976) Spray irrigation disposal of dairy factory effluent — a review of current practice in New Zealand, *N.Z. J. Dairy Sci. Technol.*, **11**, 196–205.

Parkin, M.F. and Marshall, K.R. (1981) Dairy plant effluent, in *The Waters of the Waikato*, Vol. 1, Proc. Seminar Univ. Waikato, 20–22 August 1981, Hamilton, New Zealand, 293–300.

Pengilly, A.B. (1978) Dissolved air flotation as an alternative to sedimentation, *Proc. IDF Seminar on Dairy Effluents*, Doc. 104, IDF, Brussels, 199–201.

Porges, N., Michener, T.S. Jr., Jasewicz, L. and Hoover, S.R. (1960) *Dairy Waste Treatment by Aeration — Theory, Design, Construction and Operation*, Agr. Handbook 176, Agr. Research Service, Washington, DC.

Riddle, M.J. and Chandler, W.D. (1974) Waste disposal of whey, *Proc. Whey Utilization Symposium*, FRI, DD, Canada Department of Agriculture, 94–123.

Royal, L. (1978) Reduction of milk and milk product wastage, *Proc. IDF Seminar on Dairy Effluents*, Doc. 104, IDF, Brussels, 17–27.

Salplachta, J. (1978) Study on the formation and volume of sludge in activated sludge purification plants for dairy effluents, *Proc. IDF Seminar on Dairy Effluents*, Doc. 104, IDF, Brussels, 223–234.

Scheltinga, H.M.J. (1972) Measures taken against water pollution in dairies and milk processing industries, *Pure Appl. Chem.*, **29**, 101–111.

Scheltinga, H.M.J. (1978) Hygienic aspects, fertilizing value and potential for animal feed of sludges from dairy effluent treatment, *Proc. IDF Seminar on Dairy Effluents*, Doc. 104, IDF, Brussels, 219–222.

Schraufnagel, F.H. (1957) Dairy wastes disposal by ridge and furrow irrigation, *Proc. 12th Ind. Waste Conf.*, Purdue University, Indiana, 28–49.

Schraufnagel, F.H. (1962) Ridge and furrow irrigation for industrial waste disposal, *J. Water Pollut. Control Fed.*, **34**, 1117–1132.

Scott, R. (1981) *Cheesemaking Practice*, Applied Science, London.

Seyfried, C.F. (1974) Treatment of dairy effluent by plastic medium trickling filters, *Dairy Effluent Treatment*, Doc. 77, IDF, Brussels, 101–107.

Short, J.L. (1978) Prospects for the utilization of deproteinated whey in New Zealand, *N.Z. J. Dairy Sci. Technol.*, **13**, 181–194.

Southward, C.R. and Walker, N.J. (1980) The manufacture and industrial use of casein, *N.Z. J. Dairy Sci. Technol.*, **15**, 201–217.

Stanley Associates Engineering Ltd (1979) *Biological Treatment of Food Processing Wastewater — Design and Operations Manual*, Report No. EPS 3-WP-79-7, Environmental Protection Service, Environment Canada.

Strom, A. (1974) Some parameters expressing the pollution of dairy effluent, *Dairy Effluent Treatment*, Doc. 77, IDF, Brussels, 18–23.

Svoboda, M. (1974) Waste stabilization ponds and waste stabilization basins, *Dairy Effluent Treatment*, Doc. 77, IDF, Brussels, 85–92.

Svoboda, M., Gillar, J., Hlavaka, M., Salplachta, J., Stelcova, D. and Marvan, P. (1966) Purification of dairy wastes by means of lagoons, *XVII Int. Dairy Cong. E/F*, 715–722.

Synnott, E.C., Kelly, B.F. and Moloney, A.M. (1978) Recent developments in dairy effluent treatment in Ireland, *Proc. IDF Seminar on Dairy Effluents*, Doc. 104, IDF, Brussels, 156–159.

Tanaka, T. (1974) Use of aerated lagoons for dairy effluent treatment, *Dairy Effluent Treatment*, Doc. 77, IDF, Brussels, 93–100.

Thomas, M.A. (1977) *The Processed Cheese Industry*, Department of Agriculture, New South Wales, Australia.

Thomas, T.D. (1982) Estimation of casein fines in lactic casein whey and clarifier sludge, *N.Z. J. Dairy Sci. Technol.*, **17**, 77–80.

Trebler, H.A. and Harding, H.G. (1955) Fundamentals of the control and treatment of dairy waste, *Sew. Ind. Wastes*, **27**, 1369–1382.

Tuszynski, W.B. (1978) *Packaging, Storage and Distribution of Processed Milk: Technical Requirements and their Economic Implications*, FAO Animal Production and Health Paper No. 11, FAO, Rome, Italy.

Vandamme, K. and Waes, G. (1980) Purification of dairy waste water in a two-stage treatment plant including anaerobic pretreatment, *Milchwissensch.*, **35**, 663–666.

Watson, K.S., Peterson, A.E. and Powell, R.D. (1977) Benefits of spreading whey on agricultural land, *J. Water Pollut. Control Fed.*, **49**, 24–34.

Wheatland, A.B. (1974) Treatment of waste waters from dairies and dairy product factories — methods and systems, *J. Soc. Dairy Technol.*, **27**, 71–79.

Wilcox, L.V. (1948) The quality of water for irrigation use, *US Dept Agriculture Tech. Bull.*, **962**, 1.